Signal Perception and Transduction in Higher Plants

NATO ASI Series

Advanced Science Institutes Series

A series presenting the results of activities sponsored by the NATO Science Committee, which aims at the dissemination of advanced scientific and technological knowledge, with a view to strengthening links between scientific communities.

The Series is published by an international board of publishers in conjunction with the NATO Scientific Affairs Division

A Life Sciences B Physics	Plenum Publishing Corporation London and New York
C Mathematical and Physical Sciences D Behavioural and Social Sciences E Applied Sciences	Kluwer Academic Publishers Dordrecht, Boston and London
F Computer and Systems Sciences G Ecological Sciences H Cell Biology	Springer-Verlag Berlin Heidelberg New York London Paris Tokyo Hong Kong Barcelona

Series H: Cell Biology Vol. 47

Signal Perception and Transduction in Higher Plants

Edited by

Raoul Ranjeva

Alain M. Boudet

Centre de Physiologie Végétale
Université Paul Sabatier
URA CNRS n° 241
118, Route de Narbonne
31062 Toulouse Cédex, France

Springer-Verlag Berlin Heidelberg New York
London Paris Tokyo Hong Kong Barcelona
Published in cooperation with NATO Scientific Affairs Division

Proceedings of the NATO Advanced Research Workshop on Signal Perception and Transduction in Higher Plants held in Toulouse, France, July 9–13, 1989

ISBN-13: 978-3-642-83976-4 e-ISBN-13: 978-3-642-83974-0
DOI: 10.1007/978-3-642-83974-0

Library of Congress Cataloging-in-Publication Data

Softcover reprint of the hardcover 1st edititon 1990

2131/3140-543210 – Printed on acid-free-paper

P R E F A C E

This volume summarizes the lectures and round tables of a NATO advanced workshop held in Toulouse, July 9th-13th, 1989. The workshop was held to bring together plant scientists of complementary disciplines who share a common interest in the study of signal perception and transduction in higher plants.

This volume covers only the formal proceedings but an equal length of time was devoted to discussion. Some of the main aspects will be found in the appropriate articles written by the participants who have tried to transmit a little of the flavour of the intensive exchanges and discussions. The volume reflects much of the current views and excitement in the field of signal perception and transduction in higher plants. Participants have been selected among those in the forefront of plant (and animal) physiology. They included specialists in membrane proteins, patch-clamp, cell imaging, gene cloning... to illustrate the necessity of complementary disciplines to get an integrated view.
The scientific programme was set up by the International Organizing Committee including James Morré, Andreas Sievers and Antony Trewavas.

In addition to NATO, who provided most of the funding of the workshop, thanks are also due to the Université Paul Sabatier, who was our host, the National Science Fundation, the Centre National de la Recherche Scientifique, the Ministère de l'Education Nationale, the Région Midi-Pyrénées, the Conseil Général and the Maire de Toulouse, who provided additional help. Gérard Abravanel, Béatrix Aries, Annie Boudet, Martine Charpenteau, Annick Graziana and Michèle Marolda were of valuable assistance with the day to day organization of the conference.

Raoul Ranjeva

Alain M. Boudet

CONTENTS

THE SENSITIVITY OF PLANT PROTOPLASTS TO AUXIN IS LIKELY MODULATED BY THE NUMBER OF RECEPTORS AT THE PLASMALEMMA

H. Barbier-Brygoo, G. Ephritikhine, D. Klämbt* and J. Guern
Laboratoire de Physiologie Cellulaire Végétale
C.N.R.S./I.N.R.A., Bat 15
F-91198 Gif-sur-Yvette Cedex
France

INTRODUCTION

Auxin modifies the electrical transmembrane potential of isolated protoplasts according to an inverted bell-shaped dose-response curve (Ephritikhine *et al*, 1987 ; Shen *et al*, 1988). The sensitivity of plant protoplasts to auxin can be characterized by the hormone concentration inducing the maximal variation of the transmembrane potential difference (ΔEm). On tobacco mesophyll protoplasts, the response to auxin was blocked by antibodies directed against an auxin-binding protein (ABP) from maize coleoptile membranes, or raised to the plasmalemma ATPase from yeast (Barbier-Brygoo *et al*, 1989a). The auxin-induced ΔEm thus seems to involve the occupancy of an auxin receptor at the cell surface which in turns would activate the proton-pump ATPase at the plasmalemma.

Using this membrane response on isolated protoplasts as a criterium of sensitivity to auxin, it was shown that large variations in sensitivity of plant cells to auxin can be induced by either mutagenesis on tobacco (Ephritikhine *et al*, 1987) or *Agrobacterium rhizogenes* transformation on various materials (Shen *et al*, 1988). These different materials exhibited the same sensitivity to fusicoccin as regards the hyperpolarizing effect of the toxin on the protoplasts (Barbier-Brygoo *et al*, 1989b). This, together with the fact

*Botanisches Institut der Universität Bonn, Meckenheimer Allee 170, D-5300 Bonn 1; Federal Republic of Germany

NATO ASI Series, Vol. H 47
Signal Perception and Transduction in Higher Plants
Edited by R. Ranjeva and A. M. Boudet

that the fusicoccin-induced ΔEm was not blocked by the antibody raised to the auxin receptor, but inhibited by the anti-ATPase antibody strongly suggests that the observed changes in sensitivity to auxin reflect modifications either at the reception step or at one of the early steps specifically transducing the auxin signal. In order to test the first hypothesis, the ability of the anti-ABP antibody to inhibit the auxin-induced membrane response was compared on protoplasts isolated from normal, mutant and transformed tobacco plants. Increasing the antibody concentration in the protoplast suspension led to a progressive inhibition of the response, which could allow relative immunotitrations of the functional receptor units (Barbier-Brygoo *et al*, 1989b; 1989c). These immunotitrations showed that the protoplasts from the three genotypes differed in the number of antigenic sites recognized by the anti-ABP antibody, protoplasts from the resistant mutant exhibiting less sites than wild-type protoplasts, whereas hypersensitive protoplasts from the transformed plants had more sites than their normal counterparts. According to the simplest interpretation which is to identify these sites with functional auxin receptors, the sensitivity of protoplasts to auxin thus appears to be related to the number of receptors exposed at the plasmalemma.

The present work investigates the possibility of modulating the sensitivity of protoplasts from the wild-type genotype by modifications of their number of receptors, using either the antibody directed against the auxin-binding protein or this purified receptor protein itself.

RESULTS

The sensitivity of tobacco protoplasts is decreased in the presence of monospecific immunoglobulins raised to the auxin-binding protein from maize coleoptile membranes.

Protoplasts isolated from tobacco mesophyll cells, when suspended in the culture medium depleted in auxin, exhibited a small negative inside potential difference (Em) (Figure 1A and 1C). The addition of NAA in the culture medium induced an hyperpolarization of the protoplasts which is illustrated by a shift of the Em distribution towards more negative values (Figure 1B and 1D). This auxin-induced response was previously shown to be inhibited by a preparation of total immunoglobulins (IgG) isolated from antisera of rabbits immunized with the purified auxin-binding protein from maize coleoptile membranes (Barbier-Brygoo *et al*, 1989a). In order to test the specificity of the effects observed with this crude fraction, an aliquot of the polyclonal IgG was further purified by affinity chromatography on an ABP-Sepharose column as described by Löbler and Klämbt (1985). The effects of this monospecific IgG fraction were tested on the auxin-induced hyperpolarization of protoplasts, in comparison with those of a nonimmune IgG preparation used as a negative control.

An incubation of the protoplasts either with the nonimmune IgG or with the monospecific anti-ABP IgG did not modify significantly the potential difference of the protoplasts (Figure 1A and 1C). The nonimmune IgG did not interfere with the auxin effect (Figure 1B) whereas the NAA-induced hyperpolarization was no more observed when the protoplasts had been previously treated by the anti-ABP antibody (Figure 1D). Furthermore, the monospecific anti-ABP IgG inhibited the auxin effect when applied at a concentration 100 times lower than the concentration of the crude anti-ABP IgG fraction required to produce the same inhibitory effect (Barbier-Brygoo *et al*, 1989a), thus illustrating the increased specificity of this affinity-purified anti-ABP IgG.

The dose-response curve to auxin of protoplasts pretreated with the anti-ABP antibody was compared to the one of control protoplasts (Figure 2A). In the presence of 4 nM anti-ABP IgG, which caused about 90% inhibition of the response induced by 5μM NAA (see Figure 1), the dose-response curve was shifted towards higher NAA concentrations. In this case, the maximal hyperpolarization was obtained for 50 μM NAA, a concentration 10 times higher than the optimal one observed for untreated protoplasts.

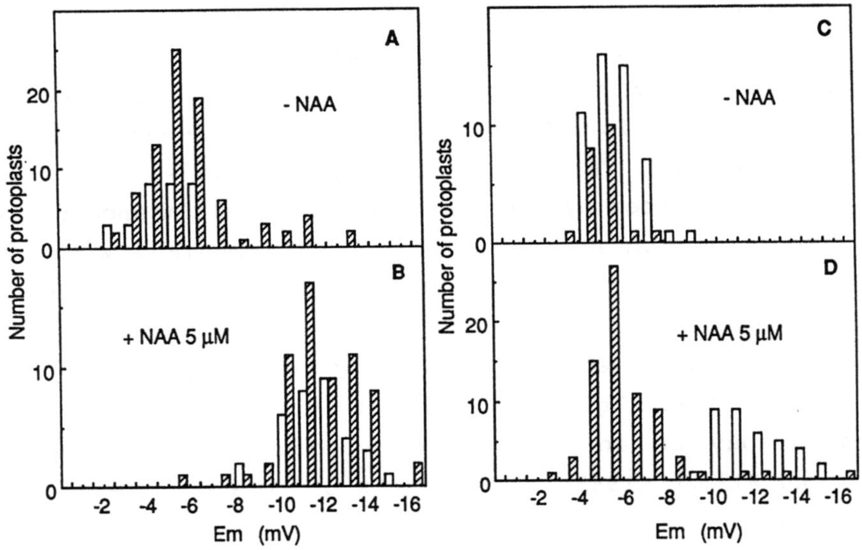

FIGURE 1. Effect of nonimmune rabbit immunoglobulins (A,B) and of monospecific immunoglobulins directed against an auxin-binding protein (ABP) from corn coleoptile membranes (C,D) on the potential difference variations induced by auxin on tobacco protoplasts.
Mesophyll protoplasts were isolated from young tobacco leaves of wild-type plants (*Nicotiana tabacum*, cv Xanthi) according to Caboche (1980). The measurements of transmembrane electrical potential difference (Em) were performed on individual protoplasts by the microelectrode technique (Barbier-Brygoo *et al*, 1989a), in the complete culture medium in the absence of auxin (A and C, white columns) or immediately after the addition of 5μM 1-naphthalene acetic acid (NAA) (B and D, white columns).
Rabbit immunoglobulins from pooled normal sera (Zymed Laboratories) were used as control nonimmune immunoglobulins. The monospecific anti-ABP immunoglobulins were purified by affinity chromatography on an ABP-Sepharose column, from the total immunoglobulin fraction isolated from antisera of rabbits previously immunized with the purified maize ABP (see Barbier-Brygoo *et al*,1989a). The protoplasts ($5x10^4$ prot.ml^{-1}) were then incubated with nonimmune IgG (0.7 μM; A and B, grey columns) or with anti-ABP IgG (4 nM; C and D, grey columns) during 1.5 h at room temperature, before the addition of NAA

The sensitivity of tobacco protoplasts can be increased by the addition of purified auxin-binding protein from maize coleoptile membranes.

The membrane-associated ABP from maize coleoptiles was purified in D. Klämbt laboratory, as already described (Barbier-Brygoo *et al*, 1989a) except that a PAA-Sepharose column was used to retain the ABP in the affinity chromatography procedure. In the presence of the purified ABP preparation (0.1 nM ABP), the dose-response curve to auxin of wild-type protoplasts was shifted towards lower concentrations by a factor of about 100 in comparison with the response of untreated protoplasts (Figure 2B).

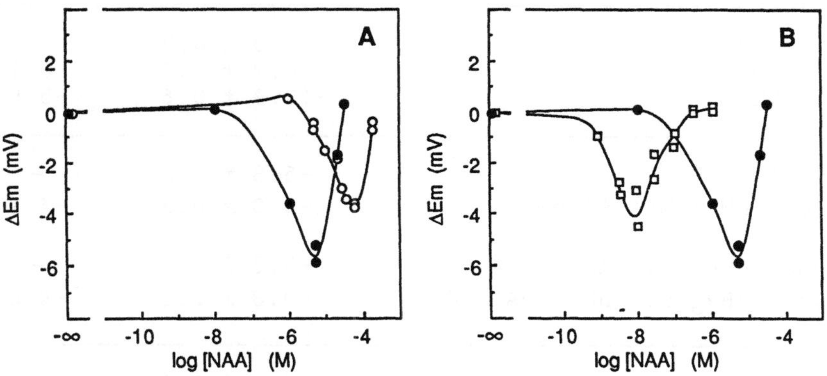

FIGURE 2. Dose-response curves for Em of tobacco protoplasts in the presence of NAA.
Mesophyll protoplasts were isolated from the wild-type genotype and Em variations (ΔEm) were measured as a function of auxin concentrations in the external medium (A and B, black symbols, one representative experiment). The protoplasts ($5x10^4$ protoplasts.ml^{-1}) were then incubated either (A) with the monospecific anti-ABP IgG (4 nM IgG, 1.5 h at room temperature), or (B) with a purified ABP preparation from maize coleoptile membranes (0.1 nM ABP, 15 min at room temperature) (see Barbier-Brygoo *et al*, 1989, for the preparation of the anti-ABP IgG and the purified ABP fractions). Dose-response curves for NAA were then established in the presence of the anti-ABP IgG (open circles, data pooled from two independant experiments) or the purified ABP (open squares, data pooled from two independant experiments)

In order to assess the role of the auxin-binding protein in this increase in sensitivity, different ABP preparations were tested for their ability to shift the sensitivity of treated protoplasts (Table 1). The concentration of 10 nM NAA was chosen for these tests as the one inducing the maximal Em variation on protoplasts treated by 0.1 nM ABP, but unaffecting control protoplasts (Figure 2B and Table 1).

TABLE 1. Effect of the treatment of tobacco protoplasts by a purified auxin-binding protein (ABP) preparation on the auxin-induced Em variations

Addition to the medium	Em ± SE (mV)	NAA-induced ΔEm (mV)
None	-7.0 ± 0.3	-
NAA 10 nM	-6.9 ± 0.7	+0.1
NAA 5 μM	-12.9 ± 0.6	-5.9
ABP_a 0.1 nM	-5.8 ± 0.3	-
ABP_a 0.1 nM + NAA 10 nM	-9.9 ± 0.3	-4.1
Dialyzed ABP_a 0.1 nM	-5.8 ± 0.3	-
Dialyzed ABP_a 0.1 nM + NAA 10 nM	-9.8 ± 0.3	-4.0
ABP_b 0.1 nM	-4.7 ± 0.3	-
ABP_b 0.1 nM + NAA 10 nM	-8.9 ± 0.4	-4.2
Boiled ABP_b 0.1 nM	-4.6 ± 0.2	-
Boiled ABP_b 0.1 nM + NAA 10 nM	-4.2 ± 0.2	+0.4

Protoplasts ($5x10^4$ protoplasts.ml^{-1}) were preincubated at room temperature for 15 min with purified ABP from two independant preparations (ABP_a and ABP_b). ABP was tested without any further treatment, after an overnight dialysis in phosphate buffered solution or after a 10 min boiling. Em measurements were performed on tobacco mesophyll protoplasts in the complete culture medium in the absence or presence of NAA. Em values are presented as mean ± SE calculated from 20 individual measurements for each experimental condition. The NAA-induced Em variations (ΔEm) were calculated from the mean control value in the absence of auxin.

Two ABP preparations (ABP_a and ABP_b) isolated from different coleoptile batches exhibited the same effect on protoplast sensitivity (Table 1). According to the purification procedure, ABP was freed from auxin after its elution from the Sepharose column with 10 mM NAA. However, to eliminate a possible residual contamination of the protein fraction by auxin the ABP preparation was further dialyzed to a potential dilution factor of 4×10^6. After this additional step, the efficiency of the ABP on the protoplasts was not modified (Table 1). In other respects, a boiled ABP preparation was no more able to shift the sensitivity of treated protoplasts (Table 1). These different results thus favour the idea that ABP itself is responsible for the increase in sensitivity to auxin.

Such an effect raises the question of the mode of interaction of ABP with the protoplasts. To investigate the stability of this interaction, protoplasts preincubated with ABP were washed prior to the evaluation of their sensitivity to auxin. Table 2 shows that the increase in sensitivity to NAA

TABLE 2. Effect of washings on protoplasts pretreated with a purified auxin-binding protein (ABP) preparation on the auxin-induced Em variations

Washing solution	Addition to the medium	Em ± SE (mV)	NAA-induced ΔEm (mV)
Culture medium	None	-5.1 ± 0.2	-
Culture medium	NAA 10 nM	-10.6 ± 0.2	-5.5
High salt medium	None	-4.6 ± 0.2	-
High salt medium	NAA 10 nM	-9.2 ± 0.2	-4.6

Protoplasts (5×10^4 protoplasts.ml^{-1}) were preincubated at room temperature for 15 min with the purified ABP_b 0.1 nM (see table 1). Protoplasts were then washed by centrifugation (100g, 5 min) either in the culture medium (Caboche, 1980) depleted in NAA or in a high-salt medium (KCl 0.3 M, $CaCl_2$ 5 mM, MES/KOH 1 mM, pH 5.7). The washing procedure led to a final dilution of the free ABP by 2×10^5. Em measurements were performed on tobacco mesophyll protoplasts in the complete culture medium in the absence or presence of NAA, as described for Table 1.

induced by ABP was equally maintained after washing in media differing by their ionic strength, the culture medium and a high-salt medium essentially composed of KCl. Preliminary results obtained in immunotitration experiments on ABP-treated protoplasts indicated that the number of titratable sites was increased by a factor of about 10 with regards to control protoplasts. Taken together, these data suggest that the ABP-induced shift in sensitivity results from the binding of ABP at the outer surface of the plasmalemma.

DISCUSSION

The results presented here show that it is possible to modulate the sensitivity of wild-type protoplasts to auxin, as judged from the position of the dose-response curves along the NAA concentration range. A partial inactivation of the auxin-binding sites by the anti-ABP antibody decreased the sensitivity, whereas a complementation of the protoplasts with purified ABP increased their responsiveness to auxin (Figure 2) and at the same time enriched the protoplast surface in antigenic sites recognized by the anti-ABP antibody. However, as the immunotitrations were performed with a polyclonal IgG preparation, they allow only a relative estimate of the abundance of sites in different situations where the sensitivity to auxin was modified. Furthermore, the system used in this study is heterologous, as tobacco protoplasts were treated by an anti-ABP IgG or a purified ABP preparations from maize. This could possibly account for the slight decrease in amplitude of the auxin-induced response after both treatments, in comparison with the control dose-response curve (Figure 2). An absolute quantification of antigenic sites could be achieved by using monoclonal antibodies directed against the auxin-binding protein. Such monoclonals raised to the maize ABP are presently developed by different groups (Napier *et al*, 1988), but the clones isolated so far exhibit very poor cross-reactivity with other species than maize.

Nevertheless, Table 3 shows that in five different situations, including the shifts in sensitivity experimentally induced here on protoplasts from wild-type tobacco but also the low and very high sensitivities to auxin exhibited by an auxin-resistant mutant and *A. rhizogenes* transformed tobacco plants, respectively, (Barbier-Brygoo *et al*, 1989b and 1989c), a good correlation was observed between the sensitivity to auxin and the amount of antigenic sites inactivated by the anti-ABP antibody. According to the assumption that such antigenic sites constitute functional auxin receptors, the simplest interpretation of these results is that the observed modulation in sensitivity is due to variations in the number of plasmalemma receptors.

TABLE 3. Correlation between the sensitivity of protoplasts to auxin and the relative abundance in antigenic sites recognized by the anti-ABP antibody

Tobacco genotype	Protoplast treatment	Sensitivity	Relative abundance of antigenic sites
Wild-type	None	++	++
Wild-type	Anti-ABP IgG	+	+
Wild-type	ABP	+++	+++
Mutant	None	+	+
A. rhizogenes transformed	None	+++	+++

The sensitivity of protoplasts to auxin was estimated by the position of the dose-response curves along the NAA concentration range. The relative abundance of antigenic sites was deduced from the amounts of anti-ABP IgG necessary to inhibit the auxin-induced membrane response.

It is not a common idea amongst plant biologists that an increase in sensitivity to an hormone could be accounted for by an increase in the number of receptors. Conversely, the idea is widely spread that the dose-response curve to an hormone should display an apparent affinity in terms of C50 closely related to the binding affinity of the receptor for the hormone. In fact, the apparent affinity displayed by the dose-response curve can differ from the affinity of the receptor for the hormone, the difference being a function of the number of receptors (Guern, 1987). This situation is encountered for example when the hormone-induced response is limited by the concentration of a secondary messenger liberated as a function of the number of occupied receptors, and when only a fraction of the receptor population needs to bind the hormone in order to liberate enough messenger to saturate the response measured (Barbier-Brygoo *et al*, 1989b).

Recently, cDNA clones derived from mRNAs encoding the maize ABP were isolated by Inohara *et al* (1989). These authors suggested, according to the amino acid sequence deduced from the nucleotidic cDNA sequence that ABP is a protein located in the ER lumen. This apparently contradicts the evidence provided here for the presence at the plasmalemma of functional auxin receptors cross-reacting with the anti-ABP antibody. Furthermore, the results showing that exogenous maize ABP is able to complement tobacco protoplasts raise many questions as to the mechanisms by which this protein is bound to the plasmalemma, to its specific role in the auxin signalling pathway and to the transmission of the hormonal signal to the cell interior, as the sequence of the mature ER protein does not exhibit any hydrophobic sequence long enough to span the membrane. We cannot exclude the possibility that the ABP preparations used contained, aside the major ER-ABP, a specific not yet characterized minor plasmalemma form. Another possibility accounting for the set of data presented here would be to consider the ER-ABP as a precursor of the ABP associated to the plasma membrane and to postulate the existence of a two-stage signalling system. In this working model, two proteins, the auxin-binding protein localized at the external surface of the plasmalemma and a putative transmembrane protein, would be associated to constitute a functional receptor unit.

REFERENCES

Barbier-Brygoo H, Ephritikhine G, Klämbt D, Ghislain M, Guern J (1989a) Functional evidence for an auxin receptor at the plasmalemma of tobacco mesophyll protoplasts. Proc Natl Acad Sci USA 86:891-896

Barbier-Brygoo H, Ephritikhine G, Shen WH, Delbarre A, Klämbt D, Guern J, (1989b) Characterization and modulation of the sensitivity of plant protoplasts to auxin. In NATO ASI Series, Transducing Pathways:Activation and Desensitization, Konijn TM ed., Springer Verlag, in press

Barbier-Brygoo H, Guern J, Ephritikhine G, Shen WH, Maurel C, Klämbt D (1989c) The sensitivity of plant protoplasts to auxins : modulation of receptors at the plasmalemma. In "Plant Gene Transfer", UCLA Symposia on Molecular and Cellular Biology, New Series, Vol. 189, eds Lamb C and Beachy R, Liss AR, Inc., New-York, in press

Caboche M (1980) Nutritional requirements of protoplast-derived, haploid tobacco cells grown at low cell densities in liquid medium. Planta 149:7-18

Ephritikhine G, Barbier-Brygoo H, Muller JF, Guern J (1987) Auxin effect on the transmembrane potential difference of wild-type and mutant tobacco protoplasts exhibiting a differential sensitivity to auxin. Plant Physiol 83:801-804

Guern J (1987) Regulation from within : the hormone dilemma. Ann Bot 60:75-102

Inohara N, Shimomura S, Fukui T, Futai M (1989) Auxin-binding protein located in the endoplasmic reticulum of maize shoots : molecular cloning and complete primary structure. Proc Natl Acad Sci USA 86:3564-3568

Löbler M, Klämbt D (1985) Auxin-binding protein from coleoptile membranes of corn (*Zea mays* L.). I Purification by immunological methods. J Biol Chem 260:9848-9853

Napier RM, Venis MA, Bolton MA, Richardson LI, Butcher GW (1988) Preparation and characterisation of monoclonal and polyclonal antibodies to maize membrane auxin-binding protein. Planta 176:519-526

Shen WH, Petit A, Guern J, Tempé J (1988) Hairy roots are more sensitive to auxin than normal roots. Proc Natl Acad Sci USA 85:3417-3421

ANTIBODIES TO THE MAIZE MEMBRANE AUXIN RECEPTOR

Michael A. Venis and Richard M. Napier
AFRC Institute of Horticultural Research, East Malling
Maidstone
Kent ME19 6BJ
United Kingdom

INTRODUCTION

Many auxin-binding systems have been studied in different laboratories but the binding sites in maize membranes are by far the best characterised. Furthermore, there is now excellent evidence, especially from recent electrophysiological data, (Barbier-Brygoo et al., 1989a, b, further elaborated by Prof. J. Guern at this Conference) that these binding sites are functionally active receptors. The binding proteins have been solubilised, purified and both polyclonal (Löbler and Klämbt, 1985; Shimomura et al., 1988; Napier et al., 1988) and monoclonal antibodies (Napier et al., 1988) produced. The receptor has been cloned and sequenced in four different laboratories (Inohara et al., 1989; Hesse et al., 1989, Tillman et al., 1989; Lazarus and Venis, unpublished).

This is the first plant hormone receptor to have been both chemically and functionally characterised. While this characterisation is still preliminary, molecular probes are now available to explore receptor structure and function in greater detail. Here, we describe some of the ways in which our antibodies are being applied and discuss the question of the cellular localisation of functional receptor.

METHODS

Receptor purification, production of monoclonal and polyclonal antibodies, SDS-PAGE, immunoblotting and auxin-binding methods have been described (Napier et al., 1988).

NATO ASI Series, Vol. H 47
Signal Perception and Transduction in Higher Plants
Edited by R. Ranjeva and A. M. Boudet

Sandwich ELISA. Plates were coated with polyclonal antibody, blocked with foetal calf serum, then purified auxin receptor added and allowed to bind. After washing away excess receptor, monoclonal antibody (mAb) was added in the presence or absence of an auxin analog. Bound mAb was detected using biotinylated second antibody and streptavidin-peroxidase.

Immunoaffinity adsorbents were prepared by coupling mAb MAC 256 or polyclonal IgG (antiserum fractionated by ammonium sulfate precipitation and DEAE-BioGel, Bio-Rad Laboratories) to Reacti-Gel 6 X (Pierce) using the manufacturers' instructions.

Immunoadsorption. Microsomal membranes were extracted at 0°C for 1-3 h in NP40 buffer (0.2% Nonidet P-40 in 5-fold diluted phosphate buffered saline, PBS, 5 mM EDTA-K_2 pH 7.3, 0.25 mM PMSF). After rapid ultracentrifugation (Beckman TL100) the clear supernatant was mixed with immunogel on a turntable for 16 h at 4°C. The gel was spun down (MSE Micro Centaur, 6500 rpm, 1 min), washed seven times in NP40 buffer, then eluted either with 0.3 mM 1-naphthaleneacetic acid, (NAA) for 10 min (MAC 256) or with 0.1 M glycine HCl pH 2.5 for 5 min (polyclonal). Eluates were precipitated with trichloroacetic acid and analysed by SDS-PAGE and blotting.

Supernatants obtained after pelleting of the microsomal fraction were precipitated at 75% ammonium sulfate saturation, taken up in a small volume of PBS/5 - 0.25 mM PMSF and dialysed against this buffer overnight. Aliquots were then taken through the immunosorption-desorption procedure described for the polyclonal matrix, followed by precipitation and analysis as above.

Immunoprecipitation. MAC 256 was incubated with post-DEAE receptor for 16 h at 4°C, then complexed with agarose-linked goat anti-rat IgG (Sigma). After centrifugation, binding of ^{14}C-NAA was measured in the supernatant and in the immunoprecipitate after release of the receptor from MAC 256 with 2 M NaCl.

Elongation tests were carried out using ten 10 mm third internode sections cut 5 mm below the hook of 6-7-day-old etiolated Alaska pea seedlings. Duplicate batches were incubated at 25°C for 16 h each in 5 ml of sterile 0.5% w/v sucrose-5 mM K maleate buffer pH 5.5 plus test substance. Final lengths were measured and growth expressed as a percentage of that in buffer controls. In most experiments, the cuticle of the epicotyls was first abraded prior to sectioning by rubbing five times with a paste of fine carborundum powder between finger and thumb (Dreyer et al., 1981).

Conjugates of 5-amino-NAA (Venis and Thomas, 1989) with bovine serum albumine (BSA) or keyhole limpet haemocyanin (KLH) were prepared by diazotisation and azo-coupling. The same ligand was also coupled to Epoxy Sepharose (Pharmacia) according to the manufacturers' recommendations. Concentrations in growth tests refer to those of the ligand.

Protoplasts were prepared from etiolated shoots of 5-day-old maize seedlings (cv. Clipper) by the method of Sarhan and Cesar (1988) with the inclusion of 1.5×10^{-5} M NAA in the isolation medium. Protoplasts were then incubated with MAC 256, polyclonal immune or pre-immune IgG fractions and washed. Bound antibody was detected with FITC-labelled second antibody and fluorescence microscopy.

RESULTS

Monoclonals

The maize membrane auxin receptor has a native M_r of ca. 44 kDa, consisting of two subunits which in our hands migrate on SDS-PAGE with an apparent M_r of 22 kDa (Napier et al., 1988). A minor band ca. 1 kDa smaller is often seen (Shimomura et al., 1986; Napier et al., 1988), and probably represents a facile cleavage product. The smaller polypeptide appears to retain full auxin-binding activity (Shimomura et al., 1986). Both polypeptides are glycosylated and treatment with endoglycosidase H indicates a single glycan chain of ca. 2 kDa

(Löbler et al., 1987; Napier et al., 1988; Inohara et al., 1989), consistent with the identification of a single glycosylation site in the cloned sequence (Inohara et al., 1989).

We have reported on the generation of five mAbs, MAC 256-260, to the receptor and their partial characterisation (Napier et al., 1988). All the mAbs recognised deglycosylated receptor as efficiently as they recognised the native form, but two - MAC 256 and MAC 259 - recognised only the major (22 kDa) polypeptide and not the 21 kDa species. These mAbs were therefore mapped to a terminal epitope within the 1 kDa region that is cleaved from the 22 kDa subunit. We have since found that the N-terminal sequences of the 22 kDa and 21 kDa forms are identical, so that MAC 256 and MAC 259 can now be mapped to the C-terminal epitope.

Further confirmation comes from FPLC chromatofocusing of the receptor on a Mono P (Pharmacia) column, which gives apparent pI values (in 6 M urea) of 4.95 and 5.1 for the 22 kDa and 21 kDa bands respectively. The 1 kDa terminus is therefore acidic and from the amino acid sequence (Inohara et al., 1989), it is clear that this must be the C-terminus. Fortuitously, this experiment yielded additional information, arising from a certain amount of degradation to a 15 kDa cleavage product. The 15 kDa band was detected by MAC 256, but not by MAC 257. The cleavage fragment therefore includes the C-terminus and hence MAC 257 can be mapped to an epitope within 7 kDa of the N-terminus.

Immunoprecipitation with either MAC 256 (Fig 1) or MAC 259 removes auxin-binding activity from partially-purified receptor preparations and some of this activity can be recovered from the precipitate. SDS-PAGE of the immune complex confirms that it is the 22 kDa polypeptide that is specifically precipitated. However, unlike the polyclonal antiserum (Venis and Napier, 1989), none of the mAbs cross-reacts strongly in other species. We consider it unlikely, therefore, that any of the mAbs is directed at an epitope within the auxin-binding site itself. Nevertheless, in the

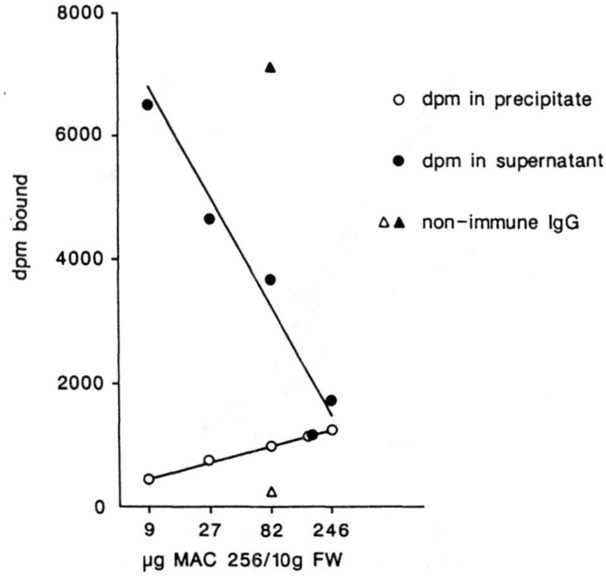

Fig 1. Specific binding of ^{14}C-NAA by supernatant and pellet fractions after immunoprecipitation of post-DEAE receptor fractions by MAC 256

sandwich ELISA assay (see Methods) with either MAC 256 or MAC 259 (Fig 2), binding of the mAb is reduced by NAA, but not by benzoic acid (BA), which is inactive as an auxin. Further assays, carried out at a fixed receptor concentration and varying concentrations of a range of auxin analogs (indoles, phenoxyacetic and benzoic acids) have shown a very good correlation between ability to depress mAb binding and in vivo auxin activity. We interpret the competition between auxins and mAb in terms of a ligand-induced conformational change, with ligand binding leading to masking of the mAb epitope. Since MAC 256 and MAC 259 recognise a C-terminal epitope, this region is implicated in the conformational change.

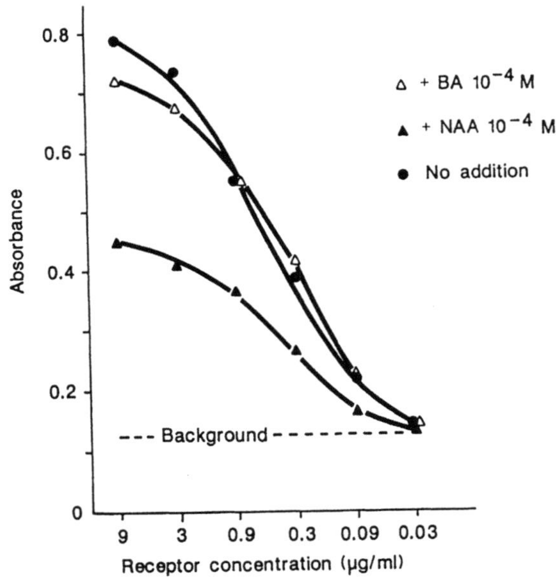

Fig 2. Competition for binding of MAC 259 in the sandwich ELISA assay at varying receptor inputs

Polyclonal

We reported previously (Venis and Napier, 1989) that the polyclonal antiserum detects homologous polypeptides in a range of species when membrane digests are run out on SDS-PAGE and immunoblotted, including the dicotyledonous species zucchini, pea and tomato. Subsequently, we have found it useful to enrich by immunosorption prior to SDS-PAGE. Unlike the monoclonal matrix, from which the maize receptor can be removed very selectively by NAA, elution from the polyclonal adsorbent requires low pH (see Methods). Using the polyclonal gel in this manner and Western blotting with the polyclonal antiserum, many more species have been shown to react positively, including rice, sorghum, barnyard grass, cotton and various members of the bean family. In some cases, reactivity has been detected only (or at least far more readily) in the supernatant fractions rather than the membranes. Whereas reactivity is frequently directed at the same size polypeptide as in maize, ie. 22 kDa, in some cases larger polypeptides are seen. The most striking example is in membranes of barnyard grass (Echinocloa crusgalli), where a strongly immunoreactive polypeptide is seen at ca. 24 kDa, rather than 22 kDa (Fig 3). Reaction is still detectable in

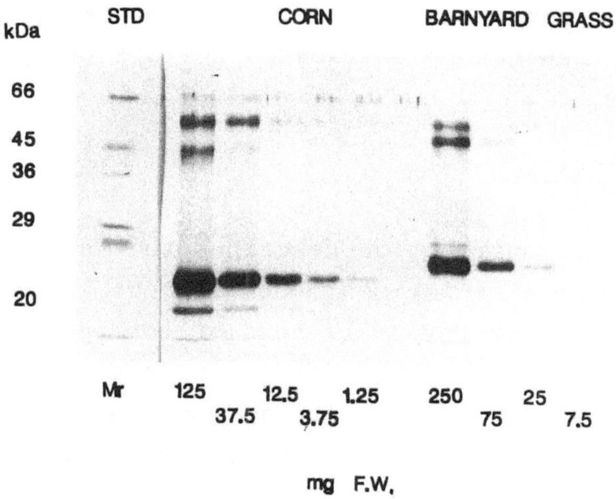

Fig 3. Titration of immunoreactive polypeptides in NP40 extracts of maize and barnyard grass after SDS-PAGE and Western blotting

maize at a loading equivalent to just over 1 mg tissue fresh weight, and in barnyard grass at 7.5 mg (on the original blot). We estimate that the abundance of the 24 kDa polypeptide in barnyard grass is 7-8-fold lower than that of the 22 kDa polypeptide in maize and this corresponds closely to the relative auxin-binding activities we have measured in microsomal preparations of the two species. There is therefore little doubt that the 24 kDa band in barnyard grass represents a receptor polypeptide homologous with the maize 22 kDa polypeptide. Furthermore, this is the most abundant receptor we have detected apart from that in maize.

Walton and Ray (1981), investigating the physiological relevance of auxin binding in maize, showed that binding decreased as sections were taken progressively further down the mesocotyl, in parallel with a reduction in the growth response to added auxin. Red light inhibition of mesocotyl growth also correlated with loss of auxin binding. Alan Jones at Chapel Hill has repeated these experiments and using our polyclonal antiserum has shown that the correlations extend to abundance of the immunologically detectable 22 kDa polypeptide (Jones et al., 1989). This is further confirmation that an auxin-binding role can be attributed to this polypeptide.

Receptor localisation

Prof. Guern's contribution describes the very elegant electrophysiological experiments carried out in his laboratory showing how the auxin responsiveness of tobacco protoplasts can be manipulated by adding maize receptors or deleting endogenous receptors using antibody from Prof. Klämbt's laboratory. Limited experiments by Barbier-Brygoo and Ephritikine at Gif using our polyclonal antibodies have provided further confirmation (Table 1), showing that auxin-induced hyperpolarisation of the membrane potential (Em) is abolished by immune, but not pre-immune IgG fractions.

The electrophysiological data show that a functional population of receptors homologous with those in maize must be accessible at the outer face of the plasma membrane. Further support for this localisation comes from two other types of evidence. First, we find that non-permeant auxin derivatives can elicit physiological responses. We have synthesised and tested three 5-substituted analogs of NAA, and shown that the 5-azido- and 5-nitro derivatives have auxin activity, while 5-amino-NAA does not (Venis and Thomas, 1989). However, after diazotisation and azo-coupling to protein, the latter analog will resemble electronically the azido derivative. When tested on growth of non-abraded pea epicotyl sections, the BSA conjugate of 5-amino-NAA is inactive (presumably because of the cuticular barrier), but shows excellent activity on sections whose cuticle has been partly removed by abrasion (Table 2). On the other hand, coupling to much larger

Table 1. Effect of $IgG_{anti-ABP}$ on the Em variations induced by NAA on tobacco mesophyll protoplasts

Medium	Em (mV ± SE)	NAA-induced EM variation (mV)
Control (C)	-4.6±0.2	-
C + NAA 5 μM	-9.7±0.4	-5.1
C + $IgG_{pre-imm}$ 4.10^{-8}M + NAA 5 μM	-9.4±0.3	-4.8
C + $IgG_{anti-ABP}$ 4.10^{-8}M + NAA 5 μM	-5.6±0.3	-1.0
Control (C)	-5.9±0.3	-
C + NAA 5 μM	-12.2±0.4	-6.3
C + $IgG_{pre-imm}$ 4.10^{-7}M + NAA 5 μM	-10.6±0.4	-4.7
C + $IgG_{anti-ABP}$ 4.10^{-7}M + NAA 5 μM	-5.1±0.2	+0.8

Protoplast suspensions were pre-incubated with the IgG fractions for 2.5 h at room temperature, before the addition of 5 μM NAA. Em values are means (± standard error) of 20-30 individual measurements in each condition. ABP = maize auxin-binding protein.

Table 2. Activity of 5-amino-NAA conjugates on growth of pea sections

Compound	Concentration μM	% of control growth	
		Abraded	Non-abraded
NAA	10	206	223
5-amino-NAA-BSA	100	209	101
"	30	152	
"	10	136	
5-amino-NAA-KLH	100	98	
5-amino-NAA-Sepharose	100	105	
BSA	0.25 mg ml^{-1}	104	

The BSA concentration is equivalent to that present at 100 μM of the 5-amino-NAA conjugate.

molecules (KLH or Sepharose) produces conjugates that are inactive (Table 2). This is entirely consistent with data on cell wall porosity showing that molecules of the size of BSA can penetrate the wall, whereas soybean agglutinin (M_r 120 kDa) cannot (Baron-Epel et al., 1988). If this interpretation is correct, we would predict that when the cell wall is removed, both BSA and KLH conjugates should be active. This is precisely what has been found when these conjugates were tested at Gif on the electrochemical response of tobacco protoplasts.

The other evidence showing that receptors are located at the plasma membrane is simply that maize protoplasts treated with monoclonal or polyclonal antibodies to the receptor fluoresce after incubation with FITC-labelled second antibodies, whereas those treated with non-immune serum do not.

DISCUSSION

We have reported on some of the ways in which our antibodies to the auxin receptor are being used. The monoclonals have enabled us to detect auxin-induced conformational change in the receptor and to localise part of the molecule involved in this change. However, we are currently attempting to generate further mAbs which we hope will include some that are directed at the auxin-binding site itself.

The polyclonal has been particularly useful in revealing homologous polypeptides in other species. In some cases these are of different M_r (eg. barnyard grass) and sometimes immunoreactivity is found mainly in the supernatant fraction rather than in the microsomes. The significance of this 'soluble' receptor population is unclear. We always use pestle and mortar rather than mechanical methods of homogenisation, but since the receptor is a peripheral rather than an integral membrane protein (as indicated by sequence, ability to be solubilised via the acetone procedure, and partial removal by high salt concentrations - unpublished data) it is possible that this provides sufficient shear to remove some of the receptor from the membrane. Comparative amino acid sequence data on soluble and membrane-bound populations from the same species might help to resolve this question.

The electrophysiological data with receptor antibodies already reported (Barbier-Brygoo et al., 1989a, b) and further results presented here by Prof. Guern and ourselves clearly show that a functional receptor population resides on the outer face of the plasma membrane. This is further supported by results with non-permeant auxin analogs and by protoplast immunofluorescence. It has always been clear that the majority of the auxin binding sites (site I) are located on the endoplasmic reticulum, ER (see Venis 1985 for a review of the earlier work), and the C-terminal tetrapeptide of the cloned receptor sequence (Inohara et al., 1989) is a consensus sequence for retention within the ER (Munro and Pelham, 1987). On the basis of immunoblotting of membrane fractions from sucrose density gradients, Shimomura et al. (1988) concluded that site II, located on the plasma membrane and/or tonoplast membrane (Venis, 1985) was immunologically distinct from site I on the ER. We have similar data (Jones et al., 1989), but disagree with the conclusion. It is clear from both sets of data that the plasma membrane regions of the gradient do contain immunoreactive polypeptide, even though reaction is strongest in the ER fractions. Furthermore, all the electrophysiological and immunofluorescence evidence establishes quite definitely that a plasma membrane receptor is recognised by polyclonal antisera that also recognise the ER polypeptide.

It is possible that the plasma membrane receptor is a minor species lacking the terminal tetrapeptide. The minor 21 kDa polypeptide does lack the C-terminus, but since only the 22 kDa band is seen when precautions are taken to minimise proteolysis (Napier et al., 1988), the smaller polypeptide may only be an artefact. Alternatively, the generalisation that polypeptides with a -lys-asp-glu-leu C-terminus are invariably retained in the ER may not be universally true. This is one of the many intriguing questions regarding the structure, operation and signal transduction pathway(s) of the auxin receptor that await resolution.

Acknowledgement: This work was partly supported under the Biotechnology Action Programme of the European Economic Communities.

REFERENCES

Barbier-Brygoo H, Ephritikine G, Klämbt D, Ghislain M, Guern J (1989a) Functional evidence for an auxin receptor at the plasmalemma of tobacco mesophyll protoplasts. Proc. Natl. Acad. Sci. USA, 86:891-895

Barbier-Brygoo H, Guern J, Ephritikine G, Shem WH, Mansel C, Klämbt D (1989b) The sensitivity of plant protoplasts to auxins: Modulation of receptors at the plasmalemma. In: Lamb C, Beachey R (eds) Plant Gene Transfer, UCLA Symposia vol 129. Alan R Liss, New York, (in press)

Baron-Epel O, Gharyal PK, Schindler M (1988) Pectins as mediators of wall porosity in soybean cells. Planta 175:389-395

Dreyer SA, Seymour V, Cleland RE (1981) Low proton conductance of plant cuticles and its relevance to the acid-growth theory. Plant Physiol 68:664-667

Hesse T, Feldwisch J, Balshuseman D, Bauw G, Puype M, Vanderkerckhove J, Löbler M, Klämbt D, Schell J, Palme K. Molecular cloning and structural analysis of a gene from Zea mays (L.) coding for a putative receptor for the plant hormone auxin. EMBO J (in press)

Inohara N, Shimomura S, Fukui T, Futai M (1989) Auxin-binding protein located in the endoplasmic reticulum of maize shoots: Molecular cloning and complete primary structure. Proc. Natl. Acad. Sci. USA, 86:3564-3568

Jones AM, Lamerson P, Venis MA (1989) Comparison of site I auxin binding and a 22 kDa auxin-binding protein in maize. Planta (in press)

Löbler M, Klämbt D (1985) Auxin-binding protein from coleoptile membranes of corn (Zea mays L.). J. Biol. Chem. 260:9848-9853

Löbler M, Simon K, Hesse T, Klämbt D (1987) Auxin receptors in target tissue. In: Fox JE, Jacobs M (eds) Molecular Biology of Plant Growth Control, UCLA Symposia vol 44. Alan R Liss, New York, pp 279-288

Munro S, Pelham HRB (1987) A C-terminal signal prevents secretion of luminal ER proteins. Cell 48:899-907

Napier RM, Venis MA, Bolton MA, Butcher G, Richardson LI (1988) Preparation and characterisation of monoclonal and polyclonal antibodies to the maize membrane auxin-binding protein. Planta 176:519-526

Sarhan F, Cesar D (1988) High yield isolation of mesophyll protoplasts from wheat, barley and rye. Physiol. Plant 72:337-342

Shimomura S, Sotobayashi T, Futai M, Fukui T (1986) Purification and properties of an auxin-binding protein from maize shoot membranes. J. Biochem. 99:1513-1524

Shimomura S, Inohara N, Tukui T, Futai M (1988) Different properties of two types of auxin-binding sites in membranes from maize coleoptiles. Planta 175:558-566

Tillman U, Viola G, Kayser B, Siemeister G, Hess T, Palme K, Löbler M, Klämbt D. cDNA clones of the auxin binding protein from corn coleoptiles (Zea mays L.): Isolation and characterization by immunological methods. EMBO J. (in press)

Venis MA (1985) Hormone Binding Sites in Plants. Longman, London

Venis MA, Napier RM (1989) Membrane-bound auxin receptors. In: Pharis RP, (ed.) Plant Growth Substances. Springer-Verlag (in press)

Venis MA, Thomas EW (1989) Synthesis and auxin activity of 5-substituted 1-naphthaleneacetic acids. Phytochemistry (in press)

Walton JD, Ray PM (1981) Evidence for receptor function of auxin binding sites in maize. Plant Physiol 68:1334-1338

ETHYLENE, FIRST OR SECOND MESSENGER?

M.A. HALL, M.H. Bell, C.P.K. Connern, N.V.J. Harpham, I. Raskin[1], I.O. Sanders, A.R. Smith and C.K. Wood
Department of Biological Sciences
University College of Wales
Aberystwyth
Dyfed SY23 3DA
Wales, U.K.

An answer to the question posed by the title of this article clearly depends on the definition adopted for both alternatives. Like the term 'hormone' the concept of a second messenger derives from endocrinology, being an intracellular signal whose formation is controlled by the interaction of a hormone with its receptor and which in turn is involved in the transduction mechanism of that hormone via interaction with another intracellular component. Since the precise applicability of the term hormone has itself been called into question in the case of the endogenous plant growth regulators (see e.g. Wareing, 1977) it would merely be an exercise in semantics to attempt to assess whether ethylene is best classified as such or given some other description. It is nevertheless understandable that, partly for historical reasons and partly because it is a gas, ethylene is seen to be different in kind to the other endogenous growth regulators and it is legitimate to raise the question as to whether that difference is real and fundamental or merely apparent, and if it is real whether some appellation needs to be coined to reflect that difference.

This article will address the matter by considering aspects of biosynthesis, metabolism and mechanism of action, particularly in relation to perception.

Biosynthesis

Probably the primary reason for even considering ethylene in a different light and perhaps as a second messenger goes back to the pioneering work of the Burgs who demonstrated that IAA

[1] I. Raskin, Analytical Laboratories, E I Du Pont de Nemours & Co., Wilmington, Delaware 19898, U.S.A.

NATO ASI Series, Vol. H 47
Signal Perception and Transduction in Higher Plants
Edited by R. Ranjeva and A. M. Boudet

could promote ethylene biosynthesis and that the ethylene so formed was at least partly responsible for the 'supraoptimal' effects of the auxin in terms of inhibition of elongation and isodiametric cell expansion (Burg & Burg, 1967). This effect has been confirmed many times since and, when the biosynthetic pathway of ethylene came to be established the metabolic basis proved to be an effect of auxin on the synthesis of the key regulatory enzyme ACC synthase (Yoshii & Imaseki, 1982) which shows a high turnover rate (Bleecker et al., 1988). Even at this point however oversimplification may be dangerous since immunological studies have indicated fundamental differences between the primary structures of auxin-induced ACC synthase and the enzyme induced by wounding (Nakagawa et al., 1988).

It is clear nevertheless that auxin as a controlling factor in ethylene biosynthesis does have profound effects upon growth and development. The work by Burg & Burg (1968) does not provide convincing evidence in this regard since the concentrations of auxin applied to intact tissue to induce enhanced ethylene biosynthesis must be regarded as unphysiological. Nevertheless, recent work by Imaseki et al. (1988) shows that application of IAA to isolated protoplasts of mung bean results in increased ethylene formation and the K_m for IAA is about 5×10^{-6}M, close to the value for the K_D of auxin binding sites (Venis, 1985) (see below). Similarly, we have found that although auxin accelerates ethylene biosynthesis in ethylene-insensitive mutants of Arabidopsis the supraoptimal auxin effect (isodiametric cell expansion) does not occur. Equally, in the tomato mutant dgt which does not show auxin-induced ethylene biosynthesis even though it contains as much IAA as the normal parent (Fujino et al., 1988), treatment with very low concentrations of ethylene restores the normal phenotype (Zobel, 1973). Recent work (Roberts et al., 1987) indicates that dgt is somehow deficient in auxin receptor(s).

On the other hand there are many instances where the control of ethylene biosynthesis by auxin appears to be absent or at best indirect in terms of the control of growth and development. As indicated above, the ACC-synthase induced in response to wounding is different to that produced in response to auxin, suggesting the existence of two separate genes controlled in different ways (Nakagawa et al., 1988). Moreover, the increase in ethylene biosynthesis in response to mechanical impedance which is very rapid and which has marked developmental effects (Goeschl et al., 1966) seems more akin to the wounding mechanism. Similarly, it appears unlikely that the phytochrome-controlled inhibition of ethylene biosynthesis by light which has dramatic and rapid effects on epicotyl hook opening is mediated by auxin (Kang & Ray, 1969). The extensive literature on the behaviour of hydrophytes and on the responses of some mesophytes to flooding indicates a more direct role of environmental factors mediated via ethylene. Thus, in deep-water rice, ethylene is trapped in submerged tissue because its diffusion from the plant into water is 10^4 times slower than from the plant to air (Metraux & Kende, 1983) and the low partial pressures of O_2 which eventuate increase the activity of ACC synthase (Cohen & Kende, 1987). The accumulation of ethylene triggers rapid elongation of the submerged internodes. Similar suggestions have been made for growth in Callitriche (Musgrave et al., 1972) and for ethylene-induced lysogenous lacuna formation in Zea (Drew et al., 1979).

There is a further aspect of the control of ethylene biosynthesis which seems to be unique to this growth regulator namely, its capacity to effect autocatalysis or autoinhibition. This process does not however appear to be of the classical feedback variety and two reports indicate that the effects are receptor-mediated. Thus, in tobacco, where ethylene inhibits its own synthesis, Ag(I) prevents autoinhibition, possibly via an effect on auxin metabolism (Aharoni, 1985) (Fig. 1); moreover Ag(I) appears to act, at least in part, by interacting with ethylene receptors (see Table 1). Equally, in carnations, where ethylene has an autocatalytic effect upon its own synthesis, norbornadiene, a specific inhibitor of ethylene responses (Sisler & Pian, 1973), abolishes the effect, apparently via a reduction in both ACC synthase and ethylene forming enzyme (Wang & Woodson, 1989) (Fig. 2). Again, norbornadiene appears to act at the receptor level (see Table 1).

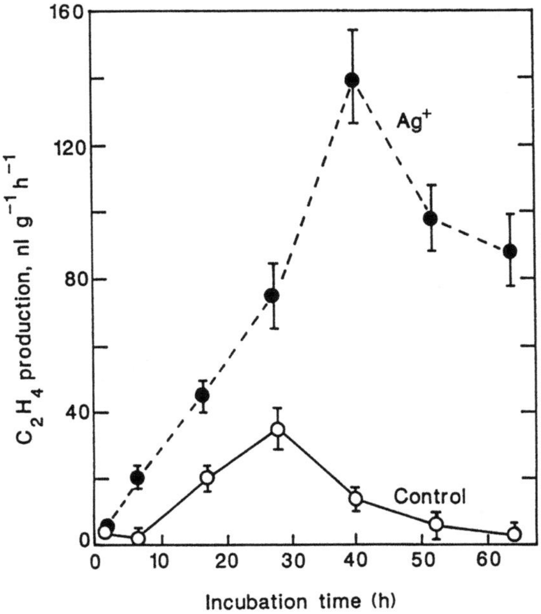

Fig. 1. Effect of Ag(I) upon auxin-induced ethylene production in *Nicotiana rustica* leaf disks. (After Aharoni, N. Physiol. Plant. 63:438-444, 1985)

Metabolism

Ethylene metabolism has also figured largely in considerations of ethylene physiology and biochemistry. In most plants investigated so far ethylene is metabolised at very low rates at physiological concentrations (for a review see Smith & Hall, 1984). Nevertheless, Beyer observed that in a number of different developmental systems the rate of ethylene metabolism was correlated with changes in sensitivity to the gas (see Beyer et al., 1984). From this arose the hypothesis that ethylene metabolising systems functioned as receptors (e.g. Sanders et al., 1986). This would certainly have set ethylene apart from the other plant growth regulators and indeed from animal hormones. However, recent work (Sanders et al., 1989 a,b) has demonstrated that not only is it unlikely that the enzyme(s) responsible for ethylene metabolism in most plants are specific for the growth

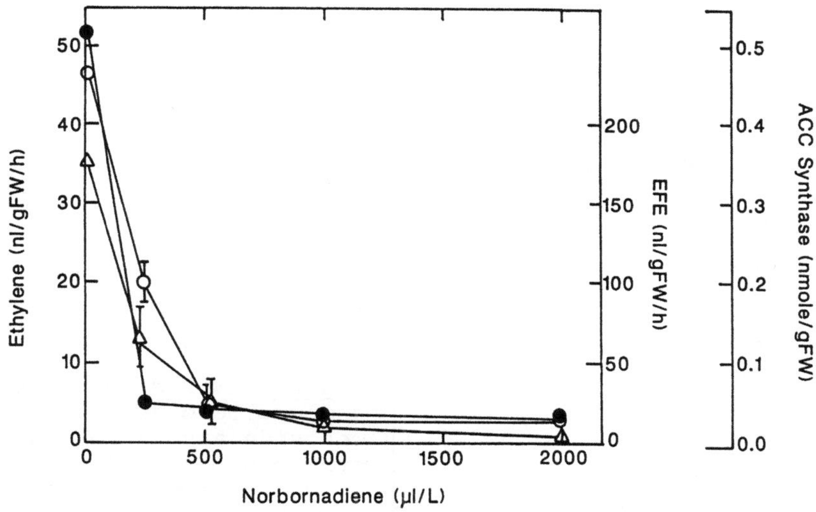

Fig. 2. Effect of norbornadiene upon autocatalytic ethylene production, ethylene forming enzyme (EFE) and ACC synthase in carnation petals. O, ethylene production; ●, ACC synthase activity; Δ, EFE activity. (Redrawn from Wang, H and Woodson, WR. Plant Physiol. 89:434-438, 1989)

regulator (indeed they show higher affinities for longer chain alkenes and can metabolise alkanes (Fig. 3)) but, in addition, inhibition of metabolism by CS_2 has no effect on the response of tissue to ethylene. Moreover, as discussed below, there is good reason to suppose that the ethylene binding sites now detected in a range of tissues qualify as ethylene receptors in the classical sense.

Ethylene responsive cells

Another feature of many ethylene effects is that they are interdependent on other growth regulators. Osborne et al. (1984) have defined three types of cell which respond to ethylene. Type I cells, such as those in epicotyls of peas, respond simultaneously to auxin and to ethylene, resulting in a cell shape different to that obtained in the presence of only one of the growth regulators. Type II cells such as those in abscission zones show en-

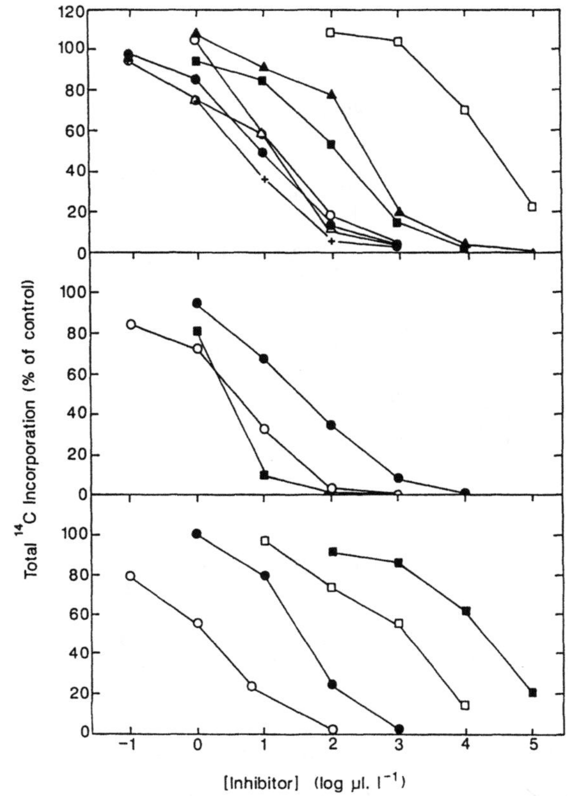

Fig. 3. Effect of aliphatic and cyclic olefins, carbon monoxide and alkanes upon ethylene metabolism by pea epicotyl tips. a) ▲, ethylene; Δ, propylene; +, butene; ●, methyl acetylene; □, carbon monoxide; O, vinyl chloride; ■, acetylene; b) ●, 2,5-norbornadiene; ■, cyclopentene; O, cis-2-butene; c) ■, ethane; □, propane; ●, pentane; O, trans-2-butene (From Sanders et al. Planta 1989, in press)

hanced expansion in the presence of ethylene and suppression of the process by auxins. Type III cells, found in the shoots of water plants, show an additive response to auxin and ethylene; a similar type of cell appears to exist in Callitriche but in this case the cooperating growth regulator is a gibberellin (Musgrave et al., 1972). The existence of these cell types suggests a more complex interaction between auxin and ethylene than that suggested by a simple primary/secondary messenger relationship.

The bases of these effects at the molecular level is unclear but some of our recent work and that in other laboratories is beginning to suggest some possible answers.

Receptors

Developing cotyledons of Phaseolus vulgaris contain a binding site whose affinities for ethylene and its structural analogues closely resemble those predicted from established developmental responses (Table 1). However, there is no known developmental response to ethylene in this tissue and moreover the rate constants of association and dissociation are very low whereas many ethylene effects show short response times. However, the relatively high abundance of the binding site (30-50 pmoles g^{-1}) and the ability to monitor purification via assessment of prebound ethylene render this a suitable system for characterisation. Equally, given the very high affinity of the site for ethylene (K_D 0.88 x 10^{-10}M at infinite dilution) and its analogues and the simplicity of the ethylene molecule it seems reasonable to assume that the binding domain itself will show strong homologies with other ethylene binding domains in the same or different species - even those with high rate constants of association and dissociation.

We have now purified this protein to homogeneity and its structure is currently being investigated. We have also raised polyclonal antibodies specific to the protein and we are developing an ELISA. These antibodies recognise homologous proteins in peas, *Arabidopsis* and tomato, although we have not yet established whether these are binding sites.

We have also developed methods for the rigorous quantitation of ethylene binding sites *in vivo* - even those with high rate constants of association and dissociation (Sanders et al., 1989b). This is possible with ethylene as distinct from other growth regulators for two reasons, firstly, because ethylene equilibrates with, and diffuses from, plant tissue within seconds; moreover its concentration in the cytosol can be accurately assessed by a knowledge of the gas phase concentration and the partition coefficient at a particular temperature. The second reason, mentioned above, is that ethylene metabolism is negligible at physiological concentrations of the gas and can in any case be wholly and specifically inhibited with CS_2 without affecting binding or developmental response.

Using these techniques we have measured ethylene binding in several species and tissues including pea and *Arabidopsis* epicotyls, rice coleoptiles and *Phaseolus* petioles and abscission zones. In all cases a common feature emerges. There appear to exist two classes of binding sites possessing high and appropriate affinities for ethylene and its analogues and responding in the expected fashion to the specific inhibitors of ethylene action Ag(I) and norbornadiene (Table 1). The only significant difference between the two classes is expressed in the rate constants of association and dissociation. One class of sites, which we have investigated in detail in pea epicotyls shows very high rate constants of association and dissociation whereas the other has low rate constants similar to those found with the binding site from Phaseolus cotyledons. In ethylene-insensitive mutants of Arabidopsis the concentration of binding sites is much lower than in the wild type. Such observations taken together with the kinetic properties of the sites make it difficult to conceive of any role for them other than as receptors.

The only qualification to this statement is that whereas the affinity of the sites is in very good agreement with the K_m for the three developmental processes involved (hook closure, inhibition of elongation and isodiametric cell expansion, (see Fig. 4)) it is difficult to see how the class with the low rate constants could be involved in two of these which have very short response times (hook closure and inhibition of elongation) unless, of course, the developmental effect is not related in a linear fashion to receptor occupancy as has been shown in a number of cases with animal hormone receptors. The dose response relationships do not however tend to support such a scenario (Fig. 4). They do on the other hand tend to support our contention that in vivo measurement of binding sites is a real reflection of the situation in the tissue. Alternatively the sites with low rate constants may represent 'down regulated' high rate constant sites or precursors of the same (see e.g. King & Cuatrecasas, 1981). Certainly the proportions of the two sites in different tissues are remarkably constant (Table 2). Moreover, such a scenario would explain the presence of the high concentrations of low rate constant sites in developing cotyledons of Phaseolus where no developmental response to ethylene is known.

Using both immunoassay and direct methods we have looked at these putative receptors in various developmental systems. In the pea epicotyl although there are marked differences in binding site concentration in different regions, with the highest values

Table 1. Properties of ethylene binding sites from various sources

	Pea epicotyl			Phaseolus		
	Fast	Slow	Growth Inhibition	Developing cotyledons[1]	petiole/ abscission zones (both sites)	Arabidopsis (both sites)
$K_D(K_m)$ Ethylene (M)	1.12×10^{-10}	1.97×10^{-10}	1.8×10^{-10}	0.88×10^{-10}	1.92×10^{-10}	3.61×10^{-10}
$K_i(K_m)$ propylene (M)	4.78×10^{-8}	1.92×10^{-7}	1.92×10^{-8}	1.49×10^{-8}	1.33×10^{-7}	
$K_i(K_m)$ carbon monoxide (M)	3.2×10^{-8}		3.6×10^{-8}	2.58×10^{-7}		
$K_i(K_m)$ vinyl chloride (M)	7.6×10^{-7}		1.45×10^{-7}	4.12×10^{-7}		
$K_i(K_m)$ acetylene (M)	3.7×10^{-7}		3.3×10^{-7}	1.02×10^{-6}		
$K_i(K_m)$ propyne (M)	1.37×10^{-6}		1.14×10^{-6}	4.23×10^{-6}		
$K_i(K_m)$ 1-butene (M)	4.66×10^{-5}		5.04×10^{-5}	6.34×10^{-5}		
K_i norbornadiene (M)	1.18×10^{-6}		1.76×10^{-6}			
K_i Ag(I) (M)	3.1×10^{-7}		6.00×10^{-7}			
k_1 $(M^{-1}s^{-1})$	10^{6}			2.97×10^{3}		
k_{-1} (s-1)	10^{-4}			1.4×10^{-6}		
k_{-1}/k_1 (M)	10^{-10}			4.7×10^{-10}		
					Fast / Slow	
$t_{1/2}$ assoc.	≃ 8 min	≃13 h	≃20 min[2]	≃13 h	≃17 min / ≃12 h	
$t_{1/2}$ dissoc.	≃32 min	≃ 22 h	≃15 min[3]	≃20 h		

All molarities quoted are those in the liquid phase

1. Calculated at infinite dilution
2. Lag time for observable response after exposure
3. Lag time for observable response after removal

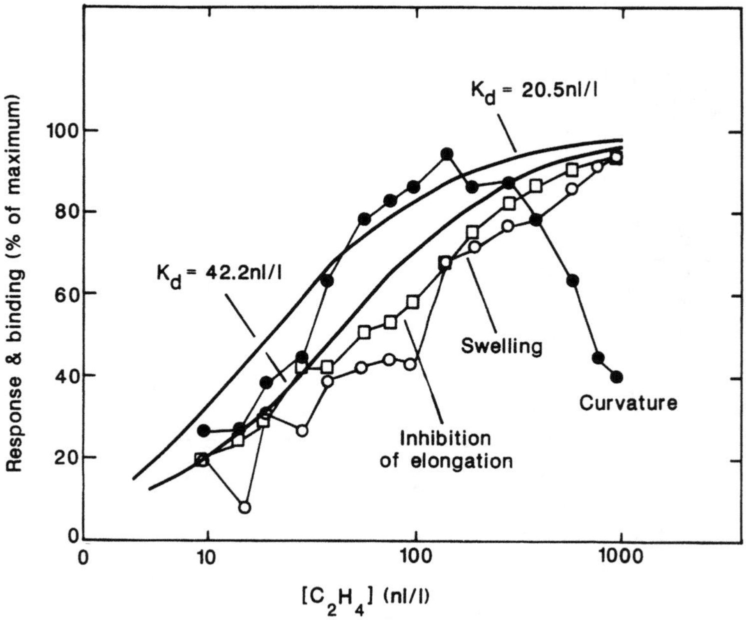

Fig. 4. Comparison of ethylene-induced developmental responses in pea epicotyls and binding site occupancy. The two solid lines represent theoretical binding site occupancy for the two classes of site present. (K_D = 20.5 nl l^{-1} fast associating site; K_D = 42.2 nl l^{-1} slow associating site)

Table 2. Properties of fast and slow associating ethylene binding sites in different plants. R_1 = fast associating site; R_2 = slow associating site

		Bn (%)	
		R_1	R_2
Pea epicotyl	Tip	23	77
	2nd internode	22	78
	1st internode	24	76
Phaseolus	petiole	19	81
	abscission zone/pulvinar	29	71
Arabidopsis	epicotyl	27	73

in the hook, on a per cell basis the number of sites of both kinds is remarkably constant whatever region is considered (Table 3). It should be borne in mind however that in this

Table 3. Ethylene binding activity in pea epicotyls. R_1 = fast associating site; R_2 = slow associating site

	R_1		R_2	
Zone	Concentration (pmoles g^{-1})	Molecules binding protein $cell^{-1}$ ($x\ 10^{-5}$)	Concentration (pmoles g^{-1})	Molecules binding protein $cell^{-1}$ ($x\ 10^{-5}$)
Tip	2.36	1.48	7.9	4.95
2nd internode	0.43	1.41	1.51	4.95
1st internode	0.33	1.33	1.07	4.31

tissue at this developmental stage even the first internode responds to ethylene, albeit not as dramatically as the hook region.

Abscission zones are well-known as ethylene-responsive tissues (Type II cells as defined by Osborne et al. (1984)) and we have looked at these in Phaseolus vulgaris. It is well established in this system that deblading and/or treatment with ethylene accelerates abscission whereas deblading and application of auxin to the end of the petiole delays the process. In the pulvinar/abscission zone area the concentration of binding sites (of both kinds) is higher than in petiolar tissue (approx. 0.9 and 0.4 pmoles g^{-1} respectively). Our preliminary conclusions using antibodies and immunogold dot-blots confirm the difference between the two tissues and further, indicate that deblading leads to an increase in binding site concentration in the abscission zone/pulvinar regions and, treatment with chlorethanephosphonic acid (Ethephon) leads to a further increase. That the situation is more complex even than that indicated above is confirmed by the observation that auxin and ethylene together can induce secondary abscission zones in Phaseolus petioles (Osborne, 1987). Such abscission zones possess the same competence in terms of response as abscission zones which develop during normal ontogeny.

Still on the subject of receptors it is worth noting that while there is no documented effect of ethylene on auxin receptors, a recent paper (Suttle, 1988) does show a pronounced inhibitory effect upon the concentration of phytotropin receptors (presumed auxin efflux carriers) with an I_{50} for ethylene of about 2.42×10^{-10}M (cf. Table 1) which is a plausible explanation for the observed effects of ethylene on auxin transport. It is interesting to note in relation to the class of sites with a low rate constant of association that this response has a lag time of ca. 8h.

Conclusions

It is clear from the foregoing that ethylene does exhibit some facets which set it apart from the other groups of endogenous plant growth regulators, at least as far as our present knowledge is concerned; equally clearly it does not correspond to the second messenger concept in animals - or indeed in plants. It is not intracellular, indeed, because of its gaseous nature, once produced it will affect more cells more rapidly than any other growth regulator synthesised in a defined locus. In this connection and confirming the foregoing it has been shown recently (Imaseki et al., 1988) that auxin-induced ethylene production in mung bean occurs in the cortex and pith whereas auxin acts primarily on epidermal cells in stimulating elongation growth - a process inhibited by ethylene - although the effect on isodiametric cell expansion is evidenced in cortical cells. Indeed, epidermal cells appear to lack ethylene-forming enzyme (Todaka & Imaseki, 1985). Equally, the fact that ethylene biosynthesis can be controlled by another growth regulator, IAA, is insufficient grounds for a distinction to be made since in animals the synthesis of many hormones is controlled by others. For example steroidigenesis is controlled both by FSH and LH in ovarian granulosa cells. There are apparent parallels with the systems we have examined in the ovarian granulosa system where 'up' or 'down' regulation of receptors may be effected either by the hormone to which they respond or by another (see e.g. Adashi et al., 1984).

The various features outlined above are summarised in Fig. 5. What this in effect reveals is a complex relationship between auxin and ethylene involving the control of biosynthesis, transport and perception. Moreover, comparisons with animal systems suggest that the scenario outlined is not likely to be in any

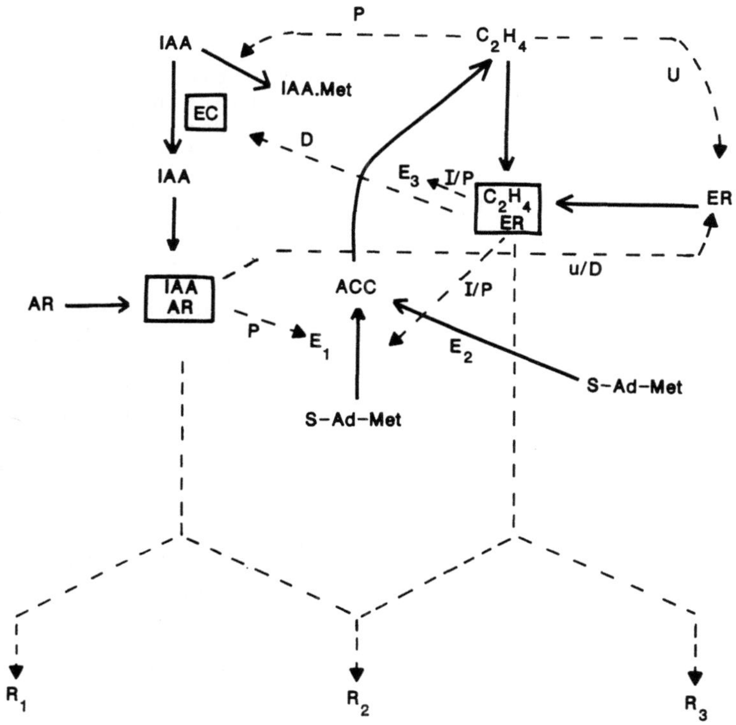

Fig. 5. Interrelationships between ethylene & auxin. AR, auxin receptor; ER, ethylene receptor; EC, auxin efflux carrier; S-Ad-Met, S-adenosyl methionine; ACC, aminocyclopropane carboxylic acid; IAA-met, auxin metabolites; E_1, auxin-induced ACC-synthase; E_2, wound-induced ACC synthase; E_3, ethylene forming enzyme; I, inhibition; P, promotion; U, up-regulation; D, down-regulation. R_1, R_2, R_3, developmental responses

sense unique and merely reflects the complexity of the control processes themselves. Indeed, it would not be incautious to predict that, as more becomes known about transduction systems, synergistic and antagonistic effects will be observed here also, as in interactions between animal hormones.

In conclusion therefore, while ethylene has its peculiarities - as indeed do the other growth regulators - nothing we know to date suggests that it should be considered and assessed in a different light.

References

Adashi EY, Hsueh AJW (1984) Hormonal induction of receptors during ovarian granulosa cell differentiation. In: Conn PM (ed) The Receptors. Academic Press, London, pp 588-636

Aharoni N (1985) Effect of silver ions and ethylene on auxin metabolism and auxin-induced ethylene production in tobacco leaf discs. Physiol. Plant 63:438-444

Beyer EM Jr, Morgan PW, Yang SF (1984) Ethylene. In: Wilkins M (ed) Advanced Plant Physiology. Pitman Press, London, pp 111-126

Bleecker AB, Robinson G, Kende H (1988) Studies on the regulation of 1-aminocyclopropane-1-carboxylate synthase in tomato using monoclonal antibodies. Planta 173:385-390

Burg SP, Burg EA (1967) Auxin stimulated ethylene formation; its relationship to auxin inhibited growth, root geotropism and other plant processes. In: Wightman F, Setterfield G (eds) Biochemistry & Physiology of Plant Growth Substances. Runge Press, Canada, pp 1275-1294

Burg SP, Burg EA (1968) Ethylene formation in pea seedlings; its relation to the inhibition of bud growth caused by indole-3-acetic acid. Plant Physiol. 43:1069-1074

Cohen E, Kende H (1987) In vivo 1-aminocyclopropane-1-carboxylate synthase activity in internodes of deepwater rice. Plant Physiol. 84:282-286

Drew MC, Jackson MB, Giffard S (1979) Ethylene-promoted adventitious rooting and development of cortical air spaces (aerenchyma) may be adaptive responses to flooding in Zea mays L. Planta 147:83-88

Fujino DW, Nissen SJ, Jones AD, Burger DW, Bradford KJ (1988) Quantification of indole-3-acetic acid in dark-grown seedlings of the Diageotropica and Epinastic mutants of tomato (Lycopersicon esculentum Mill.). Plant Physiol 88:780-784

Goeschl JD, Rappaport L, Pratt HK (1966) Ethylene as a factor regulating the growth of pea epicotyls subjected to physiological stress. Plant Physiol. 41:877-884

Imaseki H, Nakajima N, Todaka I (1988) Biosynthesis of ethylene and its regulation in plants. In: Steffens GL, Rumsey TS (eds) Beltsville Symposia in Agricultural Research 12: Biomechanisms Regulating Growth and Development. Kluwer Academic Publishers, Boston, pp 205-227

Kang BG, Ray PM (1969) Role of growth regulators in the bean hypocotyl hook opening response. Planta 87:193-205

King AC, Cuatrecasas P (1981) Peptide hormone-induced receptor mobility, aggregation and internalization. N. Engl. J. Med. 305:77-88

Metraux J-P, Kende H (1983) The cellular basis of the elongation response in submerged deep-water rice. Planta 160:73-77

Musgrave A, Jackson MB, Ling E (1972) Callitriche stem elongation is controlled by ethylene and gibberellin. Nature New Biol. 238:93-96

Nakagawa N, Nakajima N, Imaseki H (1988) Immunochemical differences of wound-induced 1-aminocyclopropane-1-carboxylic synthase from the auxin-induced enzyme. Plant Cell Physiol. 29:1255-1259

Osborne DJ, McManus MT, Webb J (1984) Target cells for ethylene action. In: Roberts JA, Tucker GA (eds) Ethylene and Plant Development. Butterworths, London, pp 197-212

Osborne DJ (1987) Physiological considerations in developmental studies: Hormones and target cells. In: Hoad GV, Lenton JR, Jackson MB, Atkin RK (eds) Hormone Action in Plant Development. Butterworths, London, pp 265-274

Roberts JA, Grierson D, Tucker GA (1987) Genetic variants as aids to examine the significance of ethylene in development. In: Hoad GV, Lenton JR, Jackson MB, Atkin RK (eds) Hormone Action in Plant Development. Butterworths, London, pp 107-118

Sanders IO, Smith AR, Hall MA (1986) Ethylene metabolism and action. Physiol Plant 66:723-726

Sanders IO, Smith AR, Hall MA (1989a) The measurement of ethylene binding and metabolism in plant tissue. Planta: in press

Sanders IO, Smith AR, Hall MA (1989b) Ethylene metabolism in *Pisum sativum* L. Planta: in press

Sisler EC, Pian A (1973) Effect of ethylene and cyclic olefins on tobacco leaves. Tobacco Sci 17:68

Smith AR, Hall MA (1984) Mechanisms of ethylene action. Plant Growth Regul. 2:151-165

Suttle JC (1988) Effect of ethylene treatment on polar IAA transport, net IAA uptake and specific binding of *N*-1-Naphthyl phthalamic acid in tissues and microsomes isolated from etiolated pea epicotyls. Plant Physiol. 88:795-799

Todaka I, Imaseki H (1985) Epidermal cells do not contribute to auxin-induced ethylene production in mung bean stem sections. Plant Cell Physiol. 26:865-871

Venis MA (1985) Hormone Binding Sites in Plants. Longman, London

Wang H, Woodson WR (1989) Reversible inhibition of ethylene action and interruption of petal senescence in carnation flowers by norbornadiene. Plant Physiol. 89:434-438

Wareing PF (1977) Growth substances and integration in the whole plant. In: Society for Experimental Botany Symposium No. 31: Integration of Activity in the Higher Plant, pp 337-365

Yoshii H, Imaseki H (1982) Regulation of auxin-induced ethylene biosynthesis. Repression of inductive formation of 1-aminocyclopropane-1-carboxylate synthase by ethylene. Plant Cell Physiol. 23:639-649

Zobel RW (1973) Some physiological characteristics of the ethylene-requiring tomato mutant *Diageotropica*. Plant Physiol 52:385-389

FUSICOCCIN RECEPTORS IN PLANTS - HOW MUCH DO WE KNOW FOR SURE ?

C.Meyer, R.Adams, E.Oelgemöller, A.Hopstock,
C.Oecking, A.Sprenger and E.W.Weiler

Dept.of Plant Physiology, Ruhr-University,
POB 102148, D 4630 Bochum, F.R.G.

INTRODUCTION

The wilt-inducing toxin of *Fusicoccum amygdali*, fusicoccin (FC) has long been used as a tool in plant physiology (Marré 1979). FC instantaneously hyperpolarizes the membrane potential and produces a strong acidification of the apoplastic space. As a result of toxin application, the transport of many metabolites and ions across the plasmalemma may change and, in accord with the acid-growth hypothesis, cell elongation may also be stimulated.

The mechanism of action of FC is still obscure, but is the subject of recently intensified research. FC may serve as a model case exemplifying how plant cells perceive a chemical stimulus and couple this to certain response elements in the cell resulting in a specific modification of cell metabolism. Thus, FC acts quite unlike the classical plant hormones which elicit multiple effects on both, cell metabolism as well as gene activity. On the other hand, the process of proton extrusion appears to be stimulated by FC in very many (and probably most, if not all) plants from the liverworts to the angiosperms, although only peach and almond are hosts to *Fusicoccum amygdali*. Under the assumption that highly conserved elements usually serve important functions, it might be surmised that the target system of FC action in the plant cell plays indeed a crucial role in cell metabolism.

NATO ASI Series, Vol. H 47
Signal Perception and Transduction in Higher Plants
Edited by R. Ranjeva and A. M. Boudet

ALTERNATIVE CONCEPTS OF FUSICOCCIN ACTION

It is a well established view that lipophilic esters, which readily penetrate the cell membranes and which are cleaved by intracellular esterases, will acidify the cytoplasm and thus stimulate the plasmalemma H^+-pumping ATPase (Hager et al.1986), the enzyme which is responsible for probably most of the proton motive force generated at this cell barrier. This mechanism can be excluded, however, for FC. Although FC is a diacetate, its metabolism in plant tissue appears to be slow (Ballio et al. 1977) and its diffusion into the cell is a slow process, as well. Using a biologically highly active, tritiated derivative of FC, (^{3}H)-9'-nor-fusicoccin-8'-alcohol (FCol), we were recently able to show that cell suspension cultures of *Corydalis sempervirens*, although responding to the compound by increased proton extrusion, took up very little (^{3}H)-FCol. Most of the radioactivity associated with the cells was found in the cell wall and microsomal membrane fraction (M.Schröder and E.W.Weiler, unpublished results). No metabolites of (^{3}H)-FCol were detected even when incubations were carried out up to 20 h. In contrast, fluorescein diacetate (FDA) was readily taken up into the cells and hydrolyzed to fluorescein, When fed at 50 - 100 nM level, FDA was ineffective in stimulating proton extrusion although it was completely hydrolyzed intracellularly. Fed at the same level, FCol was highly active biologically, but was not hydrolyzed at all. When a crude, soluble esterase preparation was used as enzyme source, FDA was hydrolyzed completely within 15 - 20 min whereas (^{3}H)-FCol was only partly hydrolyzed (5 % in ca. 50 min), to the monodeacetyl-FCol, but not to the dideacetyl-FCol. These results definitely exclude an ester hydrolysis mechanism for FC action in this tissue (M.Schröder and E.W.Weiler, unpublished results).

As early as 1977, Dohrmann et al.(1977) have demonstrated the presence, in microsomal preparations from coleoptiles, of FC-binding sites. This knowledge has since been extended considerably. FC-binding takes place at the plasmalemma of plant cells (e.g.Feyerabend and Weiler 1988) and involves a specific interaction with a high-affinity binding protein. This binding protein (FCBP) has been characterized in most detail in plasmamembranes from oat roots (e.g. Stout and Cleland 1980, de Boer et al.1989) broadbean leaves (Feyerabend and Weiler 1988, 1989) and *Arabidopsis thaliana* shoots (Meyer et al.1989). Soluble FC-binding activity was so far never detected and there is likewise no evidence for any compartment other than the plasmamembrane which might harbor the FCBP. As either an intrinsic or extrinsic-apoplastic plasmamembraneprotein, the FCBP should be synthesized at the rough ER and it might be modified by e.g. glycosylation, before reaching its target compartment. However, no data on FC-binding in ER or Golgi preparations are available although such studies are clearly needed. Most, if not all plant tissues express the FCBP, at least in *Vicia faba* (K.Waldkötter and E.W.Weiler, in preparation) and a recent extensive survey of the FCBP in plant tissue cultures including callus and suspension cultures revealed its widerspread (and, perhaps, ubiquitous ?) presence in the undifferentiated plant cell (E.Oelgemöller and E.W.Weiler, in preparation). Taxonomically, FC-binding sites can be shown from the *Marchantiales* to the monocotyledonous angiosperms (K.Waldkötter and E.W.Weiler, in preparation). A clear-cut demonstration of FC-binding sites in gymnosperms is still lacking because of technical problems with membrane preparations from these species.

PROPERTIES OF THE FC-BINDING PROTEIN

The FCBP appears to be a basic element in the plasmamembrane of land plants. In accordance with this, the best studied FCBP preparations of *Avena* (de Boer et al. 1989), *Vicia* (Feyerabend and Weiler 1988, 1989) and *Arabidopsis* (Meyer et al.1989) reveal a striking and close similarity of properties. The details need not be discussed here. Some characteristics are shown in Table 1.

Table 1. Properties of the FCBP from *V.faba* and *A.thaliana*

Parameter	*Vicia faba*	*Arabidopsis thaliana*	
	Plasmalemma	Plasmalemma	Solubilized FCBP
pH-optimum	5.5 - 6.0	6.0	8.0
Temperature optimum	25 - 30°C	20 - 25°C	30°C
t/2,association (25°C)	6 min	12 min	10 min
t/2,dissociation	120 min (25°C)	180 min (25°C)	50 min (30°C)
K_d, app. (FCol)	5 nM	1.4 nM	1.8 nM

The FC-binding site is accessible from the apoplastic space and this topology is in agreement with the pH-optimum for FC-binding. Conditions have been worked out for a highly functional solubilisation of the FCBP from *Arabidopsis* (A.Hopstock and E.W. Weiler, in preparation, cf. Table 1) as a prerequisite of reconstitution studies with the purified toxin-binding protein. The solubilized FCBP was obtained, using different techniques and tissues, as protein-detergent-lipid micelle of 60-100 kDa apparent mol . mass (Table 2). Further purification using FPLC of radioligand-FCBP complexes (Feyerabend and Weiler 1989) or affinity chromatography (de Boer et al.1989) has, for two different systems, namely oat root and broadbean leaf, resulted in the identification of two polypeptides of approx. 30 and 31-32 kDa apparent mol.mass which appear correlated with FC-binding activity in the two species. Such identification, by correlation, requires verification by another, independent technique. We have developed a photolabile FC-derivative labeled to high specific radioactivities with (^{3}H), namely 9'-nor-8'-((3,5-(^{3}H)-4-azido-benzoyl)ethylenediamine)-fusicoccin (Feyerabend and Weiler 1989). Using this photoaffinity label, a 31-32 kDa polypeptide was specifically and strongly labeled in plasmalemma preparations from *Arabidopsis* (Meyer et al.1989)

Table 2. Polypeptides associated with the FCBP. Values in brackets represent minor bands, underlined values the main band

Tissue	Method	Apparent mol. mass (kDa)	Ref.
Maize coleoptiles	Perchlorate solub., GPC	80	Pesci et al.(1979)
Oat roots	Triton X-100 solub., GPC	60-100	Stout and Cleland (1980)
Broadbean leaves	Mega-9 solubilized, GPC	80±20	Feyerabend and Weiler (1989)
Oat roots	Affinity-purified, SDS-PAGE	31.0,(29.7)	De Boer et al. (1989)
Broadbean leaves	FPLC-purified, SDS-PAGE	31.3,(29.5)	Feyerabend and Weiler (1989)
A.thaliana leaves	Photoaffinity-labeled, SDS-PAGE	(34.5),33.1 <u>31.6</u>,(30.7)	Meyer et al.1989, Weiler et al.1989

(see Table 2) and *Vicia* (data not shown). Refined purification protocols for the (^{3}H)-FCol-FCBP complexes strengthen the view that the 31-32 kDa band, and not the ~ 30 kDa band, represents the FC-binding polypeptide (C.Oecking and E.W.Weiler, in preparation). On the other hand, *in situ* cross-linking experiments with hexamethylenediisocyanate followed by photoaffinity labeling provide strong evidence that the FCBP is part of a supramolecular structure of an apparent mol. mass of over 400 kDa (A.Sprenger and E.W.Weiler, unpublished observation), part of which should be the 60 - 100 kDa core complex obtained after solubilization. This detergent extracted complex contains the 30 and 31-32 kDa polypeptides, as de Boer et al.(1989) propose, in a 1 : 2 heterotrimeric arrangement. The photoaffinity labeling experiments using *Arabidopsis* plasmalemma preparations have given some indications of additional 33 and 34-35 kDa components in the immediate vicinity of the FCBP. While pointing towards an oligomeric larger complex, part of which is the FCBP, it remains to be proven, by sequencing of the purified components, whether or not this is the case. Alternatively, proteolysis or deglycosylation during workup might result in the appearence of some of the bands observed. At present, this cannot be ruled out even more so, since the FCBP appears highly susceptible to proteolytic (e.g.Feyerabend and Weiler 1988) and cellulolytic (Meyer et al. 1989) attack.

FACTORS AFFECTING FC-BINDING

Before the structure of the FCBP can be deduced from sequence data, one has to resort to indirect structural information from, e.g., the use of reagents which affect the binding of FC to the FCBP. Additionally, biochemical data might give first hints to potential mechanisms of regulation operating at the FCBP site itself. Although this type of information has accumulated recently, the picture is still far from clear and also partly contradictory. In the absence of Mg^{2+}, FC-binding to the FCBP is greatly reduced. Ca^{2+} appears to have no effect on the *Arabidopsis* system, neither do the monovalent ions tested (K^+, Na^+). Fluoride slightly stimulates FC-binding to plasma membranes or prevents inactivation of the FCBP (Aducci et al. 1984). An additional effect of Al^{3+} plus F^- (AlF_4^-) was not observed in our own studies. We have recently discovered that nucleoside-triphosphates and, to a lesser extent, nucleoside-diphosphates strongly inhibit the binding of FC to the FCBP when given in 0.5 - 2 mM concentration. For the triphosphates, the inhibition decreases in the order ATP> GTP~CTP>UTP while for guanosine-nucleotides: GTP>GDP>GTPγS> GMP, the latter being ineffective up to 2.5 mM concentration. Inorganic phosphate is not inhibitory up to 2 mM while pyrophosphate is somewhat less effective than GDP. ATP is also effective on the solubilized FCBP of *A.thaliana*.

The action of the nucleotides appears complex. While Mg^{2+} partially compensates the inhibition of FC-binding by UTP and CTP, it is only marginally effective with GTP and ineffective with ATP. Thus, the inhibition of FC-binding by mM levels of

ATP and GTP cannot be explained on the basis of a Mg^{2+}-depletion effect brought about by the nucleotides. One of the possibilities to explain our results is that the FCBP might be inactivated through phosphorylation. However, treatment of *Arabidopsis* plasmalemma vesicles with acid phosphatase prior to the binding assay had no effect on the subsequent binding of (^{3}H)-FCol. Aducci et al. (1984) have reported on the inactivation of the solubilized FCBP from spinach leaves by treatment with acid phosphatase. These data seem to indicate that phosphorylation of the FCBP is required for FC-binding. Clearly, further experiments are needed before the question of phosphorylation of the FCBP can be answered.

The glycoprotein nature of the FCBP has also not been demonstrated unambiguously. While the solubilized spinach FCBP, after treatment with α-mannosidase, looses its ability to bind FC (Aducci et al. 1984), this was not found for the solubilized or microsomal FCBP from *Arabidopsis* (A. Hopstock, C. Meyer and E.W. Weiler, unpublished). The glycoprotein argument also rests on the observation that (a) commercial cellulase preparations strongly inactivate the FCBP and (b) that low levels of periodate (5 - 10 mM), known to modify sugars by oxidation of their vicinal diol structures drastically impair FC-binding to plasmamembrane preparations (e.g. Meyer et al. 1989). However, all commercial cellulases contain protease activity, and proteases readily degrade the FCBP (e.g. Stout and Cleland 1980, Feyerabend and Weiler 1988). The presence of FC at the binding site protects it from periodate attack. This somewhat unexpected result is difficult to explain on the basis of accepted concepts of glycoprotein structure. Furthermore, despite several attempts, it has so far proven impossible to show any effect of endoglucanases on the apparent molecular mass of the photoaffinity-labeled FCBP of *Arabidopsis* as determined by SDS-PAGE. The assumption, that the FCBP is a glycoprotein therefore requires further experimental proof.

A broad spectrum of modification reagents has been tested for effects on FC-binding in both, the broadbean and the

Arabidopsis system (Meyer et al.1989; R. Adams and E.W.Weiler, unpublished results). The FCBP from both species is affected to practically the same extent by these reagents lending further support to the notion that the FCBP from different species is very similar. Among the modification reagents which inhibit FC-binding to theFCBP are, besides periodate, chloramin T, diethylpyrocarbonate, dansylchloride, N-bromosuccimide and Koshland's reagent as well as the water-soluble carbodiimide, EDC, while DCC is ineffective. In addition, low levels of Hg^{2+} (0.5 mM) severly reduce FC-binding to the FCBP, but Hg^{2+} is ineffective on the ligand-protected FCBP (de Boer et al.1989, R.Adams and E.W.Weiler, unpublished). SH-alkylating reagents such as iodoacetamide do, however, not impair FC-binding. Modifiers of arginine, lysine and tyrosine were also found ineffective. Although none of the protein - modification reagents is absolutely specific, the information which is available allows a few suggestions for residues which might be important for FC-binding and/or for maintaining the correct structure of the FC-binding site: among these residues should be SH-groups, some of which might be located within the binding domain. The strong effects of N-bromosuccimide and Koshland's reagent suggest an important function for tryptophan residues. Ligand protection experiments in this case show that both reagents are effective in the presence of FC, thus ruling out that modification takes place within the binding domain.

HOW MUCH DO WE KNOW FOR SURE ?

The FCBP has revealed some of its secrets in the past few years. However, many important aspects remain unknown. While it seems clear now that the FCBP represents a highly conserved 31-32 kDa protein in the plasmamembrane of higher plants, its topology within the membrane remains to be elucidated. At least for the FCBP from broadbean leaves, it was shown that the FC-binding domain is exposed to the apoplastic space. Thus, the FCBP could either be an extrinsic-apoplastic or an intrinsic membrane protein. The photoaffinity-labeling and cross-linking studies available so far as well as the detergent extraction studies (see Table 2), all, suggest that other proteins are present in the vicinity of the FCBP and since cross-linking can be effected through the use of lipophilic linkers such as hexamethylenediisocyanate, at least part of the FCBP must be accessible through the lipid phase of the plasmalemma. It is thus quite likely that the FCBP is an intrinsic plasmalemma protein and part of a larger complex of still unknown composition. There is up to now no evidence that the H^+-ATPase might be closely associated with the FCBP in the plasma membrane. Such a topology, although not definitely excluded, should have resulted in at least some photoaffinity labeling of the ATPase polypeptide (100 - 105 kDa) because the azido-FC derivative used (Feyerabend and Weiler 1989) bears a relatively long and flexible spacer arm. Labeling of a 100 - 105 kDa polypeptide was, however, never seen. It is not known, if the FCBP is an integral membrane protein. Many structural features should become clearer, once the sequence of the FCBP is elaborated, and most of the immediate efforts are currently directed towards this goal. Another major blind spot is our lack of knowledge about the mechanisms of FC action and the role of the FCBP in mediating FC responses. A role of the FCBP in mediating FC action is indicated, although not proven, by the fact that a close correlation between the ability to displace (^{3}H)-FCol from the FCBP and the biological activity (stomatal opening in the dark) exists for a range of FC-derivatives in both *Vicia* (Feyerabend and Weiler 1988) as well as *Arabidopsis* (Meyer et al. 1989).

A series of plant electrophysiological experiments described by Blatt (1988,1989) leads to the conclusion that FC not only activates H^+-pumping at the plasmalemma, but changes the activity of several plasma membrane transport systems. While the toxin alters the kinetics of the H^+-Pump (presumably represented by the vanadate-sensitive H^+-ATPase) it apparently progressively inactivates K^+-channels and activates as yet unknown energy-coupled transport mechanisms (Blatt and Clint 1989). Mechanistically, FC should best bring about such an action on more than one target through a transduction (and possibly also amplification) step involving one or more "second messengers".

It will remain a challenge for the future to elucidate, step-by-step, the sequence of reactions from FC-binding to the alteration of plasmalemma transport systems. Progress on signal transduction/amplification mechanisms at the plasmalemma of plants is still slow. There is some evidence for the presence of GTP-binding proteins in the plasmalemma of *Vicia* and *Arabidopsis* (Blum et al.1988). It has also been shown recently (Palmgren et al.1988) that lyosphospholipids may activate the plant H^+-ATPase. Lysophospholipids are, in animal cells, products of phospholipases of the A-type. Some phospholipases of the A_2-type which are membrane-bound enzymes are triggered by agonists through G-proteins (Burgoyne et al.1987).

An attempt to fit all these pieces of information into a plausible working model of FC-action is the following: FC would, through the FCBP and perhaps via a G-protein-type transducing mechanism, activate a phospholipase (e.g. of the A_2-type). The products of the phospholipase reaction (e.g. lysophospholipid) would then modulate the activity of a range of plasmalemma transport systems, like the H^+-ATPase or of K^+-channels. Of course, unambiguous verification of each component of such a model will be required. Most of the model is still guesswork at present and alternative mechanisms of FC action, e.g. involving protein kinases, are perhaps equally plausible. At least this, as well as alternative models, will help us to design critical experiments which should allow us to verify or falsify our hypotheses. We are quite optimistic that within the next few years, the structure and function of the plant's fusicoccin-binding protein will be understood much better than is the case now.

ACKNOWLEDGEMENT.

Work in the author's lab was funded by the Deutsche Forschungsgemeinschaft, Bonn, and the Fonds der Chemischen Industrie, Frankfurt (literature provision).

REFERENCES

Aducci P, Ballio A, Fiorucci L, Simonetti E (1984) Inactivation of solubilized fusicoccin-binding sites by endogenous plant hydrolases. Planta 160:422-427

Ballio A, Federico R, Scalorbi D, Marrè E(1977) Metabolic stability of fusicoccin in plant tissues. Atti Accad Naz Lincei Rend Cl Sci Fis Mat Nat 63:604-608

Blatt MR (1988) Mechanisms of fusicoccin action: a dominant role for secondary transport in a higher plant cell. Planta 174:187-200

Blatt MR, Clint GM (1989) Fusicoccin activates energy-coupled K^+ uptake by stomatal guard cells. Plant Physiol 89:195

Blum W, Hinsch KD, Schultz G, Weiler EW (1988) Identification of GTP-binding proteins in the plasma membrane of higher plants. Biochem Biophys Res Comm 156:954-959

Burgoyne RD, Cheek TR, O'Sullivan AJ (1987) TIBS 12: 332-333

De Boer AH, Watson BA, Cleland RE (1989) Purification and identification of the fusicoccin-binding protein from oat root plasma membrane. Plant Physiol 89:250-259

Dohrmann U, Hertel R, Pesci P, Cocucci SM, Marrè E,Randazzo G, Ballio A (1977) Localization of *in vitro* binding of the fungal toxin fusicoccin to plasma-membrane-rich fractions from corn coleoptiles. Plant Sci Lett 9:291-299

Feyerabend M, Weiler EW (1988) Characterization and localization of fusicoccin-binding sites in leaf tissues of *Vicia faba* L. probed with a novel radioligand. Planta 174:115-122

Feyerabend M, Weiler EW (1989) Photoaffinity labeling and partial purification of the putative plant receptor for the fungal wilt-inducing toxin, fusicoccin. Planta (in press)

Hager A, Hampp R, Mehrle W (1986) Mechanism of growth induction by ester compounds. Studies with segments and protoplasts of oat coleoptiles. In: Bopp M (ed) Plant Growth Substances 1985 Springer, Berlin, Heidelberg,New York p. 284-292

Marré E (1979) Fusicoccin: a tool in plant physiology. Ann Rev Plant Physiol 30:273-288

Meyer C, Feyerabend M, Weiler EW (1988) Fusicoccin-binding proteins in *Arabidopsis thaliana* (L.)Heynh. Characterization, solubilization, and photoaffinity labeling. Plant Physiol 89:692-699

Palmgren MG, Sommarin M, Ulvskov P, Jørgensen PL (1988) Modulation of plasma membrane H^+-ATPase from oat roots by lysophosphatidylcholine, free fatty acids and phospholipase A_2. Physiol Plant 74:11-19

Pesci P, Tognoli L, Beffagna N, Marrè E (1979) Solubilization and partial purification of a fusicoccin-receptor complex from maize microsomes. Plant Sci Lett 15:313-322

Stout R, Cleland RE (1980) Partial characterization of fusicoccin binding to receptor sites on oat root membranes. Plant Physiol 66:353-359

Weiler EW, Meyer C, Oecking C, Feyerabend M, Mithöfer A(1989) The fusicoccin receptor of higher plants. In: Lamb C, Beachy R (eds) Plant Gene Transfer. Alan R Liss, New York (in press)

MODULATION OF ATPases ACTIVITY BY DIFFERENT SIGNALS

G. Zocchi

Istituto Chimica Agraria

Università degli Studi

Via Celoria 2

20133 Milano

Italy

INTRODUCTION

Membrane-bound ATPases are a very important class of enzymes which are involved in the regulation of many cellular and physiological processes. Plant membranes contain essentially two distinct classes of ATPase activities whose peculiar functions are very well differentiated.

First, there are the H^+-ATPases whose main function is the production and maintenance of an electrochemical proton gradient across membranes using the metabolic energy from ATP hydrolysis. Such an electrochemical gradient may subsequently be utilized for many physiological and biochemical purposes. The energization of secondary transport of nutrients (ions, sucrose, aminoacids, etc), osmoregulation, cell elongation, and control of intracellular and extracellular pH are certainly important characteristics linked to the activity of these enzymes (Serrano, 1985; Marre' and Ballarin-Denti, 1985).

Second, there are the Ca^{2+}-ATPases which have more likely the unique (but for this reason not less important) critical function of controlling and regulating the free cytoplasmic calcium concentration in order to keep the ion at submicromolar level (Hanson, 1984). Since a major mechanism by which external stimuli regulate cell functions is by altering Ca^{2+} influx a compensatory efflux mechanism is necessary to avoid Ca^{2+} accumulation by

NATO ASI Series, Vol. H 47
Signal Perception and Transduction in Higher Plants
Edited by R. Ranjeva and A. M. Boudet

the cell.

Clearly we might expect that the role these enzymes play in the regulation of plant development and growth makes them a possible target for modulation. It has been recognized that the response to extra- and intracellular stimuli that signals a modification of a physiological status may increase or decrease the activity of these enzymes. The aim of this paper is to put together the information now available to understand by which mechanism(s) this modulation is achieved.

Table 1

Extracellular signals affecting ATPase activities

	H^+-ATPase	Ca^{2+}-ATPase
Mechanical		
rubbing	+	-
cutting	+	-
Chemical-physical		
cold shock	+	-
pH	+	-
osmotic shock	+	-
Light		
blue light	+	-
red light	+	+
Chemical		
phytohormones	+	+
phytotoxins	+	-

EFFECT OF DIFFERENT SIGNALS ON THE ACTIVITY OF ATPases

Table 1 lists different signals that are known to affect both H^+- and Ca^{2+}-ATPase activities in plants.

Up to this time, more emphasis has been given to the H^+-ATPase activity given its earlier discovery and identifiable association with many physiological processes. Still, it has been shown that environmental and hormonal stimuli and phytotoxins directly or indirectly influence the Ca^{2+}-ATPase as well, in many plant tissues. For purpose of this paper, we will divide the signals according to the stimuli which give rise to them.

a. Mechanical and chemical-physical

Mechanical and chemical-physical stimuli such as cutting, rubbing, low temperature (cold shock), low or high pH (Gronewald and Hanson, 1980; Chastain and Hanson, 1982), osmotic shock (Rubinstein et al.,1977) give raise to a rapid blockage of electrogenic proton efflux pumping in a variety of plant tissues. It has been hypothesized that a rapid and sudden decrease in the transmembrane electrical potential may be either the cause or the effect of a rapid change in membrane permeability. In addition, Zocchi and Hanson (1982) have shown that cold shock enhanced calcium influx and phosphorylation of membrane proteins. All these effects were transient and washing of the tissue restored H^+ efflux pumping, transmembrane electrical potential and initial calcium concentration (Rincon and Hanson, 1986).

b. Light

In addition to photosynthesis, plants show a variety of responses to light, particularly in the blue and red regions.

Red irradiation of excised Phaseolus hypocotyls reduced the ATPase activity in microsomal membrane fractions by 50%. This effect was reversed if the red was followed by far-red light indicating a relationship between the form of phytochrome and ATPase activity, though the specific ATPase has not been identified (Jose, 1977). Indirect activation of ATPase by Pfr was reported in cucumber hypocotyls by Thomas and Tull (1981) and in stomatal guard cells of <u>Commelina communis</u> by Nejidat et al. (1983). Stomatal opening is greatly influenced by blue light which is postulated to activate an electrogenic H^+ -ATPase localized on the plasmalemma (Zeiger, 1983). The increased transmembrane electrical potential and K^+ influx bring about an osmotic swelling of the guard cells, and consequently, the stomata open. Guard cell protoplasts of <u>Vicia faba</u> showed a blue light dependent proton pumping in agreement with this hypothesis (Assmann et al., 1985; Shimazaki et al., 1986).

It has also been reported that Ca^{2+} transport is affected by red/far red light (Dieter and Marme', 1981; Roux et al., 1986) with the direct effect being on Ca^{2+} influx rather than efflux. As yet, however, it has not been established that Ca^{2+} pump activitys are directly influenced by phytochrome.

c. Phytohormones

With the exception of ethylene, most, if not all, of the effects exerted by plant hormones seem to be brought about by the interaction with ATPase activities. For example, activation of H^+-ATPase have been described for auxins, cytokinins and gibberellic acid (Van Steveninck, 1976; Elliott, 1986), while for abscisic acid, an inhibitory effect was detected (Lurie and Hendrix, 1979; Van Steveninck and Van Steveninck, 1983). In the case of auxins, since the development of the acid growth theory (Hager et al., 1971; Cleland, 1973), a large body of work has been done suggesting its control of H^+ active extrusion during cell elongation. Direct effects of auxins on H^+-ATPase activity <u>in vitro</u> has been shown by Scherer (Scherer, 1981; Scherer and Morre', 1978), but the difficulty of obtaining significant and repeatable stimulation might challenge these results.

Kubowicz et al. (1982) showed that preincubation of soybean hypocotyls with auxin increased the ATP dependent Ca^{2+} transport activity in a plasmalemma enriched vesicles fraction isolated from the tissue. The same effect has been detected in a membrane vesicle fraction isolated from corn coleoptiles though direct addition of auxin in vitro did not give any effect (Zocchi, unpublished). On the other hand cytokinins inhibited the ATP dependent Ca^{2+} transport in the same tissue (Kubowicz et al., 1982).

d. Phytotoxins

It is now known that some toxins produced by plant pathogens are able to interact with plasma membrane ATPase (Marre', 1980). Surely the most studied phytotoxin has been fusicoccin (a diterpene glucoside produced by the fungus Fusicoccum amygdali) largely because of its capacity to mimic some of the immediate effects of auxin (Marre', 1979). The in vivo stimulatory effect of fusicoccin on proton pumping activity has been confirmed by in vitro experiments with plasma membrane enriched fractions from spinach leaves and maize coleoptiles (Beffagna et al., 1977), from leaves of Nicotiana (Lurie and Hendrix, 1979) and more recently in purified plasma membrane vesicles isolated from germinating radish seeds (Rasi-Caldogno and Pugliarello, 1985). A stimulatory effect on ATPase activity both in vivo and in vitro has also been shown for helminthosporoside (a toxin produced by Helminthosporium sacchari) (Strobel, 1973) and for syringomycin, (a peptide toxin produced by Pseudomonas syringae) (Bidwai et al., 1987). On the contrary, toxins extracted from Cercospora beticola and Helminthosporium maydis seem to have an inhibitory effect on the ATPase activity (Marre', 1980).

POSSIBLE MECHANISMS OF ATPases MODULATION

In the last few years a large body of evidence has grown on how extracellular signals may be translated into cellular responses and consequently influence plant development and growth. The first step for signal transduction into a cellular response seems to involve the presence of a specific receptor on the external surface of the cell (most likely on the plasmalemma). Receptors have been found for light (phytochrome, blue-light receptors; Pratt, 1982; Briggs, 1980), hormones and phytotoxins (Klambt, 1987), but nothing is known about transduction for temperature, pH or mechanical signals. The means by which the information of extracellular signals flows inside and makes the cells respond are the proposed second messenger systems. In plants, this role has been recognised for calcium (Roux and Slocum, 1982; Hepler and Wayne, 1985) and for the degradation products of membrane phospholipidic components, in particular from the polyphosphoinositol breakdown (for references see Boss, 1989). Inositol 1,4,5 trisphosphate released into the cell in this way acts by mobilising calcium from internal stores (vacuole, endoplasmic reticulum) probably by opening or forming Ca^{2+}-channels. Both calcium, alone or in association with calmodulin, and diacylglycerol (DAG) activate protein kinases that, through phosphorylation of proteins, lead to physiological responses (Marme' and Dieter, 1983; Berridge and Irvine, 1984; Nishizuka, 1984; Ranjeva and Boudet, 1987). Many of the responses to external signals in plants seem to fit very well with these mechanisms. I would like to focus my attention only on the modulation which involves ATPase activities as a consequence of these responses, referring the reader to other more complete reviews for specific topics (Trewavas, 1986; Boss and Morre', 1989).

Since the paper by Zocchi et al. (1983) in which it was hypothesized that, as response to cold shock which produced rapid Ca^{2+} influx and increased membrane proteins phosphorylation, the inhibition of H^+ efflux might be the consequence of phosphorylation of the plasma membrane H^+-ATPase, only a few papers have appeared on a possible modulation of these membrame-bound enzymes by covalent modification, despite to their great importance on plant growth and development. In that paper and in a following one (Zocchi, 1985) it was shown that the ionophore-stimulated H^+-ATPase activity of membrane vesicles isolated from corn roots might be modulated by a Ca^{2+}-Cam-dependent protein kinase. In particular phosphorylation of membrane protein decreased the H^+ pumping activity in a manner that could be reversed upon phosphatase treatment (Table 2). In those papers,it could not be determined if the H^+-ATPase was the real substrate of the kinase activity or if it was other molecules, involved in modulating its activity, that were actually the targets of phosphorylation. That the H^+-ATPase is the substrate of kinase activity has since been reported by Schaller and Sussman (1988) who, however, demonstrated that the H^+-ATPase isolated from oat roots (with a MW of 100 KDa) was phosphorylated by a calcium-stimulated protein kinase both _in vivo_ and _in vitro_ in a pH-dependent manner. However, there was no indication about the activity of this H^+-ATPase upon phosphorylation. Bidwai et al. (1987) reported that

Table 2

Effect of phosphorylation-dephosphorylation treatment on ATPase activity

treatment	ATPase activity umol Pi/mg prot h		^{32}P-incorporated pmol/mg prot
	control	+CCCP	
none	4.5	6.0	-
phosphorylation	2.4	2.6	1400
phosphorylation + dephosphorylation	6.1	9.2	350

the vanadate-sensitive H^+-ATPase extracted from red beet storage tissue was stimulated by syringomycin. Stimulation was not by a direct effect of the toxin on the enzyme, but as demonstrated by a following paper (Bidwai and Takemoto, 1987) it was obtained through a protein-kinase activity which phosphorylates plasma membrane proteins. Among these proteins, a 100-KDa polypeptide, the putative H^+-ATPase catalytic subunit (Serrano, 1985), was phosphorylated in a hydroxylamine-resistant manner. The kinase activity which phosphortylated this 100-KDa polypeptide seems to have been directly stimulated by the toxin since in the presence of EGTA, phosphorylation of the peptyde was only partially affected.

Fusicoccin, the well known and characterized toxin able to stimulate H^+-ATPase both in vivo and in vitro, does not seem to behave in the same way. Recently, Tognoli et al. (Tognoli and Colombo, 1986; Tognoli and Basso, 1987) have shown that in FC-mediated phosphorylation of Acer pseudoplatanus cells, only a 33 KDa polypeptide is phosphorylated; there was no evidence of phosphorylation of the 100 KDa polypeptide. The regulatory role of this 33 KDa peptide on the H^+ -ATPase activity as a response to FC was of questionable relevance as suggested by the authors themselves.

ATPases require lipids for their activity, suggesting a possible role of these molecules in the regulation of the enzymes. Activation of H^+-ATPase by phospholipids has been reported (Cocucci and Ballarin-Denti, 1981; Cocucci and Marre', 1984; Palmgren et al., 1988). However, it is not clear whether lipids directly stimulate the enzyme activity or if the stimulation results from a modification of the enzyme micro environment that makes it more reactive (increased affinity) to its substrate. The discovery in plant tissue of calcium- and phospholipid-dependent protein kinases (Schafer et al., 1985; Elliott and Skinner, 1986; Favre and Turian, 1987) suggests that the activation of a degradation mechanism (phospholipase activity) of the phospholipidic structure of membranes in response to external stimuli might release lipids which in turn activate protein kinases. The recent papers by Scherer and co-workers seem to support this view (Scherer, 1985; Scherer and Stoffel, 1987; Scherer et al., 1988). A phospholipid very similar to the animal platelet-activating factor (PAF) was extracted from zucchini hypocotyls and shown to activate a protein kinase whose substrate is the H^+-ATPase (or an associated protein), leading after phosphorylation to the increase in the proton pumping activity. However, there was no indication if membrane proteins were phosphorylated by the kinase or if a 100 KDa polypeptide was phosphorylated.

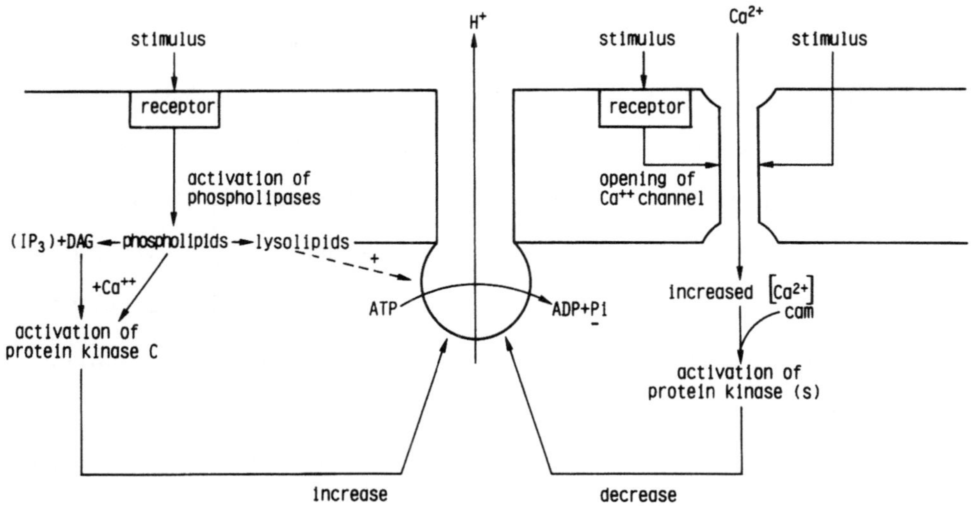

Figure 1. Hypothetical model of how extracellular signals may interact with H^+-ATPase

The supposed effect of phytohormones or light on the control of H^+-ATPase and Ca^{2+}-ATPase activities by the models depicted above is, as noted earlier, still lacking. Evidence of the transduction of their signals by the second messenger systems, of the activation of protein kinases leading to protein phosphorylation (Veluthambi and Poovaiah, 1986; Elliott, 1986; Owen, 1988; Morse et al., 1987), has been provided, although there is as yet no direct evidence that among the targets of the physiological responses are the ATPases. A tentative model on how extracellular signals may interact with H^+-ATPase is reported in Figure 1.

CONCLUSION

This brief discussion, far from be exhaustive, is intended as a contribution to the consideration of the possibility that membrane-bound ATPases might be modulated through a cascade of biochemical events by extracellular signals. In the information now arising on this specific topic there are contrasting evidences on the modulation responses to these signals. Phosphorylation-dephosphorylation is a means by which enzyme activities may be regulated (for more detail on plants see Ranjeva and Boudet, 1987). For H^+-ATPases a possible modulation by protein kinases has been proposed. It would seem that at least two classes of kinases might be involved in its modulation: a Ca^{2+} -Cam-dependent and a phospholipid-dependent Ca^{2+}-activated protein kinase (Zocchi, 1985; Schaller and Sussman, 1988; Scherer et al., 1988). The action of different kinases on the same enzyme might be a useful way to obtain differentiated and/or integrated responses to different signals. On the other hand, it is well known in mammalian cells that there are enzymes with different site of phosphorylation (multisite phosphorylations) on which different protein kinases may act (Picton et al., 1982). Another possibility is that the two classes of kinases might act on different H^+-ATPases. H^+ -ATPases from tonoplast and plasmalemma, possessing a different subunit structure (Sze, 1985), might satisfy this hypothesis. Isolation of single membrane systems and purification of their H^+-ATPase activities seems the only way to obtain clear evidence in support of these hypotheses.

Up to now, no evidence has been found of a similar control on Ca^{2+}-ATPase activities in plants. For animal cells, such a mechanism has been reported and a phosphgorylation-dephosphorylation cycle seems to modulate sarcolemma Ca^{2+}-pumping ATPase (Caroni and Carafoli, 1981). In addition, in erythrocytes, protein kinase C stimulates the activity of a plasma membrane Ca^{2+} pump presumably after catalysing the phosphorylation of the pump itself (Smallwood et al. 1988). In plants calmodulin seems to be the only system able to modulate Ca^{2+}-ATPase activity (Zocchi and Hanson, 1983; Dieter and Marme', 1980). It should be quite interesting to see if the stimulation of Ca^{2+}-pumping activity by calmodulin obtained with plant membrane preparations is a direct effect of calmodulin on the enzyme (allosteric effect) or if calmodulin activates a Ca^{2+}-Cam-dependent protein kinase which, through phosphorylation, in turn modulates the Ca^{2+}-ATPase activity.

ACKNOWLEDGEMENTS

This paper was supported in part by Consiglio Nazionale delle Ricerche, Italy, special grant "Meccanismi molecolari della trasduzione del segnale".

REFERENCES

Assmann SM, Simoncini L, Schroeder JI (1985) Blue light activates electrogenic ion pumping in guard cell protoplasts of Vicia faba. Nature 318:285-287

Beffagna N, Cocucci S, Marrè E (1977) Stimulating effect of fusicoccin on K-activated ATPase in plasmalemma preparations from higher plant tissues. Plant Sci Lett 8:91-98

Berridge MJ, Irvine RF (1984) Inositoltrisphosphate, a novel second messenger in cellular signal transduction. Nature 312:315-321

Bidwai AP, Takemoto JY (1987) Bacterial phytotoxin, syringomycin, induces a protein kinase-mediated phosphorylation of red beet plasma membrane polypeptide. Proc Natl Acad Sci 84:6755-6759

Bidwai AP, Zhang L, Bachmann RC, Takemoto JY (1987) Mechanism of action of Pseudomonas syringae phytotoxin, syringomycin. Plant Physiol 83:39-43

Boss WF (1989) Phosphoinositide metabolism: its relation to signal transduction in plants. In Boss WF, Morre' DJ (eds) "Second messengers in plant growth and development" Plant Biology Series Vol 6. Alan Liss, New York pp 29-56

Boss WF, Morre' DJ (eds) (1989) Second messengers in plant growth and development. Plant Biology Series Vol 6. Alan Liss, New York

Briggs WR (1980) A blue light photoreceptor system in higher plants and fungi. In DeGreef J (ed) "Photoreceptors and plant development" Antwerpen Univ Press, Antwerpen pp 17-28

Caroni P, Carafoli E (1981) Regulation of Ca^{2+} -pumping ATPase of heart sarcolemma by a phosphorylation-dephosphorylation process. J Biol Chem 256:9371-9373

Chastain CJ, Hanson JB (1982) Control of proton efflux from corn root tissue by an injury-sensing mechanism. Plant Sci Lett 24:97-104

Cleland RE (1973) Auxin-induced hydrogen ion excretion from Avena coleoptiles. Proc Natl Acad Sci 70:3092-3093

Cocucci M, Ballarin-Denti A (1981) Effect of polar lipids on ATPase activity of membrane preparations from germinating radish seeds. Plant Physiol 68:377-381

Cocucci MC, Marre' E (1984) Lysophosphatidylcoline-activated vanadate-inhibited, Mg^{2+} -ATPase from radish microsomes. Biochem Biophys Acta 771:42-52

Dieter P, Marme' D (1980) Calmodulin activation of plant microsomal Ca^{2+} uptake. Proc Natl Acad Sci 77:7311-7314

Dieter P, D Marme' (1981) Far-red light irradiation of intact corn seedlings affects mitochondrial and calmodulin-dependent microsomal Ca^{2+} transport. Biochem Biophys Res Commun 101:749-755

Elliott DC (1986) Calcium involvement in plant hormone action. In Trewavas AJ (ed) "Molecular and Cellular Aspects of Calcium in Plant Development" NATO ASI Series A: Life Sciences Vol 104. Plenum, New York pp 285-292

Elliott DC, Skinner JD (1986) Calcium-dependent, phospholipid-activated protein kinase in plants. Phytochem 25:39-44

Favre B, Turian G (1987) Identification of a calcium- and phospholipid-dependent protein kinase (protein kinase C) in Neurospora crassa. Plant Sci 49:15-21

Gronewald JW, Hanson JB (1980) Sensitivity of the proton and ion transport mechanisms of corn roots to injury. Plant Sci Lett 18:143-150

Hager A, Menzel H, Krauss A (1971) Versuche und hypothese zur primarwirkung des auxins beim streckungswachstum. Planta 100:47-75

Hanson JB (1984) The functions of calcium in plant nutrition. In Thinker PB, Lauchli A (eds) "Advances in Plant Nutrition" Vol 1. Praeger, New York pp 149-208

Hepler PK, Wayne WO (1985) Calcium and plant development. Annu Rev Plant Physiol 36:397-439

Jose AM (1977) Phytochrome modulation of ATPase activity in a membrane fraction from Phaseolus. Planta 137:203-206

Klambt D (ed) (1987) Plant hormone receptors. NATO ASI Series H: Vol 10. Springer-Verlag, Berlin

Kubowicz BD, Vanderhoff LN, Hanson JB (1982) ATP-dependent calcium transport in plasmalemma preparations from soybean hypocotyls. Plant Physiol 69:187-191

Lurie S, Hendrix DL (1979) Differential ion stimulation of plasmalemma ATPase from leaf epidermis and mesophyll of Nicotiana rustica. Plant Physiol 63:936-939

Marme' D, Dieter P (1983) Role of Ca^{2+} and calmodulin in plants. In Cheung WY (ed) "Calcium and Cell Functions' Vol 4. Academic Press, New York pp 263-311
Marrè E (1979) Fusicoccin: A tool in plant physiology. Annu Rev Plant Physiol 30:273-288
Marrè E (1980) Mechanism of action of phytotoxins affecting plasmalemma functions. Progress in Phytochemistry Vol 6. Pergamon Press, New York pp 253-284
Marrè E, Ballarin-Denti A (1985) Proton pumps of plasmalemma and tonoplast. J Bioenerg Biomemb 17:1-21
Morse MJ, Crain RC, Satter RL (1987) Light-stimulated phosphatidyinositol turnover in Samanea saman leaf pulvini. Proc Natl Acad Sci 84:7075-7078
Nejidat A, Itai C, Roth-Bejerano N (1983) Stomatal response to ATP mediated phytochrome. Physiol Plant 57:367-370
Nishizuka Y (1984) Turnover of inositol phospholipids and signal transduction. Science 225:1365-1370
Owen JH (1988) Role of abscisic acid in a Ca^{2+} second messenger system. Physiol Plant 72:637-641
Palmgren MG, Sommarin M, Ulvskov P, Jorgensen PL (1988) Modulation of plasma membrane H^+-ATPase from oat roots by lysophosphatidylcholine, free fatty acids and phospholipase A_2. Physiol Plant 74:11-19
Picton C, Aitken A, Bilham T, Cohen P (1982) Multisite phosphorylation of glycogen sinthase from rabbit skeletal muscle. Eur J Biochem 124:37-45
Pratt LH (1982) Phytochrome: the protein moiety. Annu Rev Plant Physiol 33:557-582
Ranjeva R, Boudet AM (1987) Phosphorylation of proteins in plants: regulatory effects and potential involvement in stimulus/response coupling. Annu Rev Plant Physiol 38:73-93
Rasi-Caldogno F, Pugliarello MC, (1985) Fusicoccin stimulates the H^+-ATPase of plasmalemma in isolated membrane vesicles from radish. Biochem Biophys Res Comm 133:280-285
Rincon M, Hanson JB (1986) Controls on calcium ion fluxes in injured or shocked corn root cells: importance of proton pumping and cell membrane potential. Physiol Plant 67:576-583
Roux SJ, Slocum RD (1982) Role of calcium in mediating cellular functions important for growth and development in higher plants. In Cheung WY (ed) "Calcium and Cell Functions" Vol 3. Academic Press, New York pp 409-453
Roux SJ, Wayne RO, Datta N (1986) Role of calcium ions in phytochrome responses: an update. Physiol Plant 66: 344-348
Rubinstein B, Mahar P, Tattar TA (1977) Effects of osmotic shock on some membrane-regulated events of oat coleoptile cells. Plant Physiol 59:365-368
Schafer A, Bygrave F, Matzenauer S, Marmé D (1985) Identification of a calcium and phospholipid dependent protein kinase in plant tissue. Fed Eur Biochem Soc Lett 187:25-28
Schaller GE, Sussman MR (1988) Phosphorylation of plasma membrane H^+-ATPase of oat roots by a calcium-stimulated protein kinase. Planta 73:509-518
Scherer GFE (1981) Auxin-stimulated ATPase in membrane fractions from pumpkin hypocotyls (Cucurbita maxima L.) Planta 151:434-438
Scherer GFE (1985) 1-O-Alkyl-2-acetyl-sn-glycero-3-phosphocholine (platelet activating factor) stimulates H^+ transport in vitro and growth. Biochem Biophys Res Commun 133:1160-1167
Scherer GFE, Morré DJ (1978) In vitro stimulation by 2,4-dichlorophenoxyacetic acid of an ATPase and inhibition of phosphatidate phosphatase of plant membranes. Biochem Biophys Res Commun 84:238-247

Scherer GFE, Stoffel B (1987) A plant lipid and the platelet-activating factor stimulate ATP-dependent H^+ transport in isolated plant membrane vesicles. Planta 172:127-130

Scherer GFE, Martiny-Baron G, Stoffel B (1988) A new set of regulatory molecules in plants: a plant phospholipid similar to platelet-activating factor stimulates protein kinase and proton-translocating ATPase in membrane vesicles. Planta 175:241-253

Serrano R (1985) Plasma membrane ATPase of plants and fungi. CRC Press, Boca Raton

Shimazaki K, Iino M, Zeiger E (1986) Blue light-dependent proton extrusion by guard cell protoplast of Vicia faba. Nature 319:324-326

Smallwood JI, Gugi B, Rasmussen H (1988) Regulation of erythrocyte Ca^{2+} pump activity by protein kinase C. J Biol Chem 263:2195-2202

Strobel AG (1973) The helminthosporoside-binding protein of sugarcane. J Biol Chem 248:1321-1328

Sze H (1985) H^+-translocating ATPases of plasma membrane and tonoplast of plant cells. Physiol Plant 61:683-691

Thomas B, Tull SE (1981) Photoregulation of K^+-ATPase in vitro by red and far red light in extracts from cucumber hypocotyls. Z Pflanzenphysiol 102:283-292

Tognoli L, Colombo R (1986) Protein phosphorylation in intact cultured sycamore (Acer pseudoplatanus) cells and its response to fusicoccin. Biochem J 235:45-48

Tognoli L, Basso B (1987) The fusicoccin-stimulated phosphorylation of a 33KDa polypeptide in cells of Acer pseudoplatanus as influenced by extracellular and intracellular pH. Plant Cell Env 10:233-239

Trewavas AJ (ed) (1986) Molecular and cellular aspects of calcium in plant development. NATO ASI Series A: Life Sciences Vol 104. Plenum Press, New York

Van Steveninck RFM (1976) Effect of hormones and related substances on ion transport. in Luttge U, Pitman MG (eds) "Encyclopedia of Plant Physiology" New Series, Vol 2B Springer-Verlag, Berlin 307-342

Van Steveninck RFM, Van Steveninck ME (1983) Abscisic acid and membrane transport. In Addicott FT (ed) "Abscisic acid". Praeger, New York pp 171-236

Veluthambi K, Poovaiah BW (1986) In vitro and in vivo protein phosphorylation in Avena sativa L. coleoptiles: Effects of Ca^{2+}, calmodulin antagonists and auxin. Plant Physiol 81:836-841

Zeiger E (1983) The biology of stomatal guard cells. Annu Rev Plant Physiol 34:441-475

Zocchi G (1985) Phosphorylation-dephosphorylation of membrane proteins controls the microsomal H^+-ATPase activity of corn roots. Plant Sci 40:153-159

Zocchi G, Hanson JB (1982) Calcium influx into corn roots as a result of cold shock. Plant Physiol 70:318-319

Zocchi G, Hanson JB (1983) Calcium transport and ATPase activity in a microsomal vesicle fraction from corn roots. Plant Cell Environ 6:203-209

Zocchi G, Rogers SA, Hanson JB (1983) Inhibition of proton pumping in corn roots is associated with increased phosphorylation of membrane proteins. Plant Sci Lett 31:215-221

PHOSPHOLIPID-ACTIVATED PROTEIN KINASE IN PLANTS: COUPLED TO PHOSPHOLIPASE A_2?

Günther Scherer
Botanical Institute
University of Bonn
Venusbergweg 22
D-5300 Bonn 1
Federal Republic Germany

INTRODUCTION

The search for second messenger systems in signal transduction mechanisms in cells of higher plants has always strongly been influenced by the concepts which had been developped for animal cells. To date, however, the only molecular signal transduction pathway providing a valid working hypothesis for signal transduction in plant cells is the phosphatidylinositol (PI) cycle which, in animals, produces two second messengers, myo-inositol,1,4,5-trisphosphate (IP_3) and diaclyglycerol (DAG) (Berridge 1987; Morse et al. 1989). The function of one of it, IP_3, is likely to be the release of Ca^{2+} from the vacuole and the ER (Schuhmaker and Sze 1987) which would be similar to its function in animal cells. The function of DAG remains unclear because evidence for a DAG-activated protein kinase in plants is still scarce and controversial (Morré et al. 1984; Elliott et al. 1988; Martiny-Baron and Scherer 1989). Our own work originated from our interest in the mechanism of proton transport stimulation in zucchini microsomes by the platelet-activating factor (PAF) (Scherer 1985) and by auxin (Scherer and Morré 1978; Scherer 1981). The at first bizarre observation of transport stimulation of plant H^+-ATPase by an animal lipid hormone lead us to discover a lipid-stimulated protein kinase in plants (Martiny-Baron and Scherer 1988; Martiny-Baron and Scherer 1989) which seems to be different from the well-known animal protein kinase C (Nishizuka 1984) in that lysophosphatidylcholine (LPC) and lysophosphatidylethanolamine (LPE) and the lysophospholipid-similar PAF activate it rather than DAG. The discovery that plants do have their own biologically active phospholipids, apart from the artificially used PAF (Scherer

NATO ASI Series, Vol. H 47
Signal Perception and Transduction in Higher Plants
Edited by R. Ranjeva and A. M. Boudet

and Stoffel 1987), lead us further to investigate their role as potential second messengers (Scherer 1989) and the role of phospholipase A_2 as a potential second messenger-generating enzyme. In this article, an auxin-activated phospholipase A_2 is described and its possible functional coupling to the lysophospholipid-activated protein kinase from plants is discussed.

MATERIAL and METHODS

Protein phosphorylation. Protein phosphorylation was done as described (Martiny-Baron and Scherer 1988). The antibody and the procedure described by Manolson et al. (1988) were used for immuno precipitation.

Phospholipase A_2 activation. Soybean cells (*Glycine max* L.) were grown in suspension culture (Nickel and Scherer 1988) and used 5 d after inoculation in the late logarithmic phase of growth. A 100 ml culture was transferred to auxin-free medium for 18 h and then 2 µCi ^{14}C-ethanolamine or ^{14}C-choline was added. After 3 h the medium was removed by gentle suction and the cells were washed with fresh medium. The cells were divided into two cultures of 40 ml each, one with $5x10^{-4}$ or 10^{-3} M hormone, one without. Samples of 5 ml were withdrawn, spun down rapidly and the cells suspended in 4 ml chloroform/methanol 1:2 (v/v) and left for 30-60 min at room temperature. Then 1 ml 0.1 M KCl was added, the resulting chloroform phase was dried *in vacuo* and dissolved in a small volume of chloroform for thin layer chromatography.

Zucchini (*Cucurbita maxima* L.) seedlings were grown for 4 d in the dark at 25-30 °C to 3-5 cm length. Pieces of 1 cm were cut and 5 g tissue incubated in 10-20 µCi ^{14}C-ethanolamine or ^{14}C-choline in 6 ml water for 4 h at room temperature in the dark. The sections were rinsed with water and membranes isolated by homogenization in 0.5 M sucrose, 1 mM EDTA, 1 mM dithioerythritol and 15 mM Tris/HCl pH 7.5. Similarly, prelabeled soybean cell culture cells were homogenized. The homogenate was centrifuged for 10 min at 10,000 g and the supernatant again for 30 min at 50,000 g. The resulting pellet was resuspended in 18 ml incubation buffer consisting of 250 mM sucrose, 250 mM MES pH 5.5 and 1.5 mM $MgCl_2$. Auxins were added from ethanolic stock solutions, ususally at $2x10^{-5}$ M. Samples were taken and lipids extracted and subjected to thin layer chromatography (Scherer and Morré 1978). An acidic solvent system was chosen. After chromatography, the radioactivity of lipids was counted.

RESULTS

Stimulation of proton transport activity by phospholipids. The stimulatory effect of PAF *in vitro* was the starting observation (Scherer 1985) which lead to the discovery of further effects of PAF on the tonoplast H^+-ATPase and the plasma membrane H^+-ATPase. The in vitro effects on tonoplast H^+-ATPase have been investigated in greatest detail (Scherer 1985; Scherer and Martiny-Baron 1987; Scherer et al. 1988; Martiny-Baron and Scherer 1988, 1989; Martiny-Baron et al. 1989). Low concentrations (5-20 µM) of PAF stimulated ATP-dependent H^+ transport activity about twofold and correlated with the tonoplast marker, nitrate-sensitive ATPase, in membrane fractionation experiments. ATPase activity, under uncoupling condititons, was also stimulated by PAF whereas membrane permeability was not affected at these concentrations. This suggested that an activation of the H^+-ATPase rather than an increase in the anion conductance was the reason for the observed transport stimulation. Similarly, LPC and LPE stimulated H^+ transport and ATPase activity in tonoplast-enriched fractions. We suggest regulatory phosphorylation of the 53 kDa subunit of the tonoplast H^+-ATPase as an activation mechanism (see below).

For the stimulation of plasma membrane H^+-ATPase in vitro by PAF and LPC our data are less complete. The observed threefold stimulation of ATPase acticity in plasma membrane-enriched fractions required higher PAF concentrations than for the tonoplast H^+-ATPase (Martiny-Baron and Scherer 1989) and transport activity was too low to be investigated. However, Palmgren and Sommarin (1989) obtained stimulation of both H^+ transport and ATPase activity in highly purified plasma membranes by LPC. We could not find a PAF-dependent 100 kDa polypeptide in but other PAF-dependent phosphorylated proteins in plasma membrane-enriched fractions. Therefore, PAF does not seem to act via phosphorylation of the ATPase itself. The lipid could either activate directly the H^+-ATPase or activate the phosphorylation of a different protein in the plasma membrane where substrates of the phospholipid-activated protein kinase have been found. Since PAF activated ATPase activity in the presence of 0.05% Triton X-100 (Martiny-Baron and Scherer 1989) we presume that LPC, its plant counterpart, should have a special role in H^+-ATPase regulation (Palmgren and Sommarin 1989).

In vivo, only PAF, but not LPC, was the biologically active

lipid (Scherer and Nickel 1988). We observed that PAF stimulated medium acidification by cultured soybean cells whereas other lipids were inactive. More recently, we demonstrated a transient hyperpolarization of the membrane potential induced within 5-10 min by PAF (Nickel, Ephretikine, Scherer and Guern, unpublished). We interprete both findings as activation of the plasma membrane pump by the lipid and the failure of LPC to stimulate in *in vivo* as a failure to reach threshhold levels due to the probably more rapid metabolism of the natural LPC as compared to the, for the plant, artificial PAF.

Additionally, by a microfluorimetric method, we could show the acidification of the vacuole of *Catharanthus roseus* protoplasts by 0.2 pH units within 20-40 min and also the acidification of isolated vacuoles (Nickel, Manigault, Scherer and Guern, unpublished). These regulatory influences of PAF in vivo show that we can successfully interfere with a suggested signal transduction chain (Fig. 10) using PAF as a tool . Filipin, used as a detergent specific for the sterol-rich tonoplast and plasma membrane (vom Dorp et al. 1986), showed no effect *in vivo* so that a trivial explanation for the PAF effects seems precluded.

Biologically active lipids from plants. A proton transport-stimulatory phospholipid has been described by us (Stoffel and Scherer 1987; Scherer 1989). In addition to this, we described the activities of LPC and LPE on *in vitro* transport stimulation and on protein kinase activity (Fig. 1; Martiny-Baron and Scherer 1989). Ongoing isolation experiments show that there is a third phospholipid, chemically similar to LPC and LPE. These latter two have also been isolated from zucchini hypocotyls and shown to stimulate H^+ transport *in vitro* with equal specific acvtivity as the commercially available substances (Schmitz and Scherer, unpublished).

Phospholipid/Ca^{2+}-activated protein kinase. Originally, we described a PAF- and Ca^{2+}-dependent protein kinase, associated with plant membranes (Martiny-Baron and Scherer 1988) but we found that it is also activated, in decreasing order of effectiveness, by LPC and by LPE (Martiny-Baron and Scherer 1989; Fig. 1). The protein kinase seems to have a preference for the choline head group since LPE but not LPS was active (Martiny-Baron, unpublished observation). The selectivity for the lipid backbone seems to be less stringent (Fig. 1) but no diacylgly-

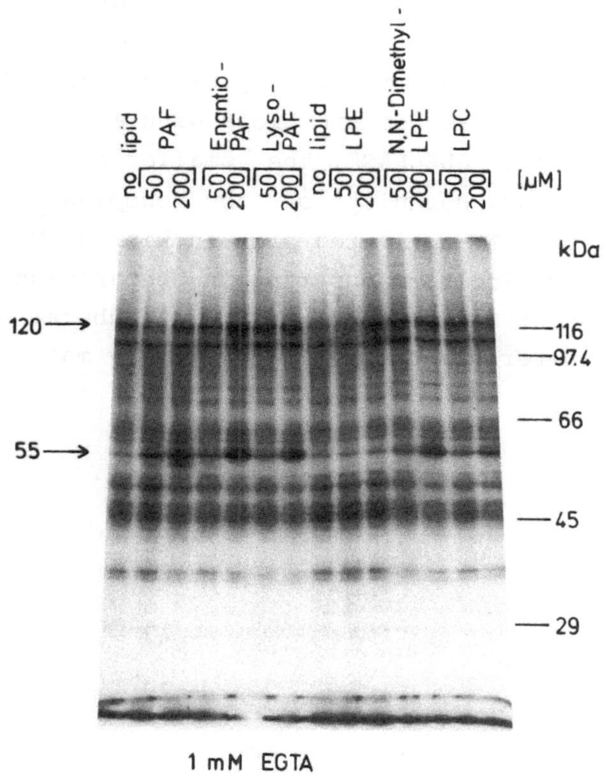

Fig. 1. Phosphorylation of polypeptides in zucchini microsomes in the absence and presence of various phospholipids. Each lipid was added at 50 and 200 µM concentration. EnantioPAF is 3-O-alkyl-2-acetyl-sn-glycero-1-phosphorylcholine, the C2-enantiomer of PAF. Note the phosphorylation of the 55 kDa polypeptide (unpublished data, Martiny-Baron and Scherer)

cerolphospholipid and no diacylglyceride or phorbol ester activated this kinase. As a second activator, submicromolar Ca^{2+} stimulated the phosphorylation of the same 55 kDa polypeptide which showed the most strongly PAF-stimulated phosphorylation. The concentration dependence suggests that mobilization of cytosolic Ca^{2+} could be physiologically relevant for the lysophospholipid-activated protein kinase.

We found several substrate polypeptides in the tonoplast and in the plasma membrane. No 100 kDa polypeptide was phosphorylated so that activation of the 100 kDa plasma membrane H^+-ATPase by the lysophospholipid-activated protein kinase is unlikely.

We could identify, by immunoprecipitation (Fig.2) and a comparison of 2-dimensional Western blots with an autoradiography of a two-dimensional SDS-PAGE, the smaller of the two large subunits ("regulatory subunit") of the tonoplast H^+-ATPase as a substrate of the PAF-activated protein kinase (Martiny-Baron, Manolson, Scherer, Poole, manuscript in preparation). Since the lipid activation of H^+ transport in vitro (Scherer et al. 1988) and in vivo (Scherer and Nickel 1988) closely matched the lipid

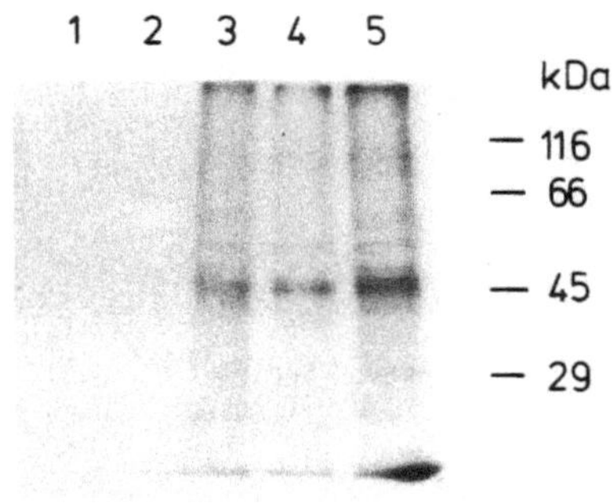

Fig. 2. Immunoprecipitation of the tonoplast H^+-ATPase small subunit from zucchini microsomes after phosphorylation. Lane 1: no microsomal protein. Lane 2: 5 mM EDTA (no $MgCl_2$). Lane 3: no PAF added. Lane 4: 10 µg/ml PAF. Lane 5: 50 µg/ml PAF. (unpublished, from Martiny-Baron et al. 1989)

specificity of the protein kinase we suggest regulatory phosphorylation of the tonoplast subunit as a mechanism of activation for tonoplast H^+-ATPase. The only PAF-dependent phosphorylated 100 kDa polypeptide that we observed was present in tonoplast-enriched fractions isolated from cultivated soybean cells

(Nickel and Scherer, unpublished) so that we have no evidence for the phosphorylation of the plasma membrane H^+-ATPase in contrast to others (Schaller and Sussman 1988).

Activation of phospholipase A_2 by auxin and generation of LPC and LPE as potential second messengers. The action of phospholipase A_2 on phosphatidylcholine (PC) and on phosphatidylethanolamine (PE), the two most abundant phospholipids in eucaryotic membranes, liberates LPC and LPE. Obviously, this would trigger, in vitro, the lysophospholipid/Ca^{2+}-activated protein kinase so that we hypothesize that this can happen also in vivo, since we found a stimulated turnover of PC and PE to LPC and LPE induced by auxin in cultered soybean cells (Fig. 3-7) and in isolated membranes (Fig. 8,9).

The time course of the turnover of radioactive PC to LPC in soybean cells prelabeled with choline is shown in Fig. 3. No apparent lag time for the action of 2,4-D was observed in this experiment but in others a 1-2 min lag time was found. This

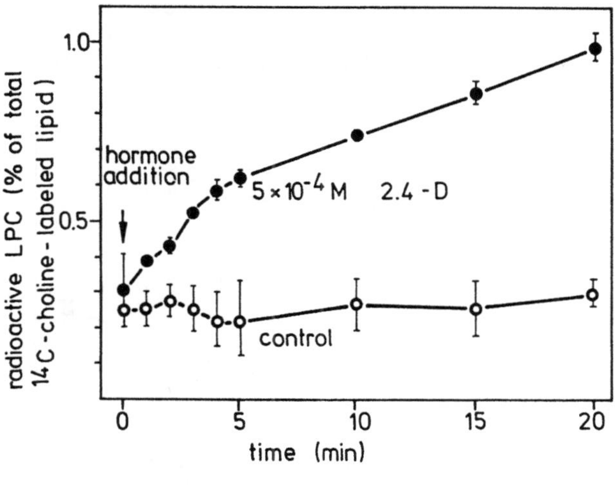

Fig. 3. Time course of redistribution of label from PC to LPC induced by the addition of $5x10^{-4}$ M 2,4-dichlorophenoxyacetic acid to cultured soybean cells. Duplicate values were taken and bars indicate the experimental error (Scherer, unpublished)

turnover reaction was auxin-specific in that active auxins, indole-3-acetic acid (IAA) (Fig. 5), α-naphthylacetic acid (α-NAA) and 2,4-dichlorophenoxyacetic acid, induced it whereas the inactive auxin β-naphthylacetic acid did not (Fig. 4,5). 2,3-

dichlorophenoxyacetic acid was relatively active although it is generally considered to be a weak auxin (Fig. 4,5). A similar turnover of PE to LPE was observed in ethanolamine-prelabeled soybean cells (Fig. 5).

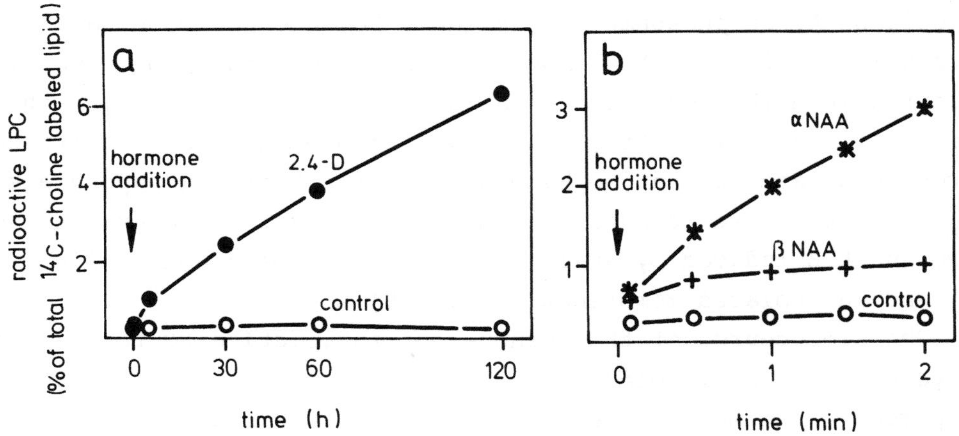

Fig. 4. Time course of turnover of radioactive PC to LPC in cultured soybean cells prelabeled with choline in the absence and presence of $5x10^{-4}$ M hormones (Scherer, unpublished)

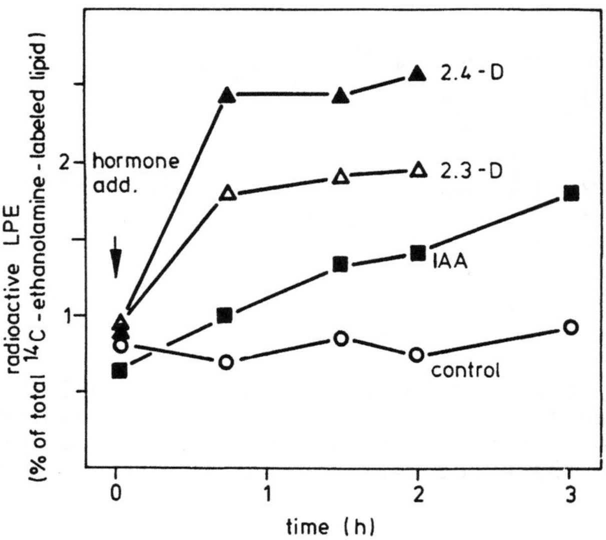

Fig. 5. Time course of the turnover of radioactive PE to LPE in cultured soybean cells prelabeled with ethanolamine in the absence and presence of $5x10^{-4}$ M hormones (Scherer, unpubl.)

The time course of auxin-induced turnover of radioactive LPC or LPE usually was steadily increasing up to two hours and then usually levelled off. Since the hormone was added in a chase and not in steady state situation for the label and since the total lipid-bound label decreased after ca. 1 h we could not decide whether or not radioactive LPC or LPE also turned over or simply accumulated. Therefore, we made an experiment where the hormone, 2,4-D, was removed from the cells after the turnover reaction had been induced for a suitable time (Fig. 6).

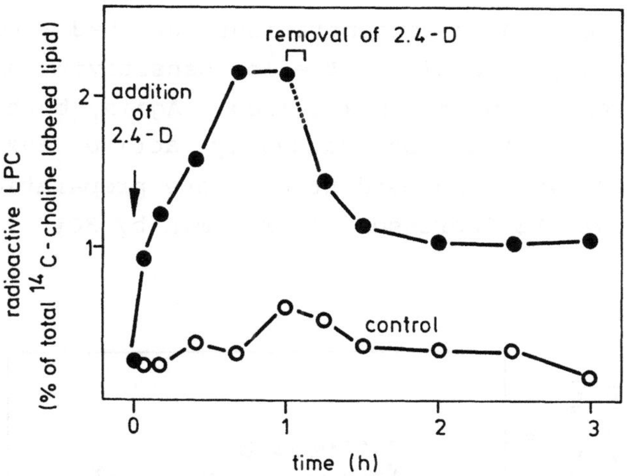

Fig. 6. Time course of radioactive turnover from PC to LPC in cultured soybean cells after the addition of $5x10^{-4}$ M 2,4-D and decrease of label in LPC after the withdrawal of hormone (Scherer, unpublished)

Clearly, the radioactive LPC declined after the withdrawal of the hormone, indicating that LPC is down-regulated. This is important if it is to serve the function of a second messenger.

We also observed that abscisic acid (ABA) antagonized the auxin-induced turnover reaction (Fig. 7). In the presence of 2,4-D, ABA decreased the accumulation of radioactive LPC whereas ABA alone had no apparent effect. However, an inhibition of phospholipase A_2 by ABA alone probably would have been overlooked due to the small amounts of radioactivity to be compared. This effect of ABA is not surprising considering the well-known antagonism of auxin and ABA.

In vitro, we observed a stimulation of phospholipase A_2 by auxins in isolated membranes from cultured soybean cells or from zucchini hypocotyls which had been prelabeled either with ethanolamine or choline. This response to auxin would explain the observed in vivo turnover of radioactive PC and PE to LPC or LPE induced by auxin. The in vitro response, even with membranes isolated from the cultured soybean cells, was much more sensitive to low auxin concentrations but had a very similar auxin specificity (Fig. 7,8). The low sensitivity of the cultured soybean cells remains unexplained. Again, both PC and PE substrates for the reaction induced by active auxins. In experiments with five independent membrane preparations, values for radioactive LPE increased after 5 min by 30% (+ 16%) above

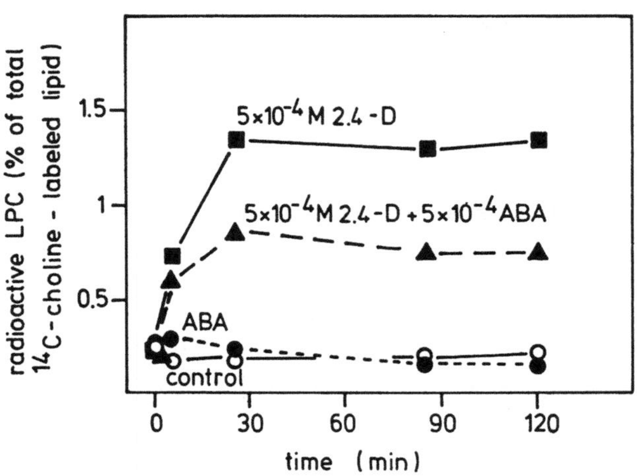

Fig. 7. Time course of radioactive turnovcer from PC to LPC in cultured soybean cells after addition of $5x10^{-4}$ M 2,4-D and/or ABA (Scherer, unpublished)

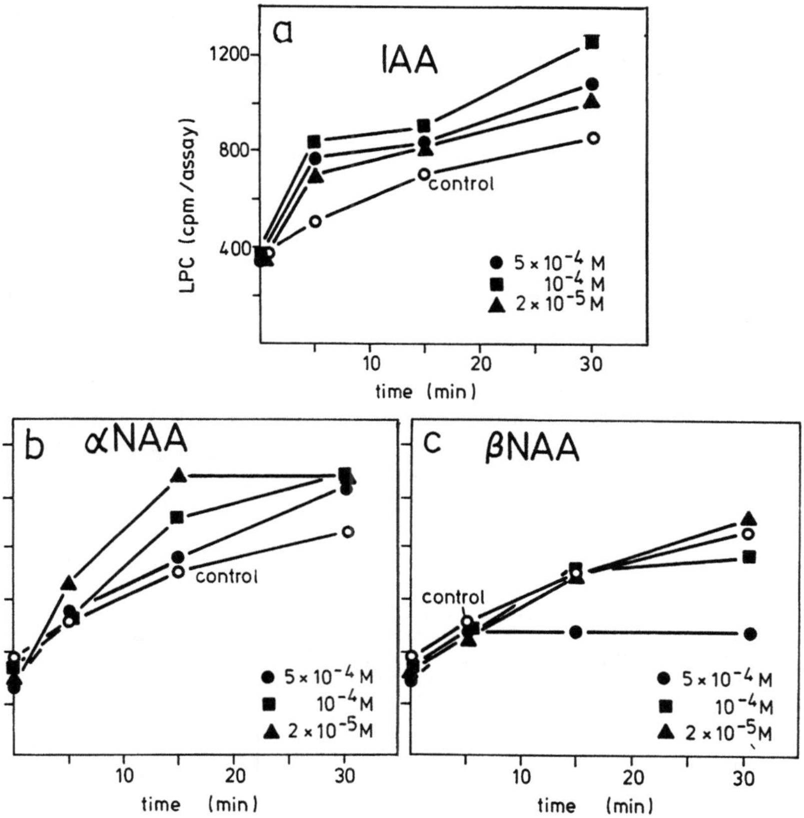

Fig. 8. Formation of lysophosphatidylcholine (LPC) in choline-prelabeled membranes isolated from cultured syobean cells. The hormone concentrations are indicated. Only active auxins induced an increase over the control (o). (André and Scherer, unpublished)

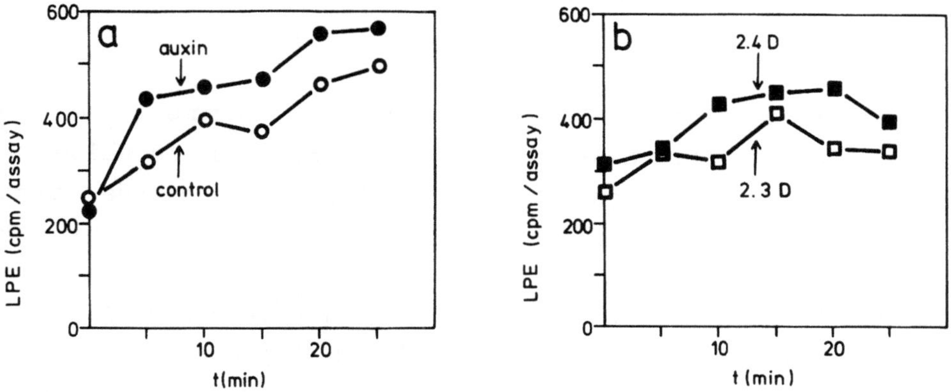

Fig. 9. Formation of lysophosphatidylethanolamine (LPE) in ethanolamine-prelabeled membranes isolated from zucchini hypocotyls. Hormones were added at $2x10^{-5}$ M (André and Scherer, unpublished)

the value at t=0 (=100%) in the controls and by 73% (± 9%) in assays with $2x10^{-5}$ M auxin, indicating again a very rapid reaction.

DISCUSSION

Our working model for a hypothetical signal transduction chain for the plant hormone auxin is shown in Fig. 10. If we can demonstrate coupling of auxin-stimulated phospholipase A_2 to lysophospholipid-activated protein kinase this would be the first complete signal transduction chain at the molecular level

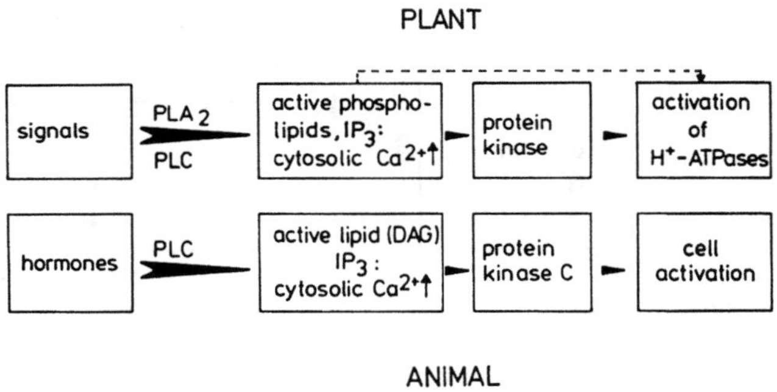

Fig. 10. Hypothetical model for a molecular signal transduction chain triggered by auxin and comparison with known animal signal transduction pathways

in plants from signal (auxin) to regulated enzymes, the H^+-ATPases. Our <u>in vitro</u> assay system (Fig. 8,9) should allow us to answer further the question of participation of the known auxin receptor (Löbler and Klämbt 1985) or of G-proteins. We have good reasons to assume that the tonoplast H^+-ATPase is regulated by phosphorylation of one of its subunits (Fig. 2; Martiny-Baron et al. 1989) whereas direct lipid activation is an alternative model for the plasma membrane H^+-ATPase (Palmgren and Sommarin 1989). We interpret previuos results on auxin stimulation of ATPase in vitro as indicative of coupling between phospholipase A_2 and lysophospholipid-activated protein kinase (Scherer and Morré 1978; Scherer 1981) whereas release of inorganic phosphate at low ATP concentrations (<10µM) stimulated by auxin probably rather indicated a futile cycle of protein kinase and phosphatase (Scherer 1984).

We did not find that fusicoccin activated phospholipase A_2 (unpublished observation) so that fusicoccin could use a different pathway for the activation of plasma membrane H^+-ATPase. Yet another tool for probing this signal transduction pathway is the application of PAF to plant cells (Scherer and Nickel; Nickel et al., unpublished results) which supports the interpretation that the PAF-activated protein kinase and/or PAF is linked to the stimulation of tonoplast and plasma membrane H^+-ATPase.

Comparing animal and plant signal transduction at the molecular level a similar function of signal-activated phospholipase C and A_2 is indicated (Berridge 1987; Morse et al. 1989; Burch et al. 1986; Bar-Sagi and Feramisco 1986; Jelsema 1987). However, the metabolites generated by these enzymes and their use as second messengers seems to be different. Diacylglycerol, generated by the action of phospholipase C on phosphatidylinositol-4,5-bisphosphate, does not stimulate the lysophospholipid/Ca^{2+}-activated protein kinase described by us (Martiny-Baron and Scherer 1989) and no activation by diacylglycerol alone of protein kinase could be convincingly demonstrated up to date (Morse et al. 1989) as it is typical for animal protein kinase C (Nishizuka 1984). The second messengers or precursors for messenger molecules released by phospholipase A_2 in animals is arachidonic acid which is not present in higher plants and a role of other fatty acids or derivatives as messengers in plants is still speculative (Anderson 1989).

Instead, we propose that lysophospholipids (LPC, LPE) constitute a major second messenger system in plant cells which are mimicked, fortitiously, by PAF (Scherer 1989) which trigger a membrane-associated protein kinase. A role of lysophospholipids in animal signal transduction is not known so that the suggested signal pathway could be unique for plants.

Acknowledgement. This work was supported by grants from the Deutsche Forschungsgemeinschaft and the Bundesministerium für Forschung und Technologie.

REFERENCES

Anderson JM (1989) In: Boss WF, Morré DJ (eds) Second Messengers in Plant Growth and Development. Alan Liss, New York, pp. 167-179

Bar-Sagi D, Feramisco JR (1986) Science 223, 1061-1068

Berridge MJ (1987) Annu. Rev. Biochem. 56, 159-163

Brightman AO, Barr R, Crane FL, Morré DJ, (1987) Plant Physiol. 86, 1264-1269

Burch RM, Luini A, Axelrod J (1986) Proc. Natl. Acad. Si. USA 83, 7201-7205

Connett RJA, Hanke DE (1986) Planta 169, 216-221

Elliott DC, Fournier A, Kokke YS (1988) Phytochemistry 27, 3725-3730

Jelsema CL (1987) J. Biol. Chem. 262, 163-168

Löbler M, Klämbt D (1985) J. Biol. Chem. 260, 9848-9853

Manolson MF, Quellette BFF, Filion M Poole RJ (1988) J. Biol. Chem. 263, 17987-17994

Martiny-Baron G, Scherer GFE (1988) Plant Cell Rep. 7, 579-582

Martiny-Baron G, Scherer GFE (1989) J. Biol. Chem., in press

Martiny-Baron G, Manolson BF, Scherer GFE, Poole RJ (1989) submitted

Morré DJ, Morré JT, Varnold RL (1984) Plant Physiol. 75, 265-268

Morse, MJ, Satter RL, Crain RC, Coté GG (1989) Physiol. Plant. 76, 118-121

Nishizuka Y (1984) Nature 308, 693-698

Palmgren MG, Sommarin M (1989) Plant Physiol., in press

Schaller GE, Sussman RM (1988) Planta 173, 509-518

Scherer GFE (1981) Planta 151, 434-438

Scherer GFE (1984) Planta 161, 394-397

Scherer GFE (1985) Biochem. Biophys. Res. Commun. 133, 1160-1167

Scherer GFE (1989) In: Boss WF, Morré DJ (eds) Second Messengers in Plant Growth and Development, Alan Liss, New York, pp. 167-179

Scherer GFE, Martiny-Baron G (1987) In: Klämbt D (ed.) Plant Hormone Receptors, NATO-ASI series H, vol 10, Springer, Berlin, Heidelberg

Scherer GFE, Martiny-Baron G, Stoffel B (1988) Planta 175, 241-253

Scherer GFE, Morré DJ (1978) Biochem. Biophys. Res. Commun. 84, 238-247

Scherer GFE, Nickel R (1988) Plant Cell Rep. 7, 575-578

Scherer GFE, Stoffel B (1987) Planta 172, 127-130

Schuhmaker KS, Sze H (1987) J. Biol. Chem.262, 3944-3946

vom Dorp B, Volkmann D, Scherer GFE (1986) Planta 168, 151-160

FUNCTIONAL RECONSTITUTION OF SOLUBILISED RECEPTORS

P. ADUCCI
Dept. of Biology
University of Rome "Tor Vergata"
Via O. Raimondo
00173 Rome, Italy

Participants:

W. MICHALKE
Dept. of Genetics
Schänzlestrasse 9-11
78100 Freiburg, FRG

M. ROSSIGNOL
Lab. Physiol. Veg.
B.V. 1540
21034 Dijon, France

M.A. VENIS
Inst. Horticultural Res.
East Malling
Maidstone, Kent ME19 6BJ, U.K.

Resolution of membrane components and their reconstitution is a classic approach to the analysis of their physiological and biological function. It has been successfully used in many animal systems and more recently its possible application to plants has emerged. In particular, the possible use of reconstitution for plant hormone receptors appears of interest as much progress in this field has been made in the last few years. It is known that receptors are, in general, proteins that specifically recognize a signal (light, hormones) and interact with one or more effector systems. The receptor is activated by the ligand and consequently triggers the effector system. This transduction step may depend on receptor changes such as conformational, chemical modification, microaggregation or internalization and/or by the interaction with transducers such as G proteins, phosphoinositides, protein kinases. For plant hormone receptors only limited evidence is available so far on the mechanisms of their activation and the possible routes of signal transduction. The elements of signal perception and transduction are in a functional membrane complex; therefore their isolation and reconstitution is an important approach to identification of the specific role of each working part of the chain. Before undertaking reconstitution experiments it is first necessary to establish suitable conditions for assay, for solubilisation and for reconstitution with preservation of function. Studies with isolated membranes and with solubilised receptors are therefore important in developing suitable conditions for reconstitution. Some examples will illustrate this approach.

NATO ASI Series, Vol. H 47
Signal Perception and Transduction in Higher Plants
Edited by R. Ranjeva and A. M. Boudet

W. Michalke. One of the elements of transduction chains, as briefly outlined above, may be G-proteins (as they are described in animal systems). Evidence for GTP binding has been reported in higher plants and is confirmed by several reports at this meeting and the following results.
In microsomes from etiolated zucchini hypocotyls, binding of guanosine triphosphate (GTP) has been demonstrated. Scatchard plots reveal more than one homogeneous class of binding sites, K_d values thus obtained being in the range from 10^{-8} to 10^{-7} mol l^{-1}. GDP is able to displace GTP from these binding sites with very similar effectivity, whereas ATP is 100 times less effective. Of the microsomal binding activity 25 to 30% appears to be located on PM vesicles according to PEG fractionation or metrizamide gradients, whereas the majority of binding is found in other membrane fractions. GTP binding is stimulated after incubation of the membrane vesicles with a raw cell wall extract from mycelia of the phytopathogenic fungus Phytophthora megasperma var. glycine. Isolation of the G-protein(s) and the receptor for the elicitor will allow reconstitution experiments to investigate the interaction of the partecipating components.
Polar auxin transport and its specific inhibition by phytotropins is another process probably mediated by an ensemble of membrane proteins, whose identity and mode of interaction has get to be established. An in vitro IAA accumulation system shows several characteristic of in vivo auxin transport. In collaboration with G. Katekar this system has been used to compare NPA binding inhibition and stimulation of IAA accumulation caused by the addition of phytotropins. The results imply that only after saturation of high affinity NPA-binding IAA accumulation is stimulated vià a site of lower affinity. From experiments with azido-NPA as photo-affinity label and chromatography on NPA affinity matrices, it appears that proteins with molecular weights of roughly 50, 60 and 75KDa are likely to be partecipating in phytotropin binding. Isolation of these proteins will allow us to test whether they can be reconstituted as phytotropin binding sites, and to search for additional components of a functional auxin transport system.
M.A. Venis. Partly purified (post-DEAE) solubilised auxin binding protein (ABP) from maize, containing ATPase activity was added to a black lipid

membrane system and effect on transmembrane current monitored. A jump in current was observed when ATP and auxin (NAA) were also present. All three components - ABP preparation, auxin and ATP - were essential, but could be added in any order. The physiologically inactive auxin analog benzoic acid was ineffective. The pH-dependence of the electrochemical response in comparison with the pH dependence of binding suggested that the current increase was due to H^+-transport. The results were consistent with auxin plus receptor stimulation of a H^+-translocating ATPase. Reconstitution experiments with more highly purified ABP preparations were less successful, probably due to elimination of one or more essential components. Recently we have begin attempts to effect more defined reconstitution using H^+-ATPase purified from maize roots and ABP of differing purities incorporated into phosphatydilcholine plus phosphatydilethanolamine liposomes by octyl glucoside dilution. So far we have been unable to detect auxin stimulation either of ATP hydrolisis or of H^+ transport by ACMA fluorescence quenching. However, we plan to investigate this further in collaboration with Rossignol and coworkers since they have been able to demonstrate auxin stimulation of H^+-translocating ATPase in plasma membrane vesicles from tobacco. This stimuation is lost when the membranes are washed in Triton X-100 which is known to remove ABP.

P. Aducci. FC receptors represent one of the best characterized example of plant receptors functionally reconstituted in a model system. These proteins are specifically recognized by the fungal toxin FC, which stimulates many physiological processes. In higher plants these effects mainly result from the activation of the plasma membrane H^+-ATPase.

In our laboratory the solubilised FC receptors either from corn or from spinach have been successfully inserted at a high yield into synthetic liposomes by the freeze-thaw method. Their orientation in proteoliposomes shows that more than 80% of sites are outside exposed. The biochemical properties of proteoliposomes have been analysed and compared with those of membrane-bound proteins. Some parameters, such as kinetics and pH-dependence, show close similarities in the two systems, suggesting that proteoliposomes are a good model for signal transduction studies.

The more recent simultaneous reconstitution into liposomes of FC receptors and the H^+-ATPase, both from corn tissues, was successfully achieved in collaboration with Rossignol and coworkers, by the freeze-thaw method. The ATP-driven H^+-pumping activity of proteoliposomes, measured by ACMA fluorescence, markedly increased upon addition of FC; the effect was not observed when receptors and ATPase preparations were separately reconstituted into liposomes. The experiments were performed with crude preparations of FC receptors and a partially purified H^+-ATPase; it is therefore impossible to establish if the signal is directly transmitted from receptors to H^+-ATPase or enters in a more complicated pathway. The model system described can however be a useful tool for future investigation of the elements involved in this transduction chain.

M. Rossignol. As pointed out by P. Aducci, reconstitution of FC-induced stimulation of the H^+-translocation activity of the plant plasma membrane H^+-ATPase can be achieved by the freeze-thaw procedure or by octyl-glucoside dilution. This latter method relies on the formation of ternary mixed micelles of lipid detergent and proteins. Optimal reconstitutions are obtained after properly adjusting the FC binding proteins-ATPase ratio, the lipid-protein ratio and the bound detergent-lipid ratio. An additional parameter is the free detergent concentration after dilution and formation of vesicles. It should be emphasized that optimal parameter values have to be determined (i) for each kind of reconstitution (nature of protein mixtures, type of lipid) and (ii) according to the effect to be studied. For instance, parameters values for maximal stimulation of H^+-translocation upon FC addition differ greatly from those for maximal H^+-translocation itself.

General discussion.

The approach of reconstitution has been, so far, very empyrical and no specific all-purpose procedure is available. Several different methods have been used such as a) spontaneous reconstitution, when a purified membrane protein is complemented to a depleted membrane b) reconstitution by sonication of organelles or membrane vesicles with or without detergent c) reconstitution with solubilised proteins. This method appears to be the most reliable in order to show a functional relationship among different membrane components because it allows reconstitution of isolated and purified components.

Some difficulties and pitfalls may be encountered with reconstituted systems such as a) the difficulty of knowing how the protein(s) really operate *in vivo* b) the failure of some proteins to be functionally reconstituted once purified c) the use of particular assays for proteoliposomes which differ in size and other properties from plasma membrane d) the simultaneous reconstitution of different components may involve conditions which are less than optimum for each individual component. On the other hand, in favourable circumstances it is possible to learn a number of things from resolution and reconstitution studies such as the number and the role of protein components, the type and function of phospholipids and the role of asimmetry of proteins and phospholipids.

In conclusion, resolution and reconstitution will be instrumental in elucidating the functional role of membrane components but need to be complemented by other approaches such as *in vivo* physiology and, of course, molecular biology.

SIGNAL TRANSDUCTION IN PLANT CELLS AND THE CALCIUM MESSAGE

M.D. Fricker, S.G. Gilroy and A.J. Trewavas
Department of Botany
The King's Buildings
Mayfield Road
Edinburgh
EH9 3JH

Introduction

In this short article we discuss the general context of signal transduction in plants. Particular reference is made to the apparent deficiency of signal transduction pathways versus the multitude of conflicting environmental and endogenous signals. Yet all these signals have to be integrated to give the appropriate developmental/ physiological response. Investigation of the individual elements of transduction events have perhaps over-emphasized simple linear pathways, whereas a network of interacting transduction elements most closely model these phenomena.

Evidence for all of the main elements of a Ca^{2+}-based transduction pathway have emerged in plant cells and these will be briefly described with emphasis on most recent advances. The information available for any individual biological system is less comprehensive and only gives equivocal support for Ca^{2+}-based transduction pathway. More detailed consideration of two putative Ca^{2+} regulated processes (abscisic acid (ABA)-induced stomatal closure and development of polarity in fucoid zygotes) will highlight some of the problems encountered when investigating signal transduction at the cellular level and underline the current paucity of experimental data.

Plant cells respond to many signals

In general, plant cells lack highly specialised cell types. Several functions are often performed by an individual cell and these cells tend to respond to a far wider range of signals than their animal equivalents. In the examples discussed below the fixation of a developmental axis in the *Fucus* zygote is sensitive to at least 14 external factors including light, temperature, electrical fields and ionic gradients (Jaffe, 1968). Stomatal guard cells respond to a wide

NATO ASI Series, Vol. H 47
Signal Perception and Transduction in Higher Plants
Edited by R. Ranjeva and A.M. Boudet

range of both environmental and endogenous factors including light, temperature, CO_2 level, abscisic acid, cytokinin, auxin, giberellin, Ca^{2+} and K^+ (Zeiger, 1983; Willmer, 1988). Response to such an extensive range of stimuli seems to be a general property of plant systems. Any process of plant differentiation which has been examined in detail (e.g. flower induction, cell division, growth, abscission, root formation or the induction and breakage of seed and bud dormancy) seems equally sensitive to a complex range of incoming signals (Trewavas 1986). Some sort of integrating transduction system must be operating to translate this multitude of stimuli to a coherent set of cellular responses. This is especially true of the examples of the polarising fucoid zygote and guard cells where the range of signals apparently lead to the same response, axis fixation and stomatal closure, respectively, *within an individual cell*. It seems unlikely that they could all operate simultaneously via the same linear transduction pathway.

Simple models to complex networks

A traditional model represents the signal transduction process as a linear train of events, with perception of a single stimulus leading, via a sequence of transduction elements, to the response. This model may be appropriate for a single stimulus considered in isolation, but is inadequate to describe a coordinated response to numerous, conflicting signals. It is important to make the distinction between the design of an experimental approach, where the number of variables is deliberately limited, and investigation of the actual control exerted by each individual factor *in vivo* as part of a network of complex interactions. The final response will be dictated by the total spectrum of signals perceived by the cell and the state of the transduction network at that time. In this context a second crucial distinction has to be drawn between responses of populations and the response of individuals within the population. Where it is possible to examine responses of individual cells it is apparent that there is variation within the population in response to the same stimulus (e.g. Omasa and Onoe, 1984). A graded response observed at the level of the population could be achieved by either gradual, continuous response of all cells, a frequency-dependent modulation for each cell or a 'switch' response with varying thresholds for each cell in the population, e.g. Trewavas (1987). The characteristics of the signal transduction pathway are radically different between these cases. Evidence at the molecular level has suggested that bistable molecular switches may occur (Blowers and Trewavas, 1988), whilst channel opening events are frequency modulated rather than amplitude modulated (e.g. Takeda, Kurkdjian and Kado, 1985). It seems that the biological variation, which as experimental scientists we try to reduce to some mean level, may be a rich source of insight in to how signal transduction phenomena interact. We require data not only about reactions of populations but of individuals within that population.

Ca^{2+}-based signal transduction in plants

The central element in the Ca^{2+}-transduction pathway is modulation of cytosolic Ca^{2+} levels in response to a signal. This may occur through increased influx or decreased efflux at the plasma membrane or shuttling of internal Ca^{2+} between cytosol and internal compartments, e.g. vacuole, endoplasmic reticulum (ER), chloroplasts and mitochondria. The consequent change in cytosolic Ca^{2+} levels may affect Ca^{2+}-dependent enzymes directly or via an intermediate Ca^{2+}-binding protein, e.g. calmodulin (Hepler and Wayne, 1985; Evans, 1988; Gilroy *et al.* 1988; Gilroy and Trewavas, 1989).

The presence of Ca^{2+}-channels in the plasma membrane has been deduced from the inhibition of various plant processes (mitosis, spore germination, polarised growth) by chemicals that antagonise the action of Ca^{2+}-channels in animal cells (Hepler and Wayne, 1985). Plant cell Ca^{2+} uptake is inhibited by the channel blocker verapamil and a verapamil binding protein has been solubilised and partially purified (Harvey *et al.*, 1989). The purified fraction exhibits Ca^{2+}-channel activity when reconstituted into black lipid bilayers (H. Harvey and M. Tester, per. comm.). The major protein present in this fraction has a molecular weight of 169 KDa similar in size to the main animal Ca^{2+}-channel protein of 165 KDa. In *Fucus* preliminary evidence suggests that LU49888 (azido desmethoxy verapamil, a photoaffinity analogue of verapamil) binds to a protein of similar molecular weight (Fricker and Trewavas, unpublished data).

A plasma membrane located Ca^{2+} pumping ATPase has been demonstrated in a range of tissues (Evans, 1988; Gilroy *et al.*, 1988). This protein cross reacts with antibodies raised to the Ca^{2+}-ATPase of erythrocyte plasma membrane, has a similar molecular weight (140 kDa), forms a phosphorylated reaction intermediate and is activated by calmodulin (Briars and Evans, 1989).

The vacuole is a major repository for Ca^{2+} in plant cells and there is evidence that vacuolar Ca^{2+} can be rapidly mobilised (Franceschi, 1989). Ca^{2+} movement appears to be mediated by a nH^+/Ca^{2+} antiport with lower affinity but greater transport capacity than the plasma-membrane Ca^{2+}-ATPase (Evans, 1988). Transport into the lumen of the ER is mediated by a Ca^{2+}-ATPase only slightly differing in properties to the plasma membrane enzyme. At present there is no direct evidence for the molecular identity of an efflux carrier or channel in the ER.

Calmodulin stimulation of the plasma membrane Ca^{2+}-ATPase has already been referred to and it seems that calmodulin is present in every plant system so far investigated. Sequences of several plant calmodulins indicate >90% homology with animal calmodulins (Roberts, Lukas, Harrington and Watterson, 1986). Recently calmodulin specifically associated with the plasma membrane has been described (Collinge and Trewavas, 1989) and there is preliminary evidence for calmodulin association with the tonoplast (Graham, Blowers and Smith, per. comm.).

A limited number of calmodulin activated enzymes have been characterised, notably Ca^{2+}-ATPase, quinate oxidoreductase, NAD kinase and protein kinases (Gilroy *et al.*, 1988). As yet only a limited number of substrates for the latter group have been determined (Blowers and Trewavas, 1989). Other Ca^{2+}-calmodulin binding proteins have been detected using gel overlay techniques but their identity is unknown (e.g. Collinge and Trewavas, 1989).

Calsequestrin is a high capacity, low affinity Ca^{2+} binding protein initially isolated from animal cells as a constituent of calciosomes (Krause *et al.*,1989). These membrane bound bodies use calsequestrin as a simple internal reservoir of Ca^{2+}. Antibodies to animal calsequestrin cross react with plant calsequestrin. Calciosomes have also been detected in, and isolated from, *Mougetia* where they are thought to be concerned with the rotation of the chloroplast in response to light (Jacobshagen *et al.*, 1985).

Numerous measurements of cytosolic Ca^{2+} have now been made in plant cells using fluorescent dyes, aequorin or Ca^{2+}-selective microelectrodes. Typically such measurements show cytosol Ca^{2+} levels are low (< 300 nM). Potentially physiologically relevant changes in cytosol Ca^{2+} are also being increasingly reported (e.g. Brownlee and

Pulsford 1988, Callaham and Hepler, 1988; Miller and Sanders, 1988; Gilroy, Hughes and Trewavas, 1987).

The potential role of phosphatidylinositols in signal transduction in plants is still under investigation. The paradigm, borrowed heavily from animal studies, revolves around the stimulus evoked activation of a phospholipase C releasing inositol-1,4,5-triphosphate (IP_3) and diacylglycerol (DAG) from phosphatidylinositol-4,5-bisphosphate (PIP_2) on the inner face of the plasma membrane. IP_3 and DAG then go on to release Ca^{2+} from the ER or activate a protein kinase C, respectively. Many of the enzymes and metabolites predicted for this system to operate have been identified, though not necessarily concurrently, in various plant cells (Boss, 1989). Notably IP_3, DAG, phospholipase C, a purported plant protein kinase C and apparent auxin-induced production of IP_3 have been reported. However, the plant plasma membrane posesses extremely small amounts of PIP_2 and plant inositol metabolism seems to differ from its animal counterpart (Boss 1989). Also, IP_3 releases vacuolar rather than ER Ca^{2+} in plants (Boudet and Ranjeva, 1989). These observations coupled with difficulties in the unequivocal identification of the component inositol lipids and phosphates have made interpretation of the role of these potential elements of plant signal transduction problematical.

Similarly the current evidence for plant G-proteins is limited. Although GTP-binding proteins have been identified in Zuchini and *Lemna paucicostata* and GTP will release Ca^{2+} from microsomal preparations (Drobak *et al.*, 1988), this is far from assigning a role to G-proteins in plant signal transduction.

The Ca^{2+} transduction pathway in specific systems

The preceding discussion has outlined the evidence for the presence of elements of the Ca^{2+} signal transduction pathway in plants, using data drawn from a wide range of disparate sources. In the next section we will concentrate on the evidence in two specific systems for a physiological role for this pathway.

Ca^{2+} and ABA induced stomatal closure

Low concentrations of external Ca^{2+} reduce stomatal opening or cause stomatal closure in epidermal strips (De Silva, Hetherington and Mansfield, 1985; Schwartz, 1985a; Schwartz, Ilan and Grantz, 1988) or

Silva *et al.*, 1985b). In similar experiments using guard cell protoplasts, however, ABA at the same concentration was effective in the absence of external Ca^{2+} (Smith and Willmer, 1988). This result does not preclude the modulation of cytosolic Ca^{2+} by mobilisation of internal Ca^{2+} stores. Attempts to measure the effect of ABA on plasma membrane Ca^{2+} fluxes also yielded contradictory data with either stimulation, inhibition or no effect (MacRobbie, 1989). Our preliminary results with the ratio photometry system have also yielded no consistant response, varying from no effect to marked oscillations of varying intensity, duration and periodicity (Figure 2).

The site of action of ABA is also unclear. Affinity cross-linking studies indicated three ABA binding proteins (M_r 20,200; 19,300 and 14,300) on the external face of the plasma membrane with low (nM) k_D binding constants, stereo-specificity but different sensitivities to pH (Hornberg and Weiler, 1984). However, patch clamp studies on excised patches of plasma membrane from guard cells indicate that ABA is effective when applied to the cytoplasmic face of the membrane (Shauf and Wilson, 1987). ABA stimulated the mean open time and burst duration of plasma membrane potassium channels.

In vivo, guard cells are known to contain large amount of ABA (Behl and Hartung, 1986; Lahr and Raschke, 1988). The sensitivity of the guard cell to ABA also increases with decreasing external pH (Paterson, Weyers and A'Brook, 1988). The k_a (4.7) for ABA indicates that a significant proportion will be protonated, uncharged and therefore membrane permeable *in vivo*. The absolute proportions are difficult to estimate as the apoplastic pH is unknown. Measurements of apoplastic hydrogen ion activity (loosely called pH: Edwards, Smith and Bowling, 1988) are difficult to relate to hydrogen ion concentrations due to the high ion exchange capacity of the wall and the consequent ion gradients in the apoplast micro-environment.

The picture is complicated further by the interaction of the ABA response with other physiologically relevant signals. ABA and CO_2 interactions are species dependent and vary with leaf/plant age (e.g. Willmer, Wilson and Jones, 1988). Differing sensitivities may result from a third interaction with IAA (Snaith and Mansfield, 1982). In addition, all these responses are influenced by KCl concentration *in vitro* (e.g. Wardle and Short, 1981). There is still controversy over the apoplastic potassium levels *in vivo*, with measurements ranging from less the equivalent processes (swelling and contraction) of guard cell protoplasts (Smith and Willmer, 1988). The k_i for Ca^{2+} inhibition of opening varies between 100-220 µM for *Commelina communis* (De Silva *et al.*, 1985a, Inoue and Katoh, 1987). The effect appears to involve trans membrane fluxes of Ca^{2+} rather than direct effects on wall components as it is stimulated by the Ca^{2+} ionophore A23187 and inhibited by the Ca^{2+} channel blockers ruthenium red, verapamil and lanthanum (De Silva, Cox, Hetherington and Mansfield, 1985b; Inoue and Katoh, 1987; Smith and Willmer, 1988).

Measurements of Ca^{2+} influx in 'isolated' epidermal peels of *C. communis* indicate the steady state flux of Ca^{2+} is fairly high (MacRobbie, 1989), similar in magnitude to rubidium (K^+) fluxes observed during stomatal opening in 'isolated' epidermis. The calculated total concentration of this presumed cytoplasmic pool was about 5.7 mM. We have measured the free cytoplasmic Ca^{2+} concentration using both fluorescence ratio imaging (Fig. 1) and fluorescence ratio photometry (Fig. 2) of guard cells microinjected with the dye Indo-1. Resting levels of Ca^{2+} were between 100 nM and 600 nM. Significant binding or sequestration of the Ca^{2+} must therefore occur.

Direct effects of external Ca^{2+} can be inhibited by the calmodulin antagonist Ophiobolin A at low concentrations, suggesting that calmodulin may be involved in transduction of the Ca^{2+} signal (Nejidat, 1987). The presence of calmodulin has not yet been investigated in guard cells. Ca^{2+} has also been shown to inhibit a possible plasma membrane proton pumping ATPase (Nejidat, Roth-Bejerano and Itai, 1986), however the k_i (c. 200 µM) is probably too high to be of physiological significance. Effects of Ca^{2+} at physiological concentrations have been observed on two types of tonoplast voltage-gated ion channels in patch clamp studies (Schroeder and Hedrich, 1989). One is activated at low (100-200 nM) Ca^{2+}, opens in response to positive membrane potentials and may be involved in anion uptake. The other operates at micromolar Ca^{2+} levels, opens at negative membrane potentials and is selective for cations over anions. The presence of other elements of the Ca^{2+} signal transduction pathway (e.g. protein kinases, inositol phosphates, G-proteins) have yet to be investigated.

The initial link between Ca^{2+} and ABA was suggested from the observed requirement of external Ca^{2+} for ABA induced stomatal closure in epidermal strips (De Silva *et al.*, 1985a). Furthermore the ABA effect was inhibited by Ca^{2+} channel blockers and calmodulin antagonists (De

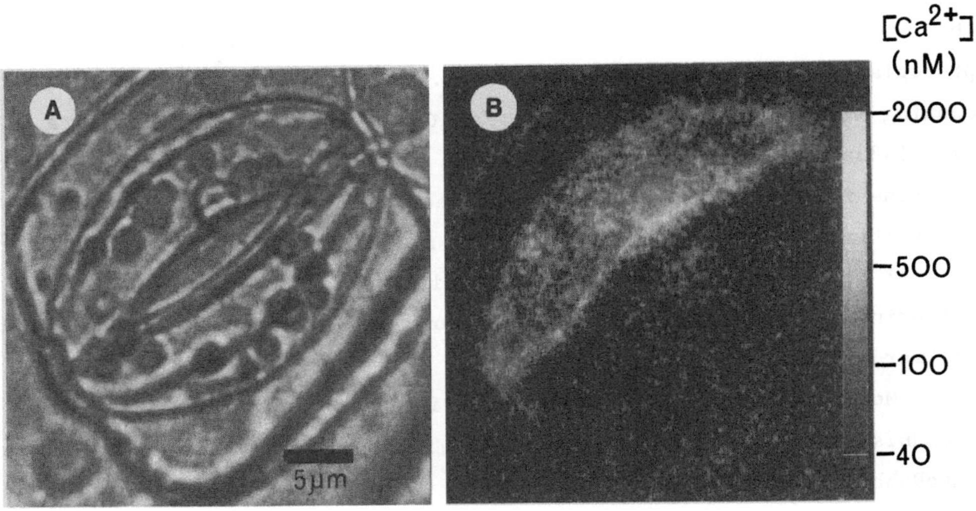

Figure 1. (A) Bright field image of a guard cell pair of *C. communis* injected with Indo-1; (B) fluorescence ratio image of the same guard cell pair. The fluorescence image at 415 nm was divided, pixel by pixel, by the image at 490 nm, yielding the 'ratio image' shown in b. The value of the ratio at each point is directly related to the free calcium concentration using a suitable calibration. The intracellular concentration of the dye was extremely low, <<10 μM. Note the exclusion of the dye (no ratio signal) from the vacuolar region

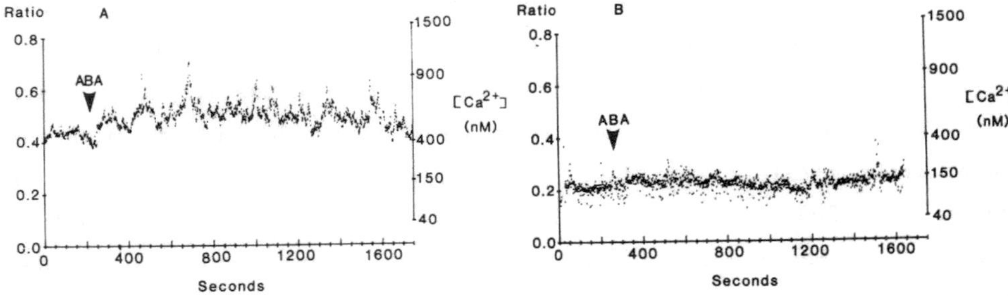

Figure 2. Fluorescence ratio analysis of the effect of 10^{-6} M ABA (added at the point indicated) on cytoplasmic free calcium concentration in a single guard cell of *Commelina communis* (conditions as for figure 1). The ratio was derived from simultaneous photomultiplier measurements of the fluorescence intensity at 405 nm and 480 nm. The response to ABA was variable, ranging from marked oscillations (A) to no apparent effect (B)

than 50 μM (Blatt, 1985) to between 3 mM and 100 mM depending on whether the stomata were open or closed (Bowling, 1987).

In conclusion, there is only weak evidence for a tight coupling between ABA, Ca^{2+} and stomatal closure. Most of the biochemical entities in the signal transduction chain remain to be detected and the crucial parallel response between Ca^{2+} and ABA stimulus is lacking.

Formation and fixation of a polar axis in developing zygotes of *Fucus* and *Pelvetia*

The unfertilised egg of members of the Fucales is spherical, apolar and lacks a cell wall. After fertilization wall material is rapidly secreted and the zygote form a developmental axis in response to external vectors, principally light. Formation of the axis occurs between 5 and 9 h post-fertilization in *Fucus*. At this stage the axis is labile and can be re-oriented. Fixation of the axis occurs between 9 and 12 h, after which rhizoid emergence and the first cell division occur. Formation of the axis is associated with an asymmetric Ca^{2+} flux entering the putative rhizoid pole and exiting the thallus pole (Robinson and Jaffe, 1975) and can be oriented by gradients of the Ca^{2+} ionophore A23187 (Robinson and Cone, 1980). Formation of the axis can be prevented by microinjection of Ca^{2+} buffers in a manner consistent with disruption of an internal Ca^{2+} gradient (Speksnijder, Weisenseel, Chen and Jaffe, 1989). Measurement of intracellular free Ca^{2+} by fluorescence ratio imaging has indicated slight Ca^{2+} gradients around 5-7 h post-fertilization but it is not clear if these are physiologically significant (Brownlee, 1989). No evidence for an internal Ca^{2+} gradient during the same period was found using the probe chlorotetracycline (CTC), which purports to measure membrane associated Ca^{2+} (Kropf and Quatrano, 1987). In addition, axis formation was not prevented by depletion of external Ca^{2+} by eight orders of magnitude (10^{-2} M to 10^{-10} M) or the presence of Ca^{2+} channel blockers (Kropf and Quatrano, 1987).

A putative Ca^{2+}-channel has been identified using a ^{3}H-photo-affinity analogue of verapamil (LU49888) followed by SDS-PAGE (Fricker and Trewavas, unpublished data). Several low molecular weight bands were labelled, but one with a molecular weight around 170 kDa, similar to the Ca^{2+}-channel mentioned above, was also detected.

Fucus zygotes contain a large number of Ca^{2+} binding proteins (9-11, Brawley and Roberts, 1989), one of which is calmodulin. There are fewer calmodulin binding proteins and their functions are not yet known. One in particular (M_r 72 kDa) may change with development.

Preliminary data indicates that protein kinase activity is present in *Fucus* eggs, but could not be localised with a particular membrane fraction after separation on sucrose gradients (Fig. 3). None of the phosphorylated proteins have been identified.

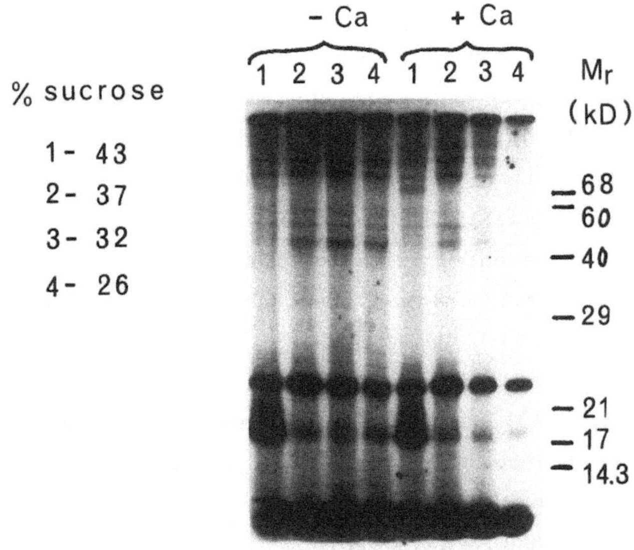

Figure 3. Autoradiography of membrane proteins from 12 h old zygotes of *Fucus serratus* after incubation with ^{32}P labeled ATP in the presence and absence of Ca^{2+}. Membrane proteins (6000 - 15000 x g pellet) were separated on a 7.5 cm^3, 10-50% linear sucrose gradient. 1 cm^3 fractions were collected, incubated with 20 nM ATP (185 TBq $mMol^{-1}$) $\pm$ 100 μM Ca^{2+} for 1 min and separated by SDS-PAGE

The situation in *Fucus* is further complicated as Ca^{2+} may be involved in a substantial number of separate physiological events during the period of early development. Golgi secretion of wall material is thought to be a Ca^{2+} dependent process (Steer, 1988) and both the F_1 and F_2 wall polymers are cross-linked by Ca^{2+}. Recycling of the plasma membrane via an endocytotic pathway may also involve Ca^{2+} (Steer, 1988) and must cause uptake of external Ca^{2+} in the lumen of the endocytotic vesicle. This Ca^{2+} may be eliminated or stored in vacuoles or calciosomes. Interestingly, this pathway would appear as Ca^{2+} uptake in flux studies but is not actually trans-membrane. Elongation of the rhizoid by tip growth commences after about 12 h and is associated with a gradient in cytosolic Ca^{2+} (Brownlee and Wood, 1986; Brownlee and Pulsford, 1988). Ca^{2+} has also been implicated in mitosis (Hepler and Callaham, 1988), which occurs after about 20 h in *Fucus*.

Conclusions and perspectives

The question remains, how can a plant cell respond to such a kaleidoscopic range of signals? Is the Ca^{2+} based transduction system sufficient to meet these needs?

The guard cell must remain sensitive to, amongst others, CO_2, light and ABA, if it is to adequately perform its role in simultaneously regulating transpiration and gaseous exchange through the stomatal pore. The observed oscillations in Ca^{2+} (figure 2) could play a role here. Their amplitude, frequency and duration could all impart information about the stimulus that evoked them. However, it still seems unlikely that a single, linear series of transduction events could support such a rich interaction of conflicting stimuli. A complex interacting network of transduction events perhaps more closely models the events in a real cell. Every element of this network interacting with, and modifying the response to, every other element to some greater or lesser degree. Thus diverse stimuli can feed in to separate points in the network but lead to apparently similar results.

Acknowledgements

We would like to thank Dr. D.P. Blowers and Dr. N.D. Read for helpful discussions and invaluable assistance. This research was funded by the Agricultural and Fisheries Research Council, Science and Engineering Research Council and the Gatsby Foundation.

References

Behl R and Hartung W (1986) Movement and compartmentation of ABA in guard cells of *Valerianella locusta*: effects of osmotic stress, external H^+ concentration and fusicoccin. Planta 168:360-368

Blatt MR (1985) Extracellular potassium activity in attached leaves and its relation to stomatal function. J Exp Bot 36:240-251

Blowers DP and Trewavas AJ (1989) Second messengers: their existence and relationship to protein kinases. In: Boss WF and Morré DJ (eds) Second messengers in plant growth and development. Alan R.Liss, New York, p 1

Boudet AM and Ranjeva R (1989) The vacuole: a potential store for second messengers in plants. In: Boss WF and Morré DJ (eds) Second messengers in plant growth and development. Alan R.Liss, New York, p 213

Boss WF (1989) Phosphoinositide metabolism: its relation to signal transduction in plants. In: Boss WF and Morré DJ (eds) Second messengers in plant growth and development. Alan R.Liss, New York, p 29

Bowling DJF (1987) Measurement of the apoplastic activity of K^+ and Cl^- in the leaf epidermis of Commelina communis in relation to stomatal activity. J Exp Bot 38:1351-1355

Brawley SH and Roberts DM (1989) Calmodulin binding proteins are developmentally regulated in gametes and embryos of fucoid algae. Dev Biol 131:313-320

Briars SA and Evans DE (1989) The calmodulin stimulated ATPase of maize coleoptiles forms a phosphorylated intermediate. Biochem Biophys Res Commun 159:185-191

Brownlee C and Wood J (1986) A gradient of cytoplasmic calcium in growing rhizoids of *Fucus serratus*. Nature 320:624-626

Brownlee C and Pulsford AL (1988) Visualisation of the cytoplasmic Ca^{2+} gradient in *Fucus serratus* rhizoids: correlation with cell ultrastructure and polarity. J Cell Sci 91:249-256

Brownlee C (1989) Visualising cytoplasmic calcium in polarizing zygotes and growing rhizoids in *Fucus serratus*. Biol Bulletin 176:14-17

Collinge M and Trewavas AJ (1989) The localisation of calmodulin in the pea plasma membrane. J Biol Chem 264:8865-8872

De Silva DLR Hetherington AM and Mansfield TA (1985a) Synergism between calcium ions and abscisic acid in preventing stomatal opening. New Phytol 100: 473-482

De Silva DLR Cox RC Hetherington AM and Mansfield TA (1985a) Suggested involvement of calcium and calmodulin in the responses of stomata to abscisic acid. New Phytol 101:555-563

Drobak BK, Allan AF, Comerford JG, Roberts K and Dawson AP (1988) Presence of guanine nucleotide-binding proteins in a plant hypocotyl microsomal fraction. Biochem Biophys Res Commun 150:899-903

Edwards MC, Smith GN and Bowling DJF (1988) Guard cells extrude protons prior to stomatal opening - A study using fluorescence microscopy and pH microelectrodes. J Exp Bot 39:1541-1547

Evans DE (1988) Regulation of cytoplasmic free calcium by plant cell membranes. Cell Biol Int Rep 12:383-396

Franceschi VR (1989) Calcium oxalate formation is a rapid and reversible process in *Lemna minor*. Protoplasma 148:130-138

Gilroy S, Blowers DP and Trewavas AJ (1988) Calcium: a regulation system emerges in plant cells. Development 100:181-184

Gilroy S, Hughes WA and Trewavas AJ (1987) Calmodulin antagonists increase free cytosolic calcium levels in plant protoplasts *in vivo*. FEBS lett 213:133-137

Gilroy S and Trewavas AJ (1989) Signal transduction across the plasma membrane. In: Larsson C and Moller IM (eds) The plant plasma membrane - structure, function and molecular biology. Springer Verlag, Berlin, in press

Harvey HJ, Venis MA and Trewavas AJ (1989) Partial purification of a protein from maize coleoptile membranes binding the calcium channel antagonist verapamil. Biochem J 257:95-100

Hepler PK and Callaham (1987) Free calcium increases during anaphase in stamen hairs of *Tradescantia*. J Cell Biol 105:2137-2143

Hepler PK and Wayne RO (1985) Calcium and plant development. Ann Rev Plant Physiol 36:397-439

Hornberg C and Weiler EW (1987) High-affinity binding sites for abscisic acid on the plasmalemma of *Vicia faba* guard cells. Nature 310: 21-324

Inoue H and Katoh Y (1987) Calcium inhibits ion-stimulated stomatal opening in epidermal strips of *Commelina communis* L. J Exp Bot 38:142-149

Jacobshagen S, Altmuller D, Grolig F and Wagner G (1986) Calcium pools, calmodulin and light regulated chloroplast movements in *Mougetia* and *Mesotaenium*. In: Trewavas AJ (ed) Molecular and cellular aspects of calcium in plant development. Plenum Press, New York, p 201

Jaffe LF (1968) Localization in the developing *Fucus* egg and the general role of localizing currents. Adv Morphog 7:295-328

Krause KH, Chou M, Thomas MA, Sjolund RD and Campbell KP (1989) Plant cells contain calsequestrin. J Biol Chem 264:4269-4272

Kropf DL and Quatrano RS (1987) Localization of membrane associated calcium during development of fucoid eggs using chlortetracycline. Planta 171:158-170

Lahr W and Raschke K (1988) Abscisic acid contents and concentrations in protoplasts from guard cells and mesophyll cells of *Vicia faba* L. Planta 173:528-531

MacRobbie EAC (1989) Calcium influx at the plasmalemma of isolated guard cells of *Commelina communis*. Planta 178:231-241

Miller AJ and Sanders D (1987) Depletion of free cytosol calcium induced by photosynthesis. Nature 326:397-400

Nejidat A, Roth-Bejerano N and Itai C (1986) K, Mg-ATPase activity in guard cells of *Commelina communis*. Physiol Plant 68:315-319

Nejidat A (1987) Effect of Ophiobolin A on stomatal movement: Role of calmodulin. Plant Cell Physiol 28:455-460

Omasa K and Onoe M (1984) Measurement of stomatal aperture by digital image processing. Plant Cell Physiol 25:1379-1388

Paterson NW Weyers JDB and A'Brook R (1988) The effect of pH on stomatal sensitivity to abscisic acid. Plant Cell Env 11:83-89

Roberts DM, Lukas TJ, Harrington HM and Watterson DM (1986) Molecular mechanisms of calmodulin action. In: Trewavas AJ (ed) Molecular and cellular aspects of calcium in plant development. Plenum Press, New York, p 11

Robinson KR and Jaffe LF (1975) Polarizing *Fucus* eggs drive a calcium current through themselves. Science 187:70-72

Robinson KR and Cone R (1980) Polarization of fucoid eggs by a calcium ionophore gradient. Science 207:77-78

Schauf C.L. and Wilson K.J. (1987) Effects of abscisic acid on K^+ channels in *Vicia faba* guard cell protoplasts. Biochem Biophys Res Comm 145:284-290

Schroeder JI and Hedrich R (1989) Involvement of ion channels and active transport in osmoregulation and signaling of higher plant cells. TIBS 14:187

Schwartz A (1985) Role of Ca and EGTA on stomatal movements in *Commelina communis* L. Plant Physiol 79:1003-1005

Schwartz A Ilan N and Grantz DA (1988) Calcium effects on stomatal movement in *Commelina communis* L. Plant Physiol 87:583-587

Smith G.N. and Willmer C.M. (1988) Effect of calcium and abscisic acid on volume changes of guard cell protoplasts of *Commelina*. J Exp Bot 39:1529-1539

Snaith PJ and Mansfield TA (1982) Control of the CO_2 responses of stomata by indol-3ylacetic acid and abscisic acid. J Exp Bot 33:360-365

Speksnijder JE, Weisenseel MH, Chen TH and Jaffe LF (1989) Calcium buffer injections arrest fucoid development by supressing calcium gradients. Biol Bulletin 176:9-13

Steer MW (1988) The role of calcium in exocytosis and endocytosis in plant cells. Physiol Plant 72:213-230

Takeda K, Kurkdjian AC and Kado RT (1985) Ionic channels, ion transport and plant cell membranes: Potential application of the patch-clamp technique. Protoplasma 127:147-162

Trewavas AJ (1986) Resource allocation under poor growth conditions. A major role for growth substances. In: Jennings DH and Trewavas AJ (eds) Plasticity in plants. CRC Press, Baton Roca, p 25

Trewavas AJ (1987) Many a mickle makes a muckle: network maxims may mitigate controversy. Bioessays 7:84-86

Wardle K and Short KC (1981) Responses of stomata in epidermal strips of *Vicia faba* to carbon dioxide and growth hormones when incubated on potassium chloride and potassium iminodiacetate. J Exp Bot 32: 303-309

Willmer CM (1988) Stomatal sensing of the environment. Biol J Linn Soc 34:205-217

Willmer CM, Wilson AB and Jones HG (1988) Changing responses of stomata to abscisic acid and CO_2 as leaves and plants age. J Exp Bot 39:401-410

Zeiger E (1983) The biology of stomatal guard cells. Ann Rev Plant Physiol 34:441-475

FUNCTION OF ION CHANNELS AND CYTOSOLIC Ca^{2+} IN THE MEDIATION OF HIGHER PLANT CELL ION TRANSPORT

Julian I. Schroeder
Department of Physiology &
Jerry Lewis Neuromuscular Research Center
UCLA School of Medicine, UCLA
Los Angeles, CA 90024, USA

Physiological processes in higher plant cells, such as osmoregulation, signal transduction, photosynthesis, movements and growth depend on membrane transport of organic and inorganic ions and on signal-induced modulation of receptor-mediated responses. New insights to the molecular dynamics of ion transport and signal transduction can now be obtained by the application of patch clamp techniques to higher plant protoplasts. This chapter will summarize new results from patch clamp studies on the identification of molecular mechanisms of ion transport and ion transport regulation in guard cells, which control the gas exchange in leaves.

NATO ASI Series, Vol. H 47
Signal Perception and Transduction in Higher Plants
Edited by R. Ranjeva and A. M. Boudet

Stomatal Movements and Signal-dependent Ion Transport

Two guard cells modulate the aperture of the pore they surround, thus regulating the gas exchange for photosynthesis, while minimizing water loss to the atmosphere. Light, among other physiological signals, causes guard cells to accumulate K^+ and anions. This increase in the osmotic potential of guard cells leads to the opening of stomatal pores (Raschke,1979). High CO_2 concentrations, darkness and the phytohormone abscisic acid mediate closure of stomata by inducing ion release. Furthermore, data indicate that Ca^{2+} plays a central role in the regulation of stomatal apertures (De Silva et al., 1985; Schwartz et al., 1988; MacRobbie, 1986). Using patch clamp techniques (Hamill et al., 1981) we have been able to identify and characterize several specialized classes of ion channels and signaling mechanisms in Vicia faba guard cells, which are capable of mediating uptake of ions during stomatal opening and ion release during stomatal closing. A schematic model of stomatal opening (Fig. 1A,B) and closing (Fig. 1C,D) summarizes these novel findings, which will be delineated throughout this chapter.

Plasma Membrane Ion Uptake During Stomatal Opening

Higher plants cells require the uptake of K^+ ions during regulation, movements, ion uptake from soils and cell expansion (Luettge & Pitman, 1976). K^+ selective channels are receiving growing attention due to their prospective importance in providing a major pathway for K^+ uptake. Upon activation, these channels open a transmembrane spanning pore which allows K^+ influx along the electrochemical gradient. It was found that hyperpolarizations of the plasma membrane to potentials more negative than -90 mV result in opening of hundreds of K^+ channels in a guard cell, which provide a pathway for K^+ influx (Fig. 1B) (Schroeder, Hedrich & Fernandez, 1984; Schroeder, Raschke & Neher, 1987).

The hypothesis that these K^+ channels serve as an important pathway during stomatal opening was supported by the following findings: The magnitude of K^+ uptake through K^+ channels in the guard cell plasma membrane is sufficiently large to account for K^+ accumulation during stomatal opening (Schroeder et al., 1984 & 1987). The permeability sequence of K^+ channels ($P_{K^+} > P_{Rb^+} > P_{Na^+} > P_{Li^+} >> P_{Cs^+}$) (Schroeder, 1988) correlates to the ability of stomata to open in the presence of various ions (Raschke, 1979). Aluminum ions block both stomatal openings (Schnabl & Ziegler, 1975) and inward conducting K^+ channels (Schroeder, 1988).

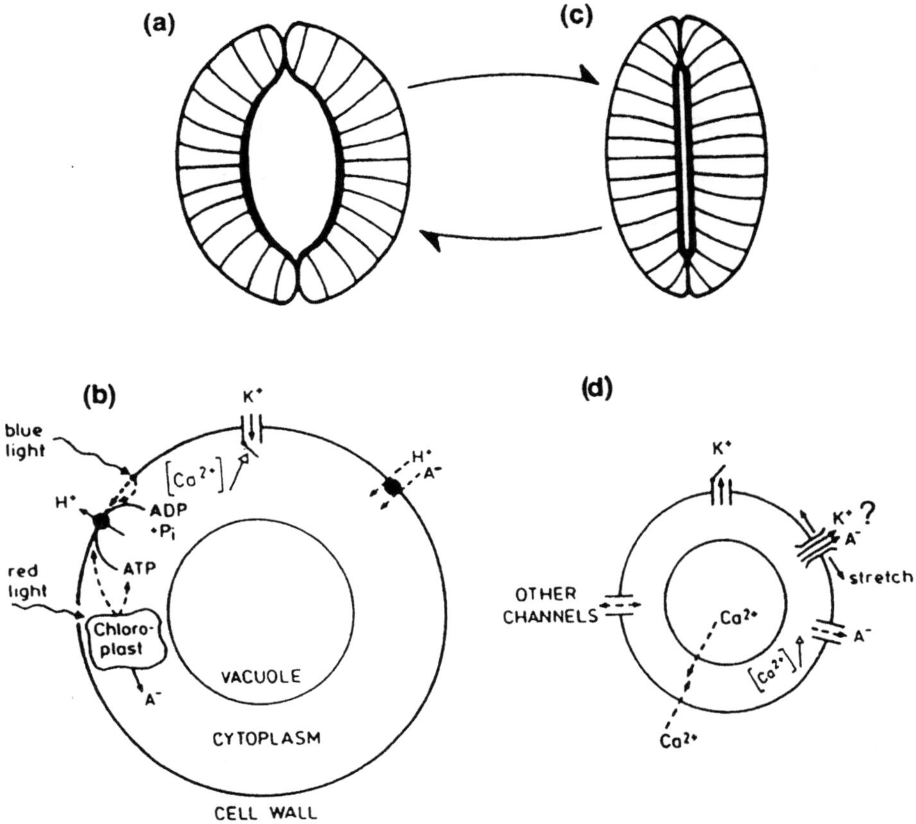

Fig. 1 Model for molecular mechanisms of plasma membrane ion transport involved in opening (A,B) and closing of stomata (C,D). **Left:** Opening of stomata (A) requires plasma membrane ion uptake (B) and malic acid (A^-) synthesis. **Right:** Stomatal closing (C) is induced by ion release (D). A^- are anions, solid lines and arrows represent ion transporters which have been directly monitored, while broken lines indicate suggested mechanisms. (modified with permission from Schroeder & Hedrich, 1989)

In detailed studies by several research groups Ca^{2+} ions have been shown to trigger stomatal closing and inhibit stomatal opening (De Silva et al., 1985; Schwartz et al., 1988). Recent results show that elevation of the Ca^{2+} concentration in the cytoplasm from the submicromolar level to the micromolar level inhibits the above mentioned inwardly conducting K^+ channels (Fig. 1B) (Schroeder & Hagiwara, 1989). These data may provide a molecular basis for Ca^{2+}-dependent inhibition of K^+ influx.

Recent studies in several laboratories show that this class of voltage-dependent K^+ uptake channels is prevalent in various types of higher plant cells (for rev. see: Hedrich & Schroeder, 1989). The ubiquity of K^+ uptake channels, as identified in guard cells, suggests that these channels may constitute general mechanisms of higher plant K^+ transport.

Channel-mediated K^+ uptake requires that a mechanism exists which can hyperpolarize the membrane to potentials more negative than -100 mV. In plants it has been clearly documented that the plasma membrane proton ATPase can polarize the membrane to values far more negative than -100 mV thus providing a driving force for solute and nutrient uptake (for reviews: Serrano, 1985; Slayman, 1987). First evidence for involvement of proton extrusion during stomatal opening was found by the recording of extracellular acidification (Raschke & Humble, 1973).

Whole-cell studies with guard cells have shown that blue light can trigger the activation of H^+ pumps (Fig. 1B). Exposure of guard cells to low fluence rates of blue light activate electrogenic ATP-dependent pumps after a delay of approximately 30 seconds (Assmann et al., 1985). Blue light-induced pump activation is not mediated by photosynthetic light absorption (Assmann et al., 1985). Extracellular acidification in response to blue light exposure supports the hypothesis that the current is generated by a proton ATPase (Shimazaki et al., 1986). Recently, a new patch clamp technique was applied to guard cells, which in contrast to whole-cell recordings does not perfuse the cytoplasm ("slow-whole-cell"; Schroeder, 1988). Using this new technique, blue light-activated H^+ pump currents in guard cells were approximately 6 times larger than in whole-cell recordings (Schroeder, 1988). These data allowed two conclusions: (1) diffusible cytoplasmic substrates together with MgATP may be required for full activation of blue light-activated electrogenic pumps. (2) The magnitude of blue light-activated pumps is sufficiently large to drive physiological fluxes of K^+ uptake through inwardly conducting K^+ channels (Schroeder, 1988).

The uptake of K^+ into the cytoplasm is balanced either by simultaneous uptake of Cl^- or by the synthesis of malic acid. In the absence of Cl^-, K^+ fluxes can be balanced by malic acid production alone (Raschke, 1979; Zeiger et al., 1987). The formation of malate in parallel to K^+ uptake during stomatal opening requires a closely coupled interrelation between ion transport and carbon metabolism (Raschke et al., 1988).

Thus voltage- and Ca^{2+}-regulated K^+ channels and proton pumps in the plasma membrane along with malate synthesis allow guard cells to increase their osmotic potential, leading to stomatal opening. An additional set of ion channels alone, may be sufficient for mediating the release of ions across the plasma membrane, required for stomatal closure.

Ion Efflux During Stomatal Closing

A separate class of K^+ channels, which allow K^+ efflux has been identified in the plasma membrane of higher plant cells (Schroeder et al., 1984 & 1987; Iijima & Hagiwara, 1987; Schauf & Wilson, 1987; Moran et al., 1988, Blatt, 1988). These outward conducting K^+ channels have been suggested to provide a pathway for K^+ release. Supportive biophysical evidence for this hypothesis was obtained in studies with guard cells (Schroeder et al., 1984; Schroeder, 1988; Blatt, 1988; Schroeder & Hagiwara, 1989). Pharmacological evidence was found in extensor and flexor cells which osmotically drive leaf movements: 10 mM TEA (tetraethylammonium) reduced both outward conducting K^+ channel currents and leaf movements (Moran, Satter et al., 1988).

These K^+ channels allow efflux of K^+ when the membrane potential is depolarized to values more positive than the equilibrium potential of K^+ (Iijima & Hagiwara, 1987, Blatt, 1988, Schroeder, 1988). Therefore, mechanisms which induce depolarization of the plasma membrane would be required to induce K^+ channel-mediated K^+ release (Schroeder et al., 1987).

The down-regulation of turgor in guard cells requires a reduction in the cellular anion content, which is partially mediated by anion efflux accompanying K^+ release, as demonstrated in tracer flux experiments with "isolated" guard cells in epidermal strips (MacRobbie, 1981). In whole-cell studies with guard cells, evidence has recently been obtained for the existence of a depolarizing voltage-dependent conductance with a permeability to anions (Schroeder & Hagiwara, 1989; Keller et al., 1989). The anion-permeable conductance in guard cells can be activated by elevation of cytosolic free Ca^{2+} to the micromolar level (Schroeder & Hagiwara, 1989). Preliminary data indicate that cytosolic Ca^{2+} may not directly modulate the anion conductance but may regulate cellular intermediates which in turn open anion channels. The activation of an anion permeability would depolarize the membrane potential, due to the electrochemical gradient of free anions in solution. This anion-dependent depolarization could , in turn, activate K^+ channels leading to a membrane potential intermediate to the equilibrium potentials of K^+ and anions. This would allow simultaneous release of K^+ and anions (Fig. 1D). The Ca^{2+}- activation of the anion conductance provides a basis for a Ca^{2+} requisite during induction of stomatal closure by abscisic acid (e.g. De Silva et al., 1985).

As the anion conductance, itself, is activated by depolarization (Schroeder & Hagiwara, 1989), a "predepolarization" may be required to trigger anion and K^+ efflux. Activation of Ca^{2+} channels and/or inhibition of H^+

pumps could induce such a predepolarization. Further research will be required to test these suggestions.

Mechanisms of Cytoplasmic Ca2+ Elevation

In living cells the resting concentration of free cytosolic Ca^{2+} usually ranges from 0.1 to 0.25 micromolar (Hepler & Wayne, 1985). Elevation of cytosolic Ca^{2+} to the micromolar level has been suggested to trigger signal-induced responses in higher plants (Hepler & Wayne, 1985). In guard cells data indicate that Ca^{2+} influx may be an initial event leading to the activation of ion release (De Silva et al., 1985; Schwartz et al., 1988; Schroeder & Hagiwara, 1989). However, signal-induced calcium influx mechanisms and the possible subsequent elevation of the cytoplasmic Ca^{2+} level have neither been clearly resolved in guard cells nor in other higher plant cells.

Ca^{2+} channel blockers inhibit stomatal closing giving rise to the hypothesis that Ca^{2+} channels may be present in the plasma membrane (De Silva et al., 1985). However, it should be noted, that the effective concentrations of Ca^{2+} channel blockers in guard cells are on the order of 10 to 100 times greater than generally employed in animal cells. At such high concentrations (10 uM to 1 mM) Ca^{2+} channel blockers have been shown to inhibit outward conducting K^+ channels in animal cells (Schlichter et al., 1986). On the other hand, modified Ca^{2+} channel blockers were recently found with a much higher affinity for carrot protoplast and maize coleoptile membranes (Graziana et al., 1988; Harvey et al., 1989). The binding of these substances correlated to a reduction of radioactively labelled Ca^{2+} influx in carrot protoplasts (Graziana et al., 1988). Whether such binding sites represent molecular components of plant Ca^{2+} channels remains to be determined.

In addition to Ca^{2+} channel-mediated influx, cytosolic Ca^{2+} may be elevated by Ca^{2+} release from intracellular organelles. Recent research has indicated that the cytosolic second messenger inositol-1,4,5-trisphosphate (IP_3) may induce an increase in the cytosolic Ca^{2+} concentration by triggering Ca^{2+} release from the endoplasmic reticulum (Berridge & Irvine, 1984) and from plant vacuoles (Shumaker & Sze, 1987). Future patch clamp studies may provide more detailed information on mechanisms which can elevate cytosolic Ca^{2+} in higher plant cells.

Ion Transport Model

The mechanisms of ion transport discussed above have been summarized in a model of ion uptake and release during guard cell movements (Fig. 1). The finding that ion channels are important for mediating long-term fluxes of ions across the guard cell plasma membrane relies on incomplete channel inactivation (see Schroeder, 1988) and the existence of mechanisms which can polarize the membrane for extended periods, thus providing a driving force for channel-mediated ion flux (e.g. H^+ pumps (Fig. 1B) or ion channels of different selectivity (Fig. 1D). The suggested model in Fig. 1 should be regarded as a simplified working hypothesis which reflects current knowledge. This model provides a basis for new conceptual insights, that may prove important for future studies. Further investigation of the above discussed ion transport mechanisms as well as of additional types of ion channels, signal receptors and active transporters should lead to extension of the proposed model in figure 1.

Recent investigations by several laboratories have shown that K^+ channels with similar properties to those identified in guard cells have been found in various higher plant cell types (Iijima & Hagiwara, 1987; Schauf & Wilson, 1987; Moran et al., 1988) (for complete references see: Tazawa et al., 1987; Hedrich & Schroeder, 1989). The similarities in basic channel properties from different cells, suggest that these ion channels may constitute general mechanisms for solute transport across higher plant membranes.

J.I.S. was supported by fellowships from NIH and the Alexander von Humboldt Foundation.

References

Assmann, S.M., Simoncini, L. & Schroeder, J.I. (1985) Blue light activates electrogenic ion pumping in guard cell protoplasts of Vicia faba Nature 318, 285- 287

Berridge, M.J. & Irvine, R.F. (1984) Inositol trisphosphate, a novel second messenger in cellular signal transduction. Nature 312, 315-321

Blatt, M.R. (1988) Potassium-dependent bipolar gating of K+ channels in guard cells. J. Mem. Biol. 102, 235-246

De Silva, D.C.R., Cox, R.C., Hetherington, A.M. & Mansfield, T.A. (1985) Synergism between calcium ions and abscisic acid in preventing stomatal opening. New Phytol. 100, 473-482

Graziana, A., Fosset, M., Ranjeva, R., Hetherington, A.M., Lazdunski, M. (1988) Ca^{2+} channel inhibitors that bind to plant cell membranes block Ca^{2+} entry into protoplasts. Biochem 27, 764-8

Hamill, O.P., Marty, A., Neher, E., Sakmann, B. & Sigworth, F.J. (1981) Improved patch-clamp techniques for high-resolution current recording from cells and cell-free membrane patches. Pflügers Archiv 391, 85-100

Harvey, H.J., Venis, M.A. & Trewavas, A.J. (1989) Partial purification of a protein from maize coleoptile membranes binding the calcium channel antagonist verapamil. Biochem.J. 275,95-100

Hedrich, R. & Schroeder, J.I. (1989) The physiology of ion channels and electrogenic pumps in higher plants. Ann. Rev. Plant Physiol. 40, 539-569

Iijima, T. & Hagiwara, S. (1987) Voltage dependent K^+ channels in protoplasts of trap-lobe cells of Dionea muscipula J. Mem. Biol. 100, 73-81

Keller, B.U., Hedrich, R. & Raschke, K. (1989) Activity of single anion channels in the plasmalemma of guard-cell protoplasts. Plant Physiol. 89(4): 886 (Abstract)

Lüttge, U. & Pitman, M.G. (Eds) (1976) Transport in Plants Vol. I,II & III Springer (Berlin, Heidelberg, New York)

MacRobbie, E.A.C. (1981) Effects of ABA in "isolated" guard cells of Commelina communis J.Exp. Bot. 32, 563-572

MacRobbie, E.A.C. (1986) in: Molecular and Cellular Aspects of Ca^{2+} in Plant Development (Trewavas, A.J. & Marme', D. eds), pp 383-384 Plenum, New York

Moran, N., Ehrenstein, G., Iwasa, K., Mischke, C., Bare, C., Satter, R.L. (1988) Potassium channels in motor cells of Samanea saman: a patch clamp study. Plant Physiol. 88, 643-648

Raschke, K. (1979) Movements of stomata in: Encyclopedia of Plant Physiology, eds. Haupt, W. & Feileib, M.E. (Springer, Berlin), Vol. 7, pp. 384-441.

Raschke, K. & Humble, G.D. (1973) No uptake of anions required during opening of stomata: guard cells release hydrogen ions. Planta 115:47-57

Raschke, K., Hedrich, R., Reckmann, U. & Schroeder, J.I. (1988) Exploring biophysical and biochemical components of the osmotic motor that drives stomatal movement Bot. Acta. 101, 283-294

Schauf, C.L. & Wilson, K.J. (1987) Properties of single K^+ and Cl^- channels in Asclepias tuberosa protoplasts. Plant Physiol. 85, 413-418

Schnabl, H. & Ziegler, H. (1975) Ueber die Wirkung von Aluminium Ionen auf die Stomata Bewegung von Vicia faba. Zur Pflanzenphysiol. 74, 394-403

Schlichter, L., Sidell, N. & Hagiwara, S. (1986) Potassium channels mediate killing by human natural killer cells. Proc. Natl. Acad. Sci. USA 83, 451-455

Schroeder, J.I. (1988) K^+ transport properties of K^+ channels in the plasma membrane of Vicia faba guard cells. Journal of General Physiology 92, 667-683

Schroeder, J.I. Hedrich, R. & Fernandez, J.M. (1984) Potassium-selective single channels in guard cell protoplasts of Vicia faba. Nature 312, 361-362

Schroeder, J.I., Raschke, K. & Neher, E. (1987) Voltage dependence of K^+ channels in guard cell protoplasts. Proc. Nat. Acad. Sci. USA 84, 4108-4112

Schroeder, J.I. & Hagiwara, S. (1989) Cytosolic calcium regulates ion channels in the plasma membrane of Vicia faba guard cells. Nature 338, 427-430

Schroeder, J.I. & Hedrich, R. (1989) Involvement of ion channels and active transport in osmoregulation and signaling of higher plant cells. Trends in Biochem. Sci. 14, 197-192

Schwartz, A., Ilan, N. & Grantz, D.A. (1988) Calcium effects on stomatal movement in Commulina communis L. Plant Physiol. 87, 583-587

Serrano, R. (1985) Plasma membrane ATPase of plants and fungi. CRC Press Inc., Boca Raton, FL

Shimazaki, K., Iino, M. & Zeiger, E. (1986) Blue-light dependent proton extrusion by guard cell protoplast of Vicia faba. Nature 319, 324-326

Shumaker K.S. & Sze, H. (1987) Inositol 1,4,5- trisphosphate releases Ca^{2+} from vacuolar membrane vesicles of oat roots. J. Biol. Chem. 262, 3944-3946

Slayman, C.L. (1987) The plasma membrane ATPase of Neurospora: a proton pumping electroenzym. J. Bioenerg. Biomembr. 19,1-20

Tazawa, M., Shimmen, T. & Mimura, T. (1987) Membrane control in the Characea. Ann. Rev. Plant Physiol. 38, 95-117

Zeiger, E., Farquhar, G.D. & Cowan, I.R. eds. (1987) Stomatal Function, Stanford University Press

Ca^{2+} AS A SIGNAL IN THE INDUCTION OF CALLOSE SYNTHESIS

H. Kauss[1], T. Waldmann[1], H. Quader[2]

[1]Fachbereich Biologie der Universität Kaiserslautern, Postfach 3049,
D-6750 Kaiserslautern, Bundesrepublik Deutschland
[2]Zellenlehre, Universität Heidelberg, Im Neuenheimer Feld 230, D-6900 Heidelberg,
Bundesrepublik Deutschland

ABSTRACT

In suspension-cultured cells and protoplasts, callose synthesis can be triggered by various biochemically unrelated substances. Laser scanning microscopy shows that the polycation chitosan is bound to the surface of protoplasts. Induction of callose synthesis by chitosan increases with the degree of chitosan polymerization up to several thousand, whereas N-acetylation of chitosan - statistically at every fourth to fifth glucosamine residue - decreases its potency. These results suggest that interaction of chitosan might occur over a large surface area with the phospholipid head groups on the plasma membrane. In contrast, the primary interaction of other callose elicitors (e.g. saponins, polyene antibiotics, acylated cyclic peptides) may occur with various constituents in the lipid phase of the plasma membrane. Taken together, signal perception for callose synthesis appears not to involve complementary receptors in the classical sense.

Common to all types of callose elicitors is the induction of a rapid K^+ efflux, which is correlated with an external alkalinization and followed temporally by net Ca^{2+} uptake. As the plasma membrane-located 1,3-ß-glucan synthase has an absolute requirement for Ca^{2+} in the μM range, it has been suggested that Ca^{2+} uptake may lead to an increase in cytoplasmic [Ca^{2+}] and thereby trigger callose synthesis. This idea is supported by the observations that external Ca^{2+} is essential for callose induction, and that inhibition of Ca^{2+} uptake by putative Ca^{2+}-channel blockers decreases callose synthesis. Increasing Ca^{2+}-uptake alone, however, appears not to be sufficient for the induction of callose formation as shown with the Ca^{2+}-ionophore A 23187 and by low doses of Amphotericin B. Plasma membrane depolarization by itself appears not to represent an additional signal since callose synthesis is increased by the hyperpolarizing toxin fusicoccin and is decreased by protonophores.

Some of the substances capable of rapidly inducing callose synthesis have also been shown to elicit the slower production of phytoalexins, suggesting that the signal transduction mechanism involved in callose synthesis may also contribute to the regulation of other metabolic pathways.

NATO ASI Series, Vol. H 47
Signal Perception and Transduction in Higher Plants
Edited by R. Ranjeva and A. M. Boudet

INTRODUCTION: PHYSIOLOGICAL ROLE OF CALLOSE

Homomeric 1,3-ß-glucan ("callose") is rarely a constituent in the regular cell wall of higher plants. Since callose can be easily recognized cytochemically by several more or less specific techniques, many examples are known where 1,3-ß-glucans are deposited in the cell wall as a result of various types of stress, often in a very localized manner (for literature see Kauss 1987, 1990). A striking example for the function of callose is the sealing of plasmadesmata or sieve pores within minutes of mechanical injury. In damaged tissues, callose deposition in the walls of surviving cells is the first event in wound-healing. Implication of callose in defense reactions against pathogens is best illustrated by the formation of "papillae". These cell wall appositions are deposited from the plasma membrane basipetal to the sites of fungal penetration. When e.g. *Phytophthora infestans* infects tubers of resistant potato cultivars, the papillae are formed earlier than in tissue from susceptible cultivars and are correlated with decreased fungal penetration (Hächler and Hohl 1984). In epidermal cells of barley coleoptiles infected with *Erysiphe graminis*, the formation of papillae begins basipetally to the appressorium about 1 h *before* fungal penetration pegs become microscopically visible in resistent lines (Gold et al. 1986). As a result, penetration is greatly suppressed. Although papillae in this (Aist et al. 1988) and other systems likely contain several chemical constituents, they are always rich in callose.

SIGNAL PERCEPTION: BINDING OF CALLOSE ELICITORS

Early experiments on the induction of callose synthesis focused on physical stresses such as bending of root hairs or compression of epidermal cells (for literature see Eschrich 1956, Kauss 1987). For biochemical studies on the mechanism of callose induction, however, a system is needed where callose synthesis in all cells of a population could be simultaneously induced. We have, therefore, searched for chemical substances able to elicit callose production in suspension cells. Chitosan, a component of fungal cell walls and of the exoskeleton of insects, was initially used as a callose elicitor (Köhle et al. 1984, 1985), and an expanded knowledge of the mechanism of chitosan binding has been obtained recently.

Binding of chitosan polymers, visualized using the confocal laser scanning microscope, appears to occur mainly at the plasma membrane surface of protoplasts (Fig. 1-3). There are always some protopasts (Fig. 2) which appear to be leaky and, therefore, also bind chitosan at intracellular membranes. This phenomenon increases with incubation time. Whether chitosan binding in these protoplasts results in metabolic changes is

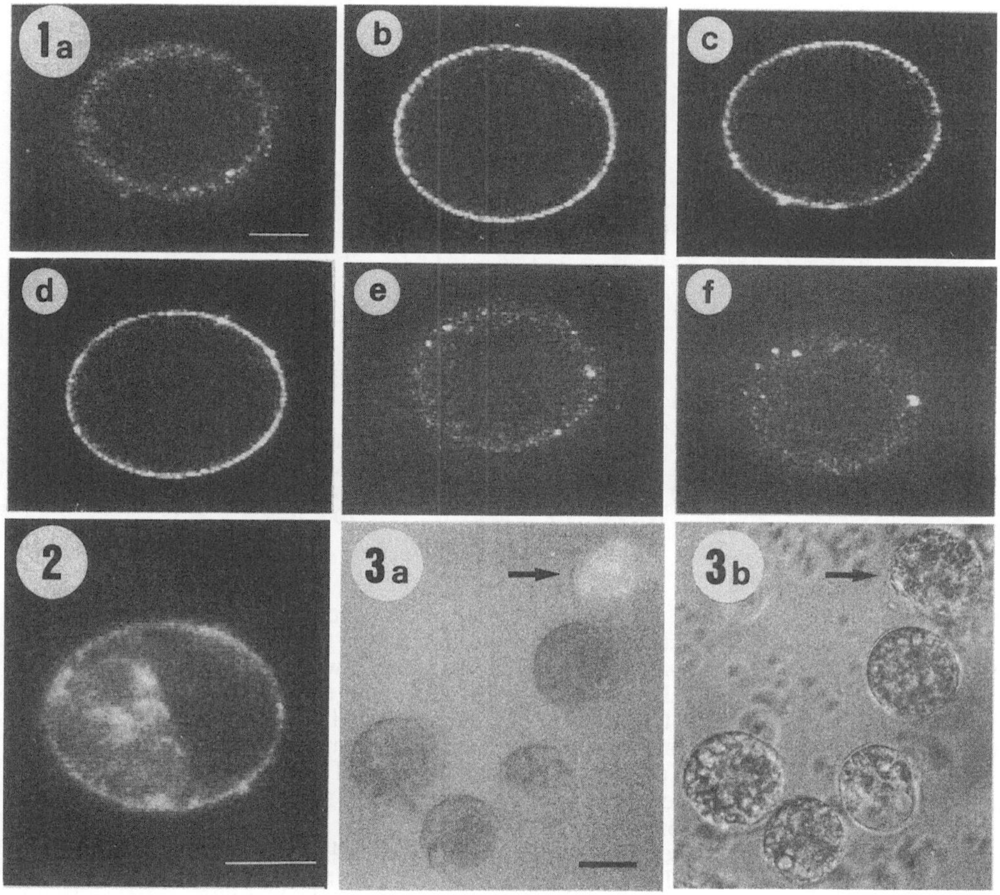

Fig. 1. Visualization of FITC-chitosan bound to the surface of intact protoplasts from suspension-cultured cells of *Catharanthus roseus*. Using the confocal laser scanning microscope (CLSM) six successive optical planes were scanned, moving down the microscopical stage (a = 3.9, b = 9.1, c = 11.7, d= 14.3, e = 19.5, f = 20.8 μm from the upper protoplast surface). The CLSM (Wild-Leitz, formerly Heidelberg Instruments, Heidelberg, FRG) was equipped with a multiline Argon-laser, excitation at 488 nm. Chitosan (DP 3420, 22.3 % N-Ac, 10 mg/ml 0.1 N acetic acid, Kauss et al. 1989) was labelled with FITC using the dialysis technique (Clark and Shepard 1963). The chitosan precipitates during the initial dialysis step at pH 9.5 but the labelling remains satisfactory. Resolubilization and determination of chitosan were performed as desribed by Kauss et al. (1989). Protoplasts (Kauss et al. 1989, $2 \cdot 10^5$ in 1 ml of suspension buffer) were supplied with 20 μg of FITC-chitosan per ml and scanned after about 20 min. Bar: 5 μm

Fig. 2. Visualization of the FITC-chitosan pattern in a presumably leaky protoplast. Same conditions as in Fig. 1

Fig. 3. (Legend see following page)

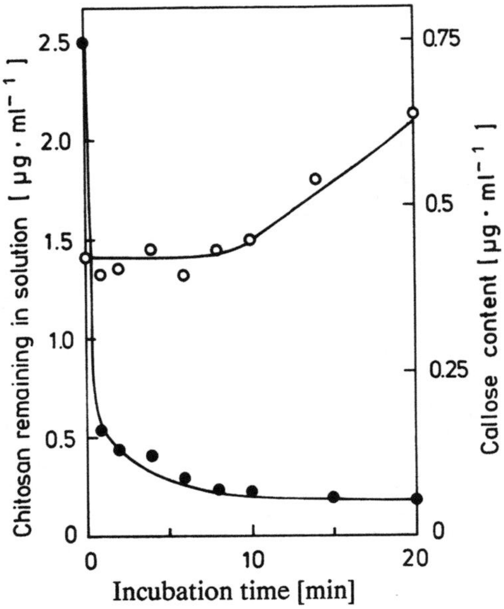

Fig. 4. Time course of callose formation (o) and of chitosan binding (•) on protoplasts. Protoplasts from suspension-cultured cells of *Catharanthus roseus* ($2 \cdot 10^5$/ml) were used under the same conditions as in Table 1. The suspension was supplied with 2.5 µg of [^{3}H]-acetyl-chitosan per ml. Aliquots of 1 ml were centrifuged for 30 sec in a Eppendorf centrifuge (about 12000xg) at the times indicated and 0.5 ml of the supernatant counted. The callose content of the protoplasts was determined fluorometrically (Kauss et al. 1989) and is expressed in µg pachyman-equivalents per ml protoplast suspension. Chitosan (DP 2500, fully N-deacetylated) was solubilized (21 mg/10 ml) in 20 mM acetic acid, freeze-dried and resolubilized in 10 ml of water. This solution (pH about 5.0) was mixed with 25 mCi of [^{3}H]-acetic anhydride. After 18 h at 4°C the chitosan was precipitated with NaOH at about pH 8.0, washed several times with water and freeze dried. A solution was dialyzed against several batches of water and cleared by centrifugation; it had a spec. activity of 12.6 µCi/mg polyglucosamine at about 0.4 % N-acetylation (calculated from spec. activity)

Fig. 3. (Legend continued from preceding page)
Protoplasts incubated with fluorescein as a control to Fig. 1. Same conditions as in Fig. 1 but fluorescein (1 µg/ml) was added instead of FITC-chitosan. Bar: 20 µm.
a: Fluorescence image. The four intact protoplasts contrast with fluorescence of the background and that of a presumably damaged protoplast (arrow) which has accumulated the dye.
b: Corresponding Nomarski-image. Note the wrinkled appearance of the damaged protoplast (arrow)

Table 1. Induction of callose formation in protoplasts by chitosan preparations exhibiting various degrees of polymerization and N-acetylation. Protoplasts ($4x10^5$) from suspension-cultured cells of *Catharanthus roseus* were suspended in 2 ml of 25 mM Bistris/Mes buffer, pH 5.8, containing 0.4 M glucose, 2 mM $CaCl_2$ and 5 mM KCl. The samples were each supplied with 5 μg of the chitosans, standardized as polyglucosamine content (for experimental details see Kauss et al. 1989). In similar experiments fully N-deacetylated chitosan oligomers (DP 8 and 14) as well as up to 20-fold higher amounts of colloidal chitin induced only traces of callose

O % N-Acetylation		23 % N-Acetylation	
DP[a]	Callose[b]	DP[a]	Callose[b]
90	1.06	82	0.82
275	2.39	190	1.09
2500	5.43	3420	2.85

[a]average degree of polymerization from viscosity measurements (Kauss et al. 1989)
[b]amount of callose is given as [μg pachyman equivalents/3 h/$4 \cdot 10^5$ protoplasts]. Controls without chitosan formed 0.07 μg callose and this value was subtracted.

difficult to access (for a discussion of similar effects in cells treated with chitosan, see Kauss et al. 1989). Binding of chitosan to protoplasts occurs rapidly (Fig. 4). After about 1 minute, most the polymer pellets with the protoplasts, but callose formation starts about 10 min later. This time gap will be discussed below.

Further conclusions toward the initial interaction of chitosan with the plasma membrane can also be drawn from the molecular properties governing the potency of chitosan as a callose elicitor. We have recently compared chitosan preparations differing in degree of polymerization and N-acetylation with regard to their efficiency in inducing callose synthesis in suspension-cultured cells and protoplasts (Kauss et al. 1989). For the comparison of different chitosan preparations, one must carefully ascertain that a concentration is choosen that lies in the region of positive slope of the dose-response curves for callose synthesis. If this requirement is fulfilled, it becomes evident (Table 1) that chitosan fragments up to DP 14 are almost inactive. The ability to induce callose formation increases with the DP of chitosan (Table 1). In similar experiments such an increase occurred up to a DP of 5000 corresponding to a linear polymer with a chain length of 2.5 μm (Kauss et al. 1989). Binding over broad areas of the membrane surface appears to be required. This binding leads to the physiological effects in a cooperative manner. At a comparable DP, fragments exhibiting about 23 % N-acetylation are less effective than fully deacetylated fragments. Similar results were found with cells, although far higher chitosan concentrations were required, presumably due to binding of some of the chitosan to the cell wall and also to exclusion effects for the largest chitosan molecules by the cell wall (Kauss et al. 1989). Taken together, these results suggest that if the row

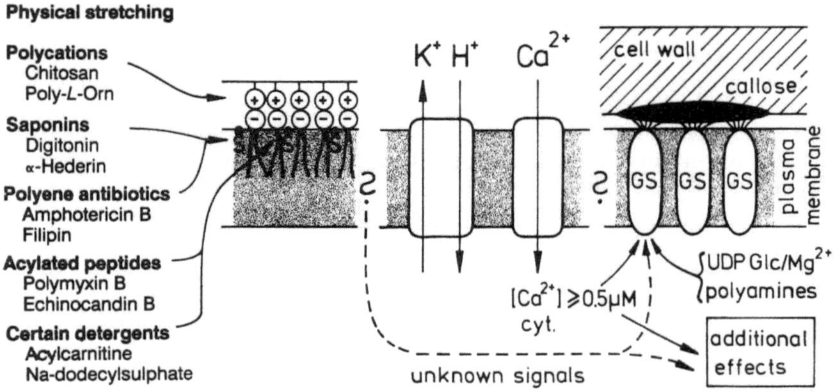

Fig. 5. Empirically found callose elicitors and schematic illustration of three aspects of plasma membrane function assumed to cooperate in callose induction. The three modules represent initial interaction, ion transport and 1,3-ß-glucan synthase activation. They should be envisaged combined in the same membrane area but are shown here side by side to point at the question how the processes are coupled to function in signal transduction. For each of the chemically different groups of callose elicitors only two typical examples are given. Note also that these substances differ considerably in effectiveness. K^+ efflux and associated external alkalinization are depicted as K^+/H^+ exchange although other explanations are equally possible.
Induction of phytoalexin synthesis and polyphenol deposition on the cell wall represent examples of additional effects induced by certain callose elicitors

S = sterols, GS = 1,3-ß-glucan synthase (GS II)

of amino groups in the polyglucosamine molecule (Fig. 5) is interrupted by N-acetyl groups, statistically at every fourth to fifth residue (2-3 nm), the binding is decreased. The points to which chitosan binds on the plasma membrane must not only be abundant but also not separated by more than a few nm. This suggests that the polycation binds by electrostatic or ternary interactions with the negatively charged head groups of phospholipids. A binding of chitosan to receptors of protein nature appears less likely as many of such molecules may not occur in such a near distance. It is easy to imagine that such a binding of rigid polymers might change overall membrane fluidity and thereby influence the phospholipid density at the borders of membrane transport proteins either functioning themselves in ion transport, or regulating the function of ion channels. This possible action mechanism bears obvious parallels to the opening of specific ion channels due to physical stretching of membranes as explored recently by patch-clamp experiments with animal (Christensen 1987) and plant (Falke et al. 1988) membranes.

In the course of several years, a number of other callose elicitors have been empirically found. Some examples are listed in Fig. 5. The substances may be taken as models for diffusable or surface-exposed (chitosan) recognition signals of pathogen origin, which can trigger defense reactions in plant cells. It should be noted that the potency of these callose elicitors varies considerably and that differences between various cell suspension cultures or strains even from the same plant species have been occasionally found. The action of the amphipathic substances appears not to result from a general detergent action as e.g. Triton-X-100, CHAPS and Octylglucoside are almost inactive. The substance groups in Fig. 5 were introduced to indicate possible modes of first interaction of the elicitors with the plasma membrane. Such an artificial system bears, however, the danger that action mechanisms suggested in the literature on the basis of hemolytic or fungistatic properties might be incorrectly implicated in callose induction. The saponins and polyene antibiotics, for instance, are widely considered to change membrane functions by their interaction with sterols, but it becomes increasingly clear that this cannot be their only action mechanism. Polyenes (Bolard 1986) and saponins (Gögelein and Hüby 1984) also change several properties of sterol-free artificial membranes, indicating that these antibiotics can also interact with the lipophilic part of phospholipids. This property is shared with acylated cyclic polypeptides (e.g. polymyxin B; Storm et al. 1977) and acylcarnitine. The latter compound, for instance, inhibits the phospholipid-activated protein kinase C, presumably by competition with the activator (Wise et al. 1982).

For induction of callose synthesis, all the substances indicated in Fig. 5 are active over a narrow concentration range, which must be experimentally determined based on cell type, cell density and composition of the suspension medium particularly with regards to divalent ion concentration. It appears that certain membrane functions need only to be slightly perturbed to trigger the metabolic effects described below. Supramaximal doses lead to severe cell damage and-or cell dead with a decreased callose production.

As discussed below, early and presumably initial effects of the callose elicitors appear to be changes in ion transport rates. One could envisage, therefore, that the low doses of amphipathic callose elicitors act by disturbing the lipid environment of membrane proteins that are involved in mediating or regulating ion transport.

Ca^{2+} UPTAKE, K^+ EFFLUX AND EXTERNAL ALKALINIZATION AS EARLY EVENTS IN SIGNAL TRANSDUCTION

Elicitation of callose synthesis in suspension cells by the substances listed in Fig. 5 is associated with typical changes in rates of ion transport. Due to the short times required for callose induction, the cells can be suspended in buffers containing essential ions at low concentrations. Specific ion-concentrations in the extracellular solution can be determined with ion-selective electrodes. The experimental protocol with chitosan as an elicitor is given as one example in Table 2.

With the cell line R of *Catharanthus roseus*, which was used in this experiment, callose synthesis was measurable after 5 min. Within this period a considerable portion of the chitosan also became bound by the cells and cellular K^+ was lost. The external [Ca^{2+}] first increased, most likely due to its ionic displacement from the cell wall (Young and Kauss 1983). After about 20 min the extracellular [Ca^{2+}] began to decrease whereas some further chitosan was bound. The uptake of Ca^{2+} into the cells now obviously dominated that portion of Ca^{2+} which was liberated from the wall. This experiment demonstrates that chitosan induced considerable Ca^{2+} uptake by the cells, at least after 20 min. The exact initial time and the amount of Ca^{2+} uptake can not, however, be determined due to the liberation of Ca^{2+} from the cell wall.

The net uptake of Ca^{2+} by cells induced to form callose can be better studied employing noncharged elicitors, which do not displace Ca^{2+} from the cell wall. For example, digitonin-elicited net Ca^{2+} uptake in soybean cells correlated in time with callose formation and was preceded by the leakage of ions, mainly K^+, from the cell (Waldmann et al. 1988). Ca^{2+} taken up by these cells would correspond to an internal increase of about 120 μM, calculated on total cell water and assuming that all cellular Ca^{2+} remains free. Such a high cytoplasmic Ca^{2+} concentration is not compatible with cellular metabolism. The only compartment that might accommodate > 100 μM Ca^{2+} is the vacuole. The movement of Ca^{2+} through the cytoplasm to the vacuole might result, in a transient increase in cytoplasmic [Ca^{2+}]. Unfortunately, the concentration of free Ca^{2+} can not yet be determined by available methods in populations of cells or protoplasts under conditions where they can be induced to form callose.

The net Ca^{2+} uptake induced by various elicitors can be inhibited by the Ca^{2+} channel blockers nifedipine (Waldmann et al. 1988) and flunarizine (Table 3), indicating that the Ca^{2+} uptake may occur through Ca^{2+}-specific channels. It was suggested above that elicitor-induced Ca^{2+} uptake may result directly from the binding of the callose elicitors to general membrane components. Alternatively, more complex mechanisms of signal transduction such as G-proteins may be involved. The considerable time gap of up to 10 min between Ca^{2+} uptake and K^+ efflux in soybean cells (Waldmann et al. 1988)

Table 2. Time course of chitosan-induced callose formation and associated changes in the external concentrations of Ca^{2+}, K^+ and soluble chitosan in suspension-cultured cells of *Catharanthus roseus*, line R. The cells (300 mg/5 ml) were suspended in 10 mM Bistris/Mes buffer, pH 5.8, containing 4 % (v/v) growth medium and 3 % (w/v) sucrose. The concentration of ions was determined with ion-selective electrodes and callose fluorometrically (Kauss et al. 1989). Chitosan in the supernatant was measured colorimetrically according to Ride and Drysdale (1972). Chitosan solution (DP 3420, 22.3 % N-Ac, 0.75 mg/g cell FW) was added at time zero. The values were corrected against controls run without chitosan for the indicated times. In similar other experiments the initial increase in [Ca^{2+}] was less pronounced and the following decrease let, consequentially, below the time zero value

Time [min]	Callose formed [μg/g FW]	External concentration of Ca^{2+} [μM]	K$^+$ [mM]	Chitosan [%]
0	0	89	1.9	100
5	6	117	2.1	61
20	12	128	2.15	48
120	169	88	2.9	17

Table 3. Callose formation and decrease in external [Ca^{2+}] between 20 and 120 min after chitosan addition to suspension cells of *Catharanthus roseus*. Same conditions as in Table 2

	Ca^{2+} [μM]	Callose [μg/2 h/g FW]
Control	- 8	- 2
Flunarizine (20 μM)	- 4	+ 3
Chitosan (0.5 mg/g FW)	- 33	+ 173
Chitosan + Flunarizine	- 13	+ 91

makes an electrical coupling by changes in the membrane potential less likely, although the methods used to determine ion fluxes may not be sensitive enough to detect small initial effects on the K^+ and Ca^{2+} transport rates. This would be necessary to decide whether threshold phenomena exist.

Callose induction by digitonin and amphotericin B also is associated with an external alkalinization, exhibiting a time course similar to K^+ efflux (data not shown). As far as can be determined for the weakly buffered suspension used in the experiments shown in Table 4, the K^+ efflux and the degree of external alkalinization also correlated in dose-response studies with amphotericin B. The phenomenon of a K^+ efflux and an associated external alkalinization has also been observed in experiments on the induction of other defense reactions, such as the induction of the hypersensitive response by bacteria (Atkinson et al. 1986) and the phytoalexin induction by glucan-type elicitors in soybean (Osswald et al. 1985). The effect appears to be associated with a decrease in intracellular pH (Ojalvo et al. 1987) and changes in plasma membrane potential (Pavlovkin et al. 1986, Pelissier et al. 1986, Mayer and Ziegler 1988). Interpretation of the data on K^+ leakage and pH-changes is still controversial. A single antiporter may mediate the coupled movement of K^+ and H^+ (Atkinson et al. 1986), or the effects may be indirectly coupled and involve the H^+-ATPase.

THE 1,3-ß-GLUCAN SYNTHASE AS A TARGET FOR Ca^{2+}

The plasma membrane-located 1,3-ß-glucan synthase (called GS II by plant cytochemists) is strictly dependent on Ca^{2+} *in vitro* (Kauss 1987, Fink et al. 1987). In the presence of 0.2 mM spermine and-or 4 mM Mg^{2+}, half-maximal activity is attained at about 0.8 μM Ca^{2+}. This type of enzyme activation is readily reversible and not dependent on nucleotides. It has been suggested that the enzyme remains latent in undisturbed cells as long as the cytoplasmic concentration of free Ca^{2+} remains at the "resting level" of about 0.1 μM (Kauss 1987). The enzyme may become activated by Ca^{2+} if the cells are damaged or rendered permeable to Ca^{2+} and may directly deposit callose from the surface of the plasma membrane in the cell wall (Fig. 5). This requires a vectorial arrangement of the enzyme. There is some indirect evidence available that the binding sites for the substrate UDP-glucose and the effector Ca^{2+} are indeed located at the cytoplasmic side of the plasma membrane (Fink et al. 1987, Kauss 1987, 1990). In plasmolyzed cells undergoing callose synthesis, polymer accumulates at the plasma membrane as clearly evident from fluorescence microscopy. The 1,3-ß-glucan synthase is present with high activity in homogenates of numerous plants but has not yet been purified. A final and direct proof for the vectorial arrangement of the enzyme awaits enzyme purification.

Ca^{2+} INFLUX IS ESSENTIAL FOR CALLOSE SYNTHESIS

Early evidence for a role of Ca^{2+} as a signal in callose synthesis by chitosan-induced suspension cells came from experiments in which the extracellular [Ca^{2+}] was lowered by the addition of EGTA. Under these conditions callose synthesis stopped (Fig. 6B), indicating that external Ca^{2+} was essential. The addition of EGTA in the presence of the elicitor, however, also appeared to increase ion leakage across the plasma membrane (Fig. 6A). External Ca^{2+} appears not only to provide a pool for Ca^{2+} functioning as second messenger but additionly may function by stabilizing membranes. The problem was

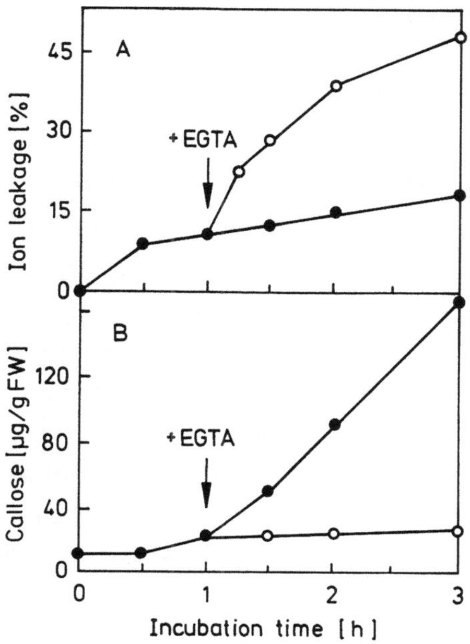

Fig. 6. Ion leakage (A) and increase in callose content (B) in soybean cells treated with chitosan (0.5 mg/g cell FW). Leakage of ions, mainly K^+, is given in % of a boiled sample. The cells were suspended in buffer at 15 μM [Ca^{2+}]. At the time indicated by the arrow, 0.5 mM EGTA (final conc.) was added to one half of the suspension to decrease [Ca^{2+}] below 0.1 μM (o), whereas the other half remained without EGTA (•). In control samples without chitosan no callose was formed and ion leakage was about 4 % after 1 h and was not considerably altered on addition of EGTA. Redrawn using data from Köhle et al. 1985

partly overcome by decreasing the external [Ca^{2+}] with ion exchange beads, which have no direct access to the membrane surface and, therefore, caused less additional ion leakage, although callose formation was greatly diminished (Köhle et al. 1985).

That Ca^{2+} uptake is required for callose synthesis is also evident from the experiments with the Ca^{2+}-channel blockers nifedipine (Waldmann et al. 1988) and flunarizine (Table 3). These substances inhibited the net Ca^{2+} uptake induced by digitonin and chitosan and in parallel also the callose synthesis.

CA^{2+} IS NOT THE ONLY SIGNAL IN CALLOSE SYNTHESIS

The above experiments indicate that Ca^{2+} influx is a prerequisite for callose synthesis. Surprisingly, however, the Ca^{2+} ionophores A 23187 and ionomycin induced little callose formation (Waldmann et al. 1988). These results were recently confirmed. At a comparable net Ca^{2+} uptake, A 23187 caused about 10 % of the callose synthesis effected by digitonin (Table 5). That Ca^{2+} cannot be the only signal in callose induction becomes also evident from dose-response studies with amphotericin B (Table 4). Half-maximal changes in K^+ and Ca^{2+} transport rates were obtained at 2 μM amphotericin B. The comparatively lower effect on pH might have resulted from the use of slightly buffered cell suspension. Rapid K^+ leakage and Ca^{2+} uptake in yeast cells were also induced by 1 μM nystatin (Eilam and Grossowicz 1982). In contrast to the effects on ion flux, callose induction at 2 μM amphotericin B was low and about 50 μM was required to fully induce callose synthesis. These results suggest that in addition to Ca^{2+} uptake, induction of callose synthesis requires triggering of another process which needs higher amphotericin B concentrations.

Table 4. Changes in ion transport rates and callose synthesis induced by various concentrations of amphotericin B in cells of *Catharanthus roseus*. The given values are expressed in % of the effects caused by 50 μM amphotericin B. Conditions as in Table 2 but buffer concentration was only 1 mM. Amphotericin B was added in dimethylsulfoxide (final conc. of the solvent 0.2 % v/v)

Amphotericin B	2 μM	5 μM	25 μM
pH increase	35	91	100
K^+ efflux	57	79	97
Ca^{2+} uptake	52	87	100
Callose synthesis	9	42	90

Table 5. K^+ release, net Ca^{2+} uptake, and callose formation caused by ionophore A 23187 and digitonin in cells of *Catharanthus roseus*. Same conditions as in Table 2. Incubation was for 2 h and the ionic changes in relation to a zero time control are reported

	Change in external		Callose
	K^+ [mM]	Ca^{2+} [μM]	formed [μg/2 h/g FW]
Control	- 0.2	0	0
+ A 23187 (10 μM)	+ 0.9	- 32	14
+ Digitonin (15 μM)	+ 1.2	- 28	154

The question thus arises, which signals may, in addition to Ca^{2+}, be required to activate the 1,3-ß-glucan synthase. Delmer (1987) has speculated that *in vivo* the callose synthase complex might be activated by a depolarisation of the plasma membrane. This appears unlikely since callose synthesis in suspension cells is increased by the presumably hyperpolarizing toxin fusicoccin and decreased by several protonophores (results not shown). One other possibility might be that the callose elicitors in addition to causing changes in the transport rates of Ca^{2+} and other ions also have a direct stimulatory effect on the 1,3-ß-glucan synthase. This enzyme requires phospholipids for activity and has been shown to be activated *in vitro* to a certain degree by low concentrations of certain free polyunsaturated fatty acids, lysophosphatidylcholine, platelet activating factor, acylcarnitine and Echinocandin B (Kauss und Jeblick 1986). The major problem with the assay of the 1,3-ß-glucan synthase is that in homogenates the enzyme is mostly present in the form of sealed right-side-out vesicles. Digitonin has, therefore, to be added to allow an access of the substrate UDP-glucose and of effectors (Mg^{2+}, spermine, cellobiose) to the cytoplasmic side. This detergent necessary to open vesicles might partly mimic the stimulation caused by the amphipathic molecules mentioned above. It was difficult up to now, therefore, to decide whether such effects contribute *in vivo* to callose synthesis in induced cells. Similarly, chitosan and other polycations also cause some stimulation of the 1,3-ß-glucan synthase in digitonin-permeabilized microsomes (Kauss und Jeblick 1985). Whether this is relevant to *in vivo* callose synthesis is not yet known. If such a direct stimulation of the 1,3-ß-glucan synthase by amphipathic elicitors contributes to callose synthesis *in vivo*, similar substances must arise endogenously when callose is induced by mechanical stress. The time gap between chitosan binding and the increase in callose content of protoplasts (Fig. 4) might be an indication that such reactions take place in the membranes before callose formation can start.

INDUCTION OF SECONDARY METABOLITES BY CALLOSE ELICITORS

Under conditions where chitosan (Köhle et al. 1984) and digitonin (Kauss 1987) induced a rapid callose synthesis in soybean suspension cells, we also observed a slower formation of the phytoalexin glyceolline. This has been confirmed by Bonhoff and Grisebach (1988) with digitonin and tomatin, using soybean roots as an experimental system. More details on the induction of secondary metabolites by chitosan have been recently reported for the synthesis of coumarin derivatives in parsley cell suspensions (Conrath et al. 1989). In conditioned growth medium, induction of callose and coumarin synthesis both increased with the degree of polymerization of chitosan, and the cells became more sensitive to chitosan with addition of 1 mM reduced glutathione. These results suggest that both processes have initial steps in common, although fully different pathways and regulatory features are involved. The induced Ca^{2+} uptake might directly represent such a common event. Alternatively or in addition, endogenous phytoalexin elicitors (oligosaccharides, free polyunsaturated fatty acids) might be formed from the cell wall or cell membrane due to necrosis, which occurs to a certain degree when the suspension cells or protoplasts are induced to form callose by the substances listed in Fig. 5.

Induction of phytoalexin synthesis by callose elicitors may reflect the fact that both processes represent defense reactions related to the phenomenon of the socalled hypersensitive response which can be triggered by a great variety of pathogens and non-pathogens exhibiting diverse chemical characteristics. Triggering of this general defense system by perturbation of basic plasma membrane properties makes, therefore, physiological sense. Accordingly, all experimental systems used up to now to study phytoalexin and callose induction do **not** reflect host/pathogen specificity (Kauss 1990).

ACKNOWLEDGEMENTS

Financial support by the Deutsche Forschungsgemeinschaft and the Fonds der Chemischen Industrie is acknowledged. We also thank T.J. Buckhout for editing the English manuscript.

REFERENCES

Aist JR, Gold RE, Bayles CJ, Morrison GH, Chandra S, Israel HW (1988) Evidence that molecular components of papillae may be involved in ml-o resistance to barley powdery mildew. Physiol Mol Plant Pathol 33:17-32

Atkinson MM, Baker CJ, Collmer A (1986) Transient activation of plasmalemma K^+ efflux and H^+ influx in tobacco by a pectate lyase isozyme from *Erwinia chrysanthemi*. Plant Physiol 82:142-146

Bolard J (1986) How do the polyene macrolide antibiotics affect the cellular membrane properties? Biochim Biophys Acta 864:257-304

Bonhoff A, Grisebach H (1988) Elicitor-induced accumulation of glyceollin and callose in soybean roots and localized resistance against *Phytophthora megasperma* f. sp. *glycinea*. Plant Sci 54:203-209

Christensen O (1987) Mediation of cell volume regulation by Ca^{2+} influx through stretch-activated channels. Nature 330:66-68

Clark HF, Shepard CC (1963) A dialysis technique for preparing fluorescent antibody. Virology 20:642-644

Conrath U, Domard A, Kauss H (1989) Correlations in the chitosan-elicited synthesis of callose and of coumarin derivatives by parsley cell suspensions. Plant Cell Reports, in press

Delmer DP (1987) Cellulose biosynthesis. Annu Rev Plant Physiol 38:259-290

Eilam Y, Grossowicz N (1982) Nystatin effects on cellular calcium in *Saccharomyces cerevisiae*. Biochim Biophys Acta 692:238-423

Eschrich W 1956 Kallose. Protoplasma 47:487-530

Falke LC, Edwards KL, Pickard BG, Misler S (1988) A stretch-activated anion channel in tobacco protoplasts. FEBS Lett 237:141-144

Fink J, Jeblick W, Blaschek W, Kauss H (1987) Calcium ions and polyamines activate the plasma membrane-located 1,3-ß-glucan synthase. Planta 171:130-135

Gögelein H, Hüby A (1984) Interaction of saponin and digitonin with black lipid membranes and lipid monolayers. Biochim Biophys Acta 773:32-38
Gold RE, Aist JR, Hazen BE, Stolzenburg MC, Marshall MR, Israel HW (1986) Effects of calcium nitrate and chlortetracycline on papilla formation, ml-o resistance and susceptibility of barley to powdery mildew. Physiol Plant Pathol 29:115-129
Hächler H, Hohl HR (1984) Temporal and spatial distribution patterns of collar and papillae wall appositions in resistant and susceptible tuber tissue of *Solanum tuberosum* infected by *Phytophthora infestans*. Physiol Plant Pathol 24:107-118
Kauss H (1987) Some aspects of calcium-dependent regulation in plant metabolism. Annu Rev Plant Physiol 38:47-72
Kauss H. (1990) Role of the plasma membrane in host/pathogen interactions. In: Larsson Ch, Møller IM (eds) The Plant Plasma Membrane - Structure, Function and Molecular Biology. Springer, in press
Kauss H, Jeblick W (1985) Activation by polyamines, polycations, and ruthenium red of the Ca^{2+}-dependent glucan synthase from soybean cells. FEBS Lett 185:226-230
Kauss H, Jeblick W (1986) Influence of free fatty acids, lysophosphatidylcholine, platelet-activating factor, acylcarnitine, and Echinocandin B on 1,3-ß-D-glucan synthase and callose synthesis. Plant Physiol 80:7-13
Kauss H, Jeblick W, Domard A (1989) Degree of polymerization and N-acetylation of chitosan determine its ability to elicit callose formation in suspension cells and protoplasts. Planta, in press
Köhle H, Young DH, Kauss H (1984) Physiological changes in suspension-cultured soybean cells elicited by treatment with chitosan. Plant Sci Lett 33:221-230
Köhle H, Jeblick W, Poten F, Blaschek W, Kauss H (1985) Chitosan-elicited callose synthesis in soybean cells as a Ca^{2+}-dependent process. Plant Physiol 77:544-551
Mayer MG, Ziegler E (1988) An elicitor from *Phytophthora megasperma* f. sp. *glycinea* influences the membrane potential of soybean cotyledonary cells. Physiol Mol Plant Pathol 33:397-407
Ojalvo I, Rokem JS, Navon G, Goldberg I (1987) ^{31}P NMR study of elicitor treated *Phaseolus vulgaris* cell suspension cultures. Plant Physiol 85:716-719
Osswald WF, Zieboll S, Elstner EF (1985) Comparison of pH changes and elicitor induced production of glyceollin isomers in soybean cotyledons. Z Naturforsch 40c:477-481
Pavlovkin J, Novacky A, Ullrich-Eberius CI (1986) Membrane potential changes during bacteria-induced hypersensitive reaction. Physiol Mol Plant Pathol 28:125-135
Pelissier B, Thibaud JB, Grignon C, Esquerré-Tugayé MT (1986) Cell surfaces in plant-microorganism interactions. VII. Elicitor preparations from two fungal pathogens depolarize plant membranes. Plant Sci 46:103-109
Ride JP, Drysdale RB (1972) A rapid method for the chemical estimation of filamentous fungi in plant tissue. Physiol Plant Pathol 2:7-15
Storm DR, Rosenthal KS, Swanson PE (1977) Polymyxin B and related peptide antibiotics. Annu Rev Biochem 46:723-763
Ullrich-Eberius CI, Pavlovkin J, Schindel J, Fischer K, Novacky A (1989) Changes in plasmalemma functions induced by phytopathogenic bacteria. In: Crane FL, Morré DJ, Löw M (eds) Plasma membrane oxidoreductase in control of animal and plant growth. Plenum Press, New York, in press
Waldmann T, Jeblick W, Kauss H (1988) Induced net Ca^{2+} uptake and callose biosynthesis in suspension-cultured plant cells. Planta 173:88-95
Wise BC, Glass DB, Jen Chou CH, Raynor RL, Katoh N, Schatzman RC, Turner RS, Kibler RF, Kuo JF (1982) Phospholipid-sensitive Ca^{2+}-dependent protein kinase from heart. J Biol Chem 257:8489-8495
Young DH, Kauss H (1983) Release of calcium from suspension-cultured *Glycine max* cells by chitosan, other polycations, and polyamines in relation to effects on membrane permeability. Plant Physiol 73:698-702

INORGANIC IONS IN MORPHOGENETIC RESPONSES OF PLANTS

M. Thellier*, M.O. Desbiez**, C. Ripoll*, M. Demarty* and A. Monnier*
*Faculté des Sciences, URA CNRS 203,
BP118, F76134 Mont-Saint-Aignan Cedex, France
**Phytomorphogenèse, URA CNRS 45,
Université Blaise Pascal, 4 rue Ledru, F63038 Clermont-Ferrand, France

Signal transduction (i.e. the chain of events between reception and expression), depends on the interaction of organic molecules (e.g. hormones) with stereospecific cellular sites. However, inorganic ions also may be involved (Davies, 1987b ; Desbiez and Thellier, 1987).

MORPHOGENETIC EVENTS INVOLVING MINERAL IONS

Ionic effects related to mineral nutrition

A plant reacts against an ionic deficiency (or excess) by increasing (or reducing) the rate of absorption of the unbalanced ion. This was observed initially with monocellular species and small water weeds (Yamamoto and Segel, 1966 ; Vallée and Jeanjean, 1968 ; Thoiron et al., 1970 ; Thellier et al., 1971 ; Falkner et al., 1984). In these cases, the cells suffering from the deficiency (or excess) were also those responding to it by a modification of their mechanisms of cellular transport. With terrestrial higher plants, deficiency (or satiety) signals are set forth by the shoots, and they are forwarded to the root where they cause specific changes of the absorption rates, even if the root does not suffer by itself from the deficiency or the excess of the element under consideration (Cogliatti and Clarkson, 1983 ; Clarkson et al., 1983 ; Drew and Saker, 1984). This mechanism of information transfer from shoot to root provides us with an interesting model for the study of the reception and of the transduction of signals in a plant (Claude Grignon, personal communication).

Ionic effects related to signal reception and transduction, or to message migration in morphogenetic responses

Independent of any problem with the mineral nutrition, the ionic status of plant cells, or of subcellular compartments, depends on the balance between passive leaks through ionic channels and active transports via primary or secondary processes. We are going to recall cases where the ionic (hence the electrical) characteristics, and the way how they may vary, are correlated with characteristic events of the plant morphogenesis.

NATO ASI Series, Vol. H 47
Signal Perception and Transduction in Higher Plants
Edited by R. Ranjeva and A. M. Boudet

Ca^{2+} acts as an intermediate messenger in the control of a number of cellular processes, including polarity, secretion, growth, division, gene expression (Trewavas, 1986), apical dominance (Tamas, 1987) and gravitropism (Pickard, 1985). In some cases changes of cytosolic pH occur simultaneously with the changes of cytosolic Ca^{2+}. For more details about Ca^{2+}, see the chapter by Trewavas in the same volume.

The promotion of cell-growth by auxin or fusicoccin begins with the activation of a H^+/K^+ pump ; in various processes involving growth promotion or inhibition (axillary-bud and tuber dormancy, seed germination, enlargement of roots, hypocotyls, cotyledons and leaves, geotropism), H^+ and K^+ exchanges have been observed (Marrè, 1979). The hypersensitive reaction of tobacco to infection with *Pseudomonas* involves an efflux of K^+ and an equimolar influx of H^+ (Atkinson et al., 1985), while fungal elicitors depolarize the plasmalemma of plant roots (Pelissier et al., 1986). In circumnutating bean shoots, there is a longitudinal gradient of K^+ with a maximum at the level of the curved zone ; moreover, K^+ is more concentrated, while the turgor potential is lower, on the concave than on the convex side (Millet et al., 1987). With *Bidens pilosa*, the effect of cotyledon prickings on the growth of the hypocotyl, and on the breaking of the symmetry of the axillary-bud growth, depends on the ionic composition (K^+, Na^+, pH) of the nutrient medium of the plants (Desbiez and Thellier, 1987).

The action potentials occurring in the seismonastic responses of "excitable" plants involve cellular efflux of Cl^- and K^+ and influx of Ca^{2+} ; however, action potentials occur also in normal plants (Davies, 1987a). The propagation of morphogenetic messages in a plant has been shown to be associated with an electric depolarization wave (Frachisse et al., 1985), but the ions involved are still not known.

Although Li^+ is practically absent from plants under natural conditions, the addition of Li^+ often causes morphogenetic disturbances. Such effects have been observed in the breaking of the symmetry of bud growth, thigmomorphogenesis, circumnutation, root orthogravitropism, flowering processes, and hypocotyl growth ; K^+, but not Na^+, counterbalances the effect of Li^+ (Desbiez and Thellier, 1987).

Occurrence of current loops in regenerating or growing cells and tissues

According to measurements performed with vibrating electrodes, natural electric currents, corresponding to steady fields of a few mV (Jaffe, 1981), seem to occur systematically during differentiation and growth of cells and tissues (Weisenseel and Jaffe, 1976 ; Brawley et al., 1984 ; Miller et al., 1986 ; Gorst et al., 1987 ; Rathore et al., 1988). The current leaves from the mature areas, while it enters the main elongation zones and the zones of branching. Miller et al. (1986) have underlined that fungal pathogens, myccorhizal symbionts and nitrogen fixing bacteria infect the root regions known to be the most electrogenically active. In clusters of carrot suspension cultures, electrical polarity was established even when the cells were grown under conditions when they do not form any embryo (Gorst et al., 1987) : this means that these cells somehow possess a stored determination for embryogenesis even when they cannot express it. In several cases, it appeared that the outward currents were corresponding to an active H^+ efflux, and the inward currents to a passive K^+ influx (Weisenseel and Jaffe, 1976 ; Rathore et al., 1988).

PHYSICO-CHEMICAL PROCESSES LIKELY TO INTERPRET THE INTERFERENCE OF PLANT MORPHOGENESIS BY MINERAL IONS

It is assumed most often that the distribution of the mineral ions is homogeneous in any aqueous subcellular compartment. Moreover, one considers mainly two mechanisms according to which an ion may interfere with cellular processes : i) the ion may be engaged in a binding reaction with other soluble molecules (including proteins) within a subcellular compartment, in which case the important parameter is the chemical potential of the ion in this compartment, or ii) the ion may be involved in a membrane process (ion channelling, co or countertransport, primary active transport), and the important parameter then is the electrochemical potential difference of the ion between the two compartments bathing the membrane. We are going to see that the reality of the ionic interactions may be more complicated.

Equilibrium processes

Reversible binding of an ion with a stereospecific site of a macromolecule. This is the most simple way for an ion to act as an intermediate messenger. Consider a protein controlling one step of plant morphogenesis. Assume that the protein can bind an ion, for instance a cation, M^{z+} (in which case the protein is termed P^{z-}), according to $P^{z-} + M^{z+} \rightarrow PM$, and that either only the free protein, P^{z-}, or only the bound protein, PM, is catalytically active. Mass-action law makes it possible that the cation triggers the activity of the protein. However two conditions have to be fulfilled for that. First, the process must be a real binding reaction, characterized by a dissociation constant : since the anionic sites of the proteins are mainly carboxylic groups which form actual bonds with calcium and protons, $-(COO)_2Ca$ and -COOH, whereas the interaction with alkaline cations is more like a statistical electrical balance, $-COO^- Na^+$ or $COO^- K^+$, mainly calcium and possibly protons may serve as "second messengers" with such processes, but probably neither sodium nor potassium. Secondly the cells must be able to change their concentration of the messenger cation very much, so that the equilibrium of the reaction shifts from a situation where there is practically no bound protein to another where almost all of the protein is bound, and conversely.

A typical example is that of the interaction of calcium with calmodulin and with various cellular enzymes. According to Marmé (1982) this is written

$$4\,Ca^{2+} + CAM \rightleftharpoons Ca_4CAM \qquad (1)$$

$$Ca_4CAM + Enz(actI) \rightleftharpoons Ca_4CAMEnz(actII) \qquad (2)$$

where CAM holds for calmodulin, and where the activity of the free enzyme, Enz(actI), is different (either increased or decreased) from that of the complex Ca_4CAMEnz(act II). Under non stimulated conditions, the Ca^{2+} concentration of the cytosol is low, and the equilibrium of reactions (1) and (2) is in favour of the free enzymes ; when the cell is excited (for instance by light or hormones), the Ca^{2+} concentration raises up due to Ca^{2+} transport from the exterior and from cell organeles, and reactions (1) and (2) shift to the right ; when excitation ceases, Ca^{2+} is pumped out of the cytosol and the enzyme complex dissociates. Calcium-activated ion channels, whose presence is well known in animal cells, have been shown to exist in the tonoplast of plant cells too (Hedrich and Neher, 1987). It seems likely (Begenisich, 1988) that, at least for a part, the effect of Ca^{2+} is via a specific binding with a subunit of the channel protein.

Changes in pH are frequently observed following application of plant hormones or other active substances. The mechanism of action may be via the reversible protonation of a specific site of a protein, while either only the protonated (PH) or only the non protonated (P^-) protein would be biologically active. For instance, the switch of a system of orthophosphate transport, from a low activity at wall pH > 6 to a high activity at pH < 4, is likely to be due to the protonation of a group with pK 5.5 (Sentenac and Grignon, 1985). In the experiments by Ojalvo et al. (1987), addition of an elicitor (glucan) to bean cells affected the vacuolar and cytoplasmic pH's in less than 10 min, and finally caused the production of the phytoalexin "phaseollin" to increase. This might mean that pH acts as a second messenger in this process. However, since the recorded pH-changes were corresponding to an increase of the H^+ activity of a factor of only 3 in the vacuole and of only 1.5 in the cytoplasm, it is unlikely that this was enough to shift a site from a completely non-protonated to a completely protonated form. In this case, the role of the pH-change as an internal signal factor has probably to be explained in another way than by application of mass-action law to the proton-concentrations in the bulk of the cell compartments.

"Ionic cloud" and "ionic diffuse layers". Close to any electrical charge, there is a spherical domain, termed "ionic cloud", where the concentration of ions with an opposite sign is statistically larger than that of the ions with the same sign. Although the ionic cloud has no clear-cut limit, one may characterize it by a "radius", r, which is the distance from the charge under consideration where the radial charge density of the ionic cloud reaches a maximum : the reciprocal of r is proportional to the square root of the ionic strength of the bulk (r ranging from 1 to 3 nm when the ionic strength ranges from 100 to 10 mM). The ionic cloud exerts a screening effect which abates the intensity of the electrostatic interactions between the charges.

Consider a macromolecule bearing fixed charges, in a solution of small ions. The ionic clouds of the fixed charges more or less overlap one another. Therefore, if the saline concentration increases (or decreases), as a result of ion pumping or channeling, or of loss of water, the screening effect increases (or decreases), and the interactions between the fixed charges decrease (or increase). The conformation of the macromolecule thus may change a great deal according to the ionic strength of the medium. For a protein, this might correspond to a transition between an active and an inactive form.

A similar reasoning applies to a surface bearing a distribution of ionizable sites, such as a negatively charged cell membrane, except that the concept of "ionic cloud" is replaced by that of "ionic diffuse layer". Therefore, any change of the saline composition in a cell compartment will change the screening effect of the diffuse layer, and thus might change the accessibility of charged solutes to the membrane, significantly. This corresponds to a functional transition where a membrane protein (enzyme, or transport system) may be reached, or not by its substrate. For instance (Begenisich, 1988), most of the effects of Ca^{2+} on ion channel function have long been described only by Ca^{2+} screening the surface charges, thus causing an apparent voltage shift of channel voltage-dependent parameters. This might seem contradictory with the hypothesis of a specific binding of Ca^{2+} with a subunit of the channel proteins, as indicated above. In fact, it might be that Ca^{2+} acts simultaneously via both mechanisms, specific binding to the channel, and unspecific voltage shift. In the present reasonings, we have postulated implicitely that the membrane surface charge was uniform. This seems to be actually the case, resulting from identical local charge densities above the lipidic and proteic parts of the membrane (Gibrat et al., 1985).

Donnan effect. Within a cellular compartment, the electrostatic interactions between the mobile ions and the non-diffusible charges are responsible for Donnan effect (co-ion exclusion and counter-ion accumulation). In biological media, the ideal Donnan theory has to be sophisticated by taking into account all of the interactions (ion/ion, ion/fixed charge) due to the nature of the polyelectrolyte (Demarty et al., 1978). Attenuation of Donnan exclusion in the cell wall, due to the protonation of fixed groups, might again change the accessibility of charged substrates (e.g. $H_2PO_4^-$) to the cell membrane (Sentenac and Grignon, 1985).

...

The "condensation" effect. Consider a linear polyelectrolyte, for instance a polyanion. One may characterize this polyelectrolyte by the dimensionless parameter, $\xi = e\beta/\epsilon kT$ where k is Boltzmann constant, ϵ the dielectric constant of the medium, e the elementary charge, T the temperature (K), and β the linear charge density. According to Manning (1969), when the ξ value is not too large, there is an atmosphere of mobile cations close to the polyelectrolyte. However if $\xi > z^{-1}$ (where z is the charge of the mobile cations), the atmosphere is unstable and the ions with the highest value of z condense along the polyelectrolyte. As a result ξ decreases until $\xi = z^{-1}$. Such a behaviour, which does not obey mass action law, is well suited for acting as a secondary messenger. A possible mechanism is as follows. Consider the neutralization curve of the polyelectrolyte (Fig.1). In certain domains of pH, a small increase in pH corresponds to a large increase of the linear density of charged sites, β, hence to a large increase of the value of parameter ξ. This promotes sudden condensation of mobile ions beginning by the divalent. Calcium-dependent processes are likely to be affected primarily in this case, and possibly also Magnesium-dependent processes. If the condensation process extends to the monovalent cations, the functionning of K^+ or Na^+ activated enzymes might also be affected.

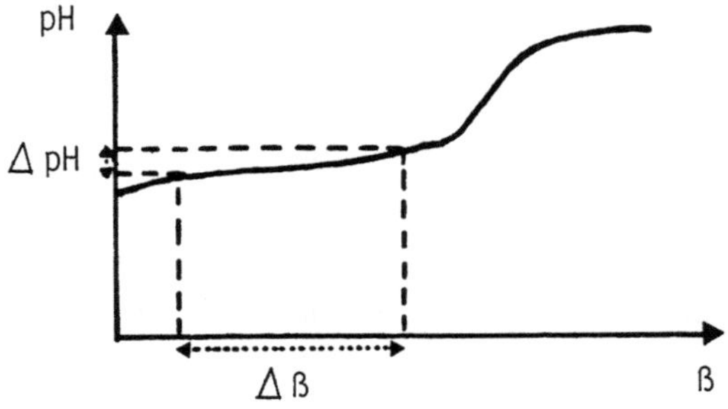

Fig. 1. Neutralization curve of a polyelectrolyte. In certain parts of the curve, a small change in pH corresponds to a huge change of the linear density of charged sites, β, on the polyelectrolyte

Strictly speaking, condensation occurs only with linear polyelectrolytes. Therefore, up to now, ion condensation has been studied mainly with the uronic acids of the cell wall : under the usual conditions, Ca^{2+} (and possibly other ions) are condensed in the cell wall (Demarty et al., 1980 and 1984). The pumping of protons out of the cytoplasm (following auxin or other treatment) might protonate pectin carboxylic groups, thus decreasing the charge density and the amount of condensed ions, finally changing the concentrations of free ions in the wall very significantly. This might be non negligible for the induction of cell growth.

Even with spherical macroions, a situation approximating "condensation of counter-ions" may occur, although, in this case, there is no critical ξ value (Oosawa, 1971). With charged surfaces (i.e. cell membranes), a Stern layer may occur, and have a role equivalent to that of ionic condensation. As a consequence, transition effects equivalent to those which we have described in the cell wall may occur also in the cytoplasm.

Non-equilibrium processes

Due to the existence of primary and secondary active transmembrane ion transports, many ions are likely not to be at chemical equilibrium. This will be responsible for ionic fluxes and electric fields. These processes may be stationary, or non-stationary, for instance oscillatory.

Sustained ionic gradients perpendicularly to a membrane. Since diffusion equilibrium is not immediate, when cells perform an active pumping of protons towards the exterior a proton gradient is maintained in the wall normally to the plasmalemma. The more active the pump and the less intense the leak, the steeper is the proton gradient. The charge distribution at the surface of the plasmalemma, the accessibility of the solutes and the activity of the membrane proteins depend on the value of the local pH thus created by the H^+ pump close to the plasmalemma, which is clearly different from the pH that would exist at equilibrium with the bulk. A slight change of the rate of the proton pump or of the leak thus might dramatically change the plasmalemma functions, consequently reorienting the cell metabolism and growth. It has been shown i) that H^+ excretion actually caused the surface pH of corn root cells to be different from the pH of the bulk medium (Sentenac and Grignon, 1987), and ii) that the H^+-cotransports (phosphate, nitrate, malate, K^+) were related to the surface pH and not to the pH of the bulk medium (Thibaud et al., 1988).

The same reasoning which we have made with H^+ may be extended to any other mineral ion (Ca^{2+}, K^+, etc.) exchanged by cells. In a recent contribution by Alexandre et al. (1989) it was shown i) that IP_3 opens Ca^{2+} channels in the tonoplast, ii) that the corresponding efflux of Ca^{2+} causes non specific membrane channels to close, but iii) that the latter effect is caused by the local concentration increase of Ca^{2+} close to the vacuole and not by the Ca^{2+} concentration in the bulk of the bathing medium.

One may build artificial systems of membrane-enzymes, and of primary or secondary active transports, i) by inserting enzymes in a gel slab used to separate two aqueous solutions from one another, and ii) by imposing across the gel slab a gradient of an ion interfering with the catalysed reaction (Vincent and Thellier, 1983 ; Vincent et al., 1988a and b). Even when simple Michaelian enzymes have been inserted in the gel slab, the observed kinetics may be hyperbolic, plurihyperbolic or sigmoïdal depending on the structural characteristics (mainly the ionic gradient) imposed to the system. Therefore, in the highly compartmented cellular medium with many ionic gradients existing across the membranes or close to them, it is likely that changes of the ionic status (following activation of pumps or channels) may affect the kinetics of protein-catalysed processes very much.

Ricard and Noat (1986) have shown how such mechanisms may be applied to a better interpretation of plant cell growth. The cell wall is postulated to be equivalent to a set of two antagonistic enzyme reactions, i) the first one conditioned by pectin methyl esterase which increases the fixed charge density of the wall and which is inhibited by protons in the pH range 4-8, and ii) the second one corresponding to a number of "growth enzymes", incorporating uncharged carbohydrates in cell-wall material (thus decreasing the wall charge density), and activated by protons in the same pH range. Such a system may display a very high co-operativity of its response (i.e. a dramatic increase or decrease of the wall charge density) to slight changes of pH. Ca^{2+} may diminish or even suppress this abrupt charge transition. This may help to understand the regulation of cell wall growth : with a high charge density, the pH is low and the building up of the wall is active ; as a consequence, the charge density decreases, the pH increases and the methyl-esterases become active, and so on.

Tangential non-equilibrium cases. Apart from the polarized ionic exchanges of the growing cells or tissues, adult characean cells exhibit alkaline bands along their wall, corresponding to specialized membrane areas for OH^- efflux (Lucas and Dainty, 1977). Therefore, the segregation of the various ionic fluxes on the cell surface, inducing current loops, may be a general feature of the plant cells. In fact, even when no current loops are detectable in the external medium, they may exist in the cell wall. According to Lucas (personal communication), the segregation of the ionic fluxes would provide the plant cell with a sensor of its ionic exchanges, via the interference of the electric momentum of membrane proteins by the local electric field (which would not be the case if the various ionic fluxes were intermingled with one another over the entire cell surface). In a multicellular organism, such as a higher plant, the different cells have to concert with one another in some way. The diffusion of signal molecules is a powerful but slow method for intercellular regulation. The network of current loops may contribute to immediate synchronization of distant cells. To our knowledge such a possibility has not been considered, yet, with plant systems, but there is at least one example (Lassalles et al., 1980 and 1981) with an animal epithelium.

Membrane electrophoresis. According to Jaffe (1977) and Poo and Robinson (1977) an electric potential difference of 1 mV across a cell of 30 μm in diameter, may be responsible for a difference of 10 to 50% in the density of the charged membrane proteins floating in the bilayer, between the two sides of the cell. For instance, this "lateral electrophoresis" can move hormone receptors or ion channels in the plane of the membrane (Brawley et al., 1984). According to Raven (1979), such a longitudinal gradient of electrical potential difference (base positive to apex) would tend to accumulate IAA^- uniporters on the basal side of plant cells : this would ultimately explain why auxin migrates from tip to base of the plants, with the well-known morphogenetic consequences of this polarity of movement. The proposed mechanism is as follows. For any given cell of the plant, since the cytoplasm is more alkaline than the wall, the equilibrium of the dissociation reaction, IAA/IAA^-, is more in favor of IAA^- in the cytosol than it is in the wall; but, since the cytoplasm is electrically negative with respect to the wall, there is a net driving force on IAA^- tending to push it out of the cell. There is thus a net efflux of IAA^-, occuring mainly at the basal end of the cell where the density of IAA^- carriers is higher. As a consequence, the concentration of the total auxin in the cell wall is higher at the basal than at the apical end of the cell, which causes a slight increase of the influx of IAA into the apical end of the next cell down from the apex, and so on from cell to cell. Moreover, the asymmetry of the auxin distribution at the level of each cell would tend to activate the plasmalemma H^+-efflux more at the basal than at the apical end of the cell, thus maintaining and reinforcing the longitudinal electric potential difference of the cell.

Dissipative structurations. The forces (reaction affinities, gradients of electrochemical potentials, etc.) drive fluxes (scalar reaction fluxes, vectorial fluxes of transport of substances, etc.).

The fluxes, J, are functions, f, of the forces, X : J = f(X). Close to thermodynamic equilibrium, f is a linear function and the process is always stable. Farther from thermodynamic equilibrium, when f is no longer a linear function, and especially if feed-back loops are present in the system, stability is no longer compulsory. A random fluctuation can then cause the system to shift to a different behaviour. The simplest example is that described in Fig. 2 : a reaction depends non linearly on the concentration of a single given ion, for instance H^+, which is supplied by diffusion (with a linear dependence of the diffusion flux on the concentration).

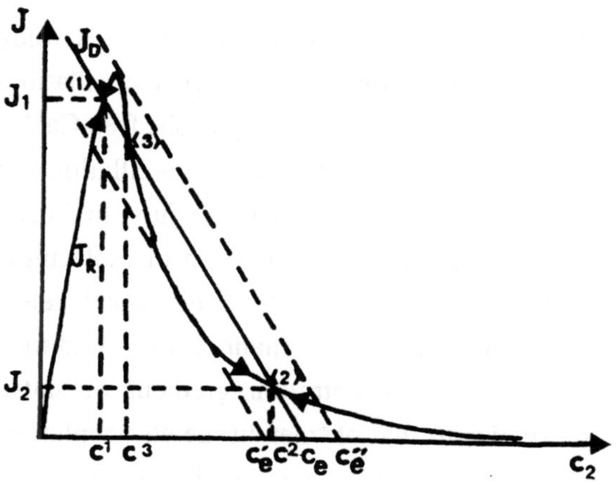

Fig. 2. Reaction/diffusion coupling. In a cell domain, i (membrane or sub-cellular compartment), a reaction, R (enzymic or transmembrane transport), consumes a given ion which is supplied via diffusion from the exterior, e ; C_i and C_e are the concentrations of this ion in i and e ; J_R and J_D are the reaction and diffusion fluxes

For appropriate values of C_e, there are three steady points of functioning, (1), (2) and (3), i.e. points such that $J_R = J_D$. It is clear on the figure that (1) and (2) are stable steady states (a small fluctuation spontaneously tends to vanish), while (2) is unstable.

- Consider the system functionning at high rate, J_1, and low concentration, C^1. If some external signal produces an increase of C_i from C^1 to C^3 (even a transient one), the system will shift to the low rate, J_2, high concentration, C^2, steady state.

- If the external concentration, C_e, is changed from a value slightly below $C_e^{'}$ to a value slightly above $C_e^{''}$ ($C_e^{'}$ and $C_e^{''}$ corresponding to the diffusion curve being tangent to the reaction curve) the system again shifts from the (J_1,C^1) to the (J_2,C^2) steady state.

- Similar effects (not represented on the figure) are obtained when the diffusion coefficient (i.e. the slope of the diffusion line) is changed, following a given signal (opening or closing channels for instance).

- Even more sophisticated effects are expected when considering populations of cells, either in steady-state (1) or (2). In any case, such a process may clearly act as an intermediate messenger in the transduction of a signal.

When at least two diffusible substances are involved, the system may shift from a steady to an oscillatory behaviour following a perturbation. This is termed a "dissipative structuration" (Nicolis and Prigogine, 1977). Although temporal oscillations occur frequently in living systems, their biological advantage compared to steady processes is not clear yet. An oscillatory process may be more easily controlled than a steady one. As a matter of fact, the mechanism of cell synchronisation reminded above (Lassalles et al., 1980 and 1981) occurred in cells whose electric and ionic parameters were oscillating. Addition of auxin triggers oscillations of the membrane electric potential, of the cytosolic pH and of the cytosolic free Ca^{2+} in epidermal cells of maize coleoptiles : it is still not clear whether the oscillations are relevant to auxin action on growth regulation or whether they are just an intriguing side effect (Felle, 1988). Spatial oscillations, i.e. self-organization of molecules in an originally homogeneous system, may be of more immediate biological relevance. According to Fromherz (1988) electrically charged ion channels in a fluid membrane may form dissipative structures driven by a concentration gradient of salts. Such an effect is an interpretation alternative to that of "membrane electrophoresis" for explaining the segregation of the ionic fluxes over a cell membrane by an actual segregation of the ion carriers. However, Lucas and Dainty (1977) have suggested that the OH^- efflux transporters of *Chara* cells would be distributed over the entire membrane surface, but at any one time the transporters would be operational only in a few membrane bands. In such a case, a dissipative structuration of effectors of the OH^- carriers might be the explanation for the observed phenomenon.

Non-equilibrium non-steady processes. Mineral ions may be engaged in at least two different non-equilibrium non-steady processes. The first type corresponds to a dissipative structuration inducing a temporal oscillation. In such cases, the interpretation of the oscillation does not require to postulate any variation of the ionic conductivity of the cell membranes, but it depends only on the shape of the kinetic equations and on the limiting conditions imposed to the system. Spontaneous or induced waves of electric depolarization may also be observed, corresponding to transient changes of the conductivity of ionic channels in the cell membrane. Pricking the cotyledons of plantlets induce correlations between the cotyledonary buds, while they initiate the immediate transmission of a wave of electric depolarization ; it has been suggested (Frachisse et al., 1985) that the electrical event was corresponding to the migration of the morphogenetic message.

METHODOLOGICAL APPROACHES

The quantitative microlocation of the chemical elements, and the discrimination of isotopes in view of ion-exchange measurements are a requisite for studying the interference of plant morphogenesis by mineral ions. References about classical and modern relevant methods are given in a previous review (Thellier, 1984).

A few methods are utilizable for *in vivo* measurements. Coloured or fluorescent dyes, or the emission of light by aequorin, serve to measure local Ca^{2+} concentrations. The methods for intracellular pH measurements (weak acids, weak bases, H^+-selective microelectrodes, ^{31}P-NMR) have been reviewed (Kurkdjian and Guern, 1989). Specific microelectrodes and NMR may be used also for other ions.

However, most physical methods of detection apply to histological sections. Autoradiography is the most popular of such methods ; however, it is practically restricted to the use of a small number of low-energy β-emitters, radioisotope discrimination is difficult, and the time-lapse for obtaining the autoradiographs is usually long. The electron-microprobe is also a popular method for mapping elements : the resolving power is of the order of 1 μm ; the method has been used widely for middle-mass elements, but it is not utilizable for low-mass elements ; it does not recognize the different isotopes of a given element. The PIXE method (proton induced X-ray emission) is an alternative to the electron-microprobe, with a better sensitivity. With the "Electron energy-loss analysis" the spatial resolution is remarkable (in the magnitude of 0.5 nm), and the method is well adapted to detect light elements (Bendayan et al., 1989), but it does not discriminate isotopes.

In the "Secondary ion emission microanalysis" a beam of accelerated primary ions is projected onto the sample which it erodes progressively. Part of the matter thus ejected comes out under an ionized form. These secondary ions may be treated according to the stigmatic image mode or to the scanning mode. The method can discriminate any isotope of any element of the periodic classification, and, especially in the scanning mode, there is practically no risk of interferences. The sensitivity ranges from ppm to ppb. The resolving power on the sample surface is 0.2 μm at the best. Profiles of the distribution in depth of any isotope of any element may be obtained with a resolution as good as a few nm. It is possible to image the distribution of as many different isotopes of as many different elements as one may wish at the same place of a specimen. The time lapse per measurement is of a few minutes. The NCR method ("Neutron capture radiography") is based upon using neutron, e.g. (n,α) or (n,p), nuclear reactions and appropriate "dielectric detectors", for identifying and imaging a few stable isotope (mainly ^{6}Li, ^{10}B, ^{14}N, ^{17}O). The resolving power is close to 1 μm. The handling of the method is very simple. The main disadvantage is in the small number of nuclides which the method can detect and map. Using proton or deuteron nuclear reactions makes it possible to detect most isotopes of most elements. The sensitivity is close to ppm. The resolving power is of the order of magnitude of 1 μm. By adjusting the energy of the bombarding particles and measuring that of the collected ones, one may draw the profile of the distribution in depth of the nuclide under study. In the "Nuclear microprobes" a computer reconstitutes the image of the distribution of each nuclide.

DISCUSSION AND CONCLUSIONS

In the cellular medium, the distribution of the mineral ions is fairly complex, since it depends on equilibrium interactions with other cell constituents and on the dynamics of the non-equilibrium processes which are going on. As a consequence, ionic gradients may occur normally or laterally to the membranes, and the ionic concentrations close to the membranes are probably quite different from what they are in the bulk of the aqueous compartments.

The characteristics of the ionic distributions may display a high cooperativity of their responses to slight perturbations, or even abrupt changes due to instability processes. This means that the mineral ions may act as "intermediate messengers" or contribute to relay-and-amplification mechanisms, in many different ways, apart from being engaged in association/dissociation reactions. In fact, even when an ion is engaged in an association/dissociation reaction, one must not forget that this ion is also subjected to all the intricate physico-chemical interactions which we have considered.

The ion-transport dynamics may have unexpected effects on the segregation of important membrane proteins (such as ion channels or hormone receptors), or on the segregation of the membrane areas where these proteins are active, via "membrane electrophoresis" or dissipative structurations. This may be responsible for inducing and maintaining cell or tissue polarity, or for triggering symmetry-breaking effects. Since feed-back loops are involved, these processes may be subjected to instability and amplification. According to Almirantis and Nicolis (1987), when a two-variable reaction-diffusion system displays a symmetry breaking bifurcation, even a slight initial asymmetry can give rise to a marked asymmetry in the resulting dissipative structure.

Davies (1987b) has already underlined that "ions (...) *are likely to play a major role, especially as local (intracellular) signals*", and that *"electrical signals (...) need to attain respect from a greater number of plant researchers, as candidates for major intercellular communication devices"*. Our present study fully support these views. In the past, the coupling of ATP synthesis with exergonic cell-processes was first interpreted only by postulating the existence of specialized coupling molecules ; then it became apparent that ion gradients played a major role too. Now, we might face a similar situation concerning the transduction of morphogenetic signals in plants.

Alexandre J., Lassalles J.P., Kado R.T. (1989) IP_3 opens calcium channels of the red beet vacuole tonoplast. 8th international workshop on plant membrane transports. Elsevier, Amsterdam (in press).

Almirantis Y., Nicolis G. (1987) Morphogenesis in an asymmetric medium. Bull. Math. Biol. 49:519-530.

Atkinson M.M., Huang J.S., Knopp J.A. (1985) The hypersensitive reaction of tobacco to *Pseudomonas syringae* pv. pisi. Plant Physiol. 79:843-847.

Begenisich T. (1988) The role of divalent cations in potassium channels. Trends in Neurosciences 11:270-273.

Bendayan M., Barth R.F., Gingras D., Londono I., Robinson P.T., Alam F., Adams D.M., Mattiazzi L. (1989). Electron spectroscopic imaging for high resolution immunocytochemistry. J. Histochem. Cytochem. 37:573-580.

Brawley S.H., Wetherell D.F., Robinson K.R. (1984) Electrical polarity in embryos of wild carrot precedes cotyledon differentiation. Proc. Natl. Acad. Sci. USA 81:6064-6067.

Clarkson D.T., Smith F.W., Van Den Berg P.J. (1983) Regulation of sulphate transport in a tropical legume, *Macroptilium atropurpureum*, cv. Siratro. J. Exp. Bot. 34:1463-1483.

Cogliatti H., Clarkson D.T. (1983) Physiological changes in, and phosphate uptake by potato plants during development of, and recovery from phosphate deficiency. Physiol. Plant. 58:287-294.

Davies E. (1987a) Action potentials as multifunctional signals in plants : a unifying hypothesis to explain apparently disparate wound responses. Plant Cell Environment 10:623-631.

Davies E. (1987b) Plant responses to wounding. In : Davies D.D. (ed). The biochemistry of plants, vol. 12, Acad. Press, New York, p. 243.

Demarty M., Morvan C., Thellier M. (1978) Exchange properties of isolated cell walls of *Lemna minor* L. Plant Physiol. 62:477-481.

Demarty M., Ripoll C., Thellier M. (1980) Ion exchanges in plant cell wall. In : Spanswick R.M. et al. (eds) Plant membrane transports. Elsevier, Amsterdam, p.39.

Demarty M., Morvan M., Thellier M (1984) Calcium and the cell wall. Plant Cell Environment 7:441-448.

Desbiez M.O., Thellier M. (1987) Rôle des ions minéraux dans la morphogenèse végétale. In : le Guyader H. (ed.) Le développement des végétaux : aspects théoriques et synthétiques. Masson, Paris, p. 173.
Drew M.C., Saker L.R. (1984) Uptake and long-distance transport of phosphate, potassium and chloride in relation to internal ion concentrations in barley : evidence of non-allosteric regulation. Planta 160:600-607.
Falkner G., Strasser P., Graffins D. (1984) Phosphate uptake by blue green algae *in vitro* and in a lake during an algal bloom : useful application of a flow-force relationship. Hydrobiol 108:265-271.
Felle H. (1988) Auxin causes oscillations of cytosolic free calcium and pH in *Zea mays* coleoptiles. Planta 174:495-499.
Frachisse J.M., Desbiez M.O., Champagnat P., Thellier M. (1985) Transmission of a traumatic signal via a wave of electric depolarization. Physiol. Plant. 64:48-52.
Fromherz P. (1988) Self-organization of the fluid mosaic of charged channel proteins in membranes. Proc. Natl. Acad. Sci. USA 85:6353-6357.
Gibrat R., Grouzis J.P., Rigaud J., Grignon C. (1985) Electrostatic characteristics of corn root plasmalemma : effect on the Mg^{2+}-ATPase activity. Biochim. Biophys. Acta 816:349-357.
Gorst J., Overall R.L., Wernicke W. (1987) Ionic currents traversing cell clusters from carrot suspension cultures reveal perpetuation of morphogenetic potential as distinct from induction of embryogenesis. Cell Differentiation 21:101-109.
Hedrich R., Neher E. (1987) Cytoplasmic calcium regulates voltage-dependent ion channels in plant vacuoles. Nature 329:833-836.
Jaffe L.F. (1977) Electrophoresis along cell membranes. Nature 265:600-602.
Jaffe L.F. (1981) The role of ionic currents in establishing developmental pattern. Phil. Trans. R. Soc. Lond. B295:553-566.
Kurkdjian A. Guern J. (1989) Intracellular pH : measurement and importance in cell activity. Ann. Rev. Plant Physiol. Biochem. (in press).
Lassalles J.P., Hartmann A., Thellier M. (1980) Oscillation of the electric potential of frog skin under the effect of Li^{+} : experimental approach. J. Membrane Biol. 56:107-119.
Lassalles J.P., Hyver C., Thellier M. (1981) Oscillation of the electrical potential of the frog skin under the effect of Li^{+} : theoretical formulation. Biophys. Chem. 14:65-80.
Lucas W.J., Dainty J. (1977) Spatial distribution of functional OH^{-} transporters along a characean internodal cell, determined by the effect of cytochalasin B on $H^{14}CO_3^{-}$ assimilation. In : Thellier M., Monnier A., Demarty M., Dainty J. (eds) Transmembrane ionic exchanges in plants. Editions du CNRS, Paris, Éditions de l'Université, Rouen, p.289.
Manning G.S. (1969) Limiting laws and counterion condensation in polyelectrolyte solutions : I colligative properties. J. Chem. Phys. 51:924-933.
Marmé D. (1982) The role of Ca^{2+} in signal transduction of higher plants. In : Wareing P.F. (ed) Plant Growth substances. Acad. Press, New York, p. 419.
Marrè E. (1979) Fusicoccin : a tool in plant physiology. Ann. Rev. Plant Physiol. 30:273-288.
Miller A.L., Raven J.A., Sprent J.I., Weisenseel M.H. (1986) Endogenous ion currents traverse growing roots and root hairs of *Trifolium repens*. Plant Cell and Environment 9:79-83.
Millet B., Melin D., Badot P. (1987) Circumnutation : a model for signal transduction from cell to cell. In : Wagner E. et al. (eds). The cell surface in signal transduction. NATO ASI Series H12, Springer-Verlag, Berlin, p. 169.
Nicolis G., Prigogine I. (1977) Self-organization in nonequilibrium systems. John Wiley & sons, New York.
Ojalvo I., Rokem J.S., Navon G., Goldberg I. (1987) ^{31}P NMR study of elicitor treated *Phaseolus vulgaris* cell suspension cultures. Plant Physiol., 85:716-719.
Oosawa F. (1971) Polyelectrolytes. Marcel Dekker, New York.

Pelissier B., Thibaud J.B., Grignon C., Esquerré-Tugayé M.T. (1986) Cell surfaces in plant microorganism interaction. VII elicitor preparations from two fungal pathogens depolarize plant membranes. Plant Science 46:103-109.
Pickard B.G. (1985) Early events in geotropism of seedling shoots. Ann. Rev. Plant Physiol. 36:55-75.
Poo M., Robinson K.R. (1977) Electrophoresis of concanavalin A receptors along embryonic muscle cell membrane. Nature 265:602-605.
Rathore K.S., Hodges T.K., Robinson K.R. (1988) Ionic basis of currents in somatic embryos of *Daucus carota*. Planta 175, 280-289.
Raven J.A. (1979) The possible role of membrane electrophoresis in the polar transport of IAA and other solutes in plant tissues. New Phytol. 82:285-291.
Ricard J., Noat G. (1986) Electrostatic effects and the dynamics of enzyme reactions at the surface of plant cells : 1 a theory of the ionic control of a complex multi-enzyme system. Eur. J. Biochem. 155:183-190.
Sentenac H., Grignon C. (1985) Effect of pH on orthophosphate uptake by corn roots. Plant Physiol. 77:136-141.
Sentenac H., Grignon C. (1987) Effect of H^+ excretion on the surface pH of corn root cells evaluated by using weak acid influx as a pH probe. Plant. Physiol. 84:1367-1372.
Tamas I.E. (1987) Hormonal regulation of apical dominance. In : Davies P.L. (ed) Plant hormones and their role in plant growth and development. Martinus Nijhoff Pub., Dordrecht, p. 393.
Thellier M., Thoiron B., Thoiron A. (1971) Electrokinetic formulation of overall kinetics of *in vivo* processes. Physiol. Veg. 9:65-82.
Thellier M. (1984) From classical to modern methods for element localization in plant systems. Physiol. Veg. 22:867-886.
Thibaud J.B., Davidian J.C., Sentenac H., Soler A., Grignon C. (1988) H^+-cotransports in corn roots as related to the surface pH shift induced by active H^+-excretion. Plant Physiol. 88:1469-1473.
Thoiron A., Thoiron B., Thellier M. (1970) Absorption du sulfate par la *Riccia fluitans* : effet des conditions antérieures de nutrition en sulfate. C.R. Acad. Sc. Paris, série D, 270:328-330.
Trewavas A.J. (1986) Molecular and cellular aspects of calcium in plant development. NATO ASI Series, Series A : Life Sciences, Vol. 104, Plenum Press, New York.
Vallée M., Jeanjean R. (1968) Le système de transport de SO_4^- chez *Chlorella pyrenoidosa* et sa régulation : II. Recherches sur la régulation de l'entrée. Biochim. Biophys. Acta 150:607-617.
Vincent J.C., Thellier M. (1983) Theoretical analysis of the significance of whether or not enzymes or transport systems in structured media follow Michaelis-Menten kinetics. Biophys. J. 41:23-28.
Vincent J.C., Alexandre S., Thellier M. (1988a) How a soluble enzyme can be forced to work as a transport system : description of an experimental design. Arch. Biochem. Biophys. 261:405-408.
Vincent J.C., Alexandre S., Thellier M. (1988b) How a soluble enzyme can be forced to work as a transport system : theoretical interpretation. Bioelectrochem. Bioenergetics 20:215-222.
Weisenseel M.H., Jaffe L.F. (1976) The major growth current through lily pollen tubes enters as K^+ and leaves as H^+. Planta 133:1-7.
Yamamoto L.A., Segel I.H. (1966) The inorganic sulphate transport system of *Penicillium chrysogenum*. Arch. Biochem. Biophys. 114:523-538.

ACKNOWLEDGEMENTS :
We are thankful to Claude Grignon for his help in the discussion of the effects of ionorganic ions in morphogenetic responses of plants.

THE IN-VIVO APPROACH: CELL IMAGING

J. TEISSIE B. GABRIEL M.H. MONTANE D. ROUAN
Centre de Recherches de Biochimie et de Génétique Cellulaires du CNRS
118 Route de Narbonne
31062 Toulouse Cédex
FRANCE

Movements of ions in cells are considered as a driving forces in the metabolism. Most of the studies have been dealing with flows across the plasma membranes. For such purposes, studies are operated on a large number of cells and an value averaged on the population is obtained. Exchange can be quantified by using radioactive labelled compounds or analogs, flame photometry or the use of ion sensitive probes. In this last case, fluorescence is a very sensitive method. Recently, the evaluation of ionic exchange has been extended to the single cell level. This was made possible by use of fluorescence microscopy and by the synthesis of new and very sensitive probes. As far as this second point is concerned in the case of Ca^{++}, the work of the group of Tsien should be acknowledged because it made commercially available different Ca^{++} specifically sensitive fluorescent dyes which have been proved suitable for studies at the single cell level.

Working at the single cell level showed that in many cases the distribution of ions and of other charged species was not uniform in the cytoplasm. From a biophysical point of view, this was predictable if one took into account the electrical properties of biological membranes. They are charged interfaces due to the presence of many ionizable groups on their surfaces (phospholipids, proteins). As such, the distribution of ions in their neighbourhood obeys the Gouy-Chapman relationship. In other words, counter ions are going to accumulate close to the interface creating a gradient arising from the surface. Another ionic anisotropy is associated to transient changes which occur either along active transport or with granules release. The local ionic increase would then dissipate through passive diffusion in the cytoplasm.

Single cell observation made possible the monitoring of these local changes. They can indeed be quantitated by use of video monitoring associated to the computer assisted handling of the video signal.

NATO ASI Series, Vol. H 47
Signal Perception and Transduction in Higher Plants
Edited by R. Ranjeva and A.M. Boudet

In this work, we would describe briefly the methods which are used for such purposes. Many informations are indeed already available (Inoue, 1986; Ploem, 1989; Tsien, 1989). As such, by taking advantage of the troubles we have been facing, our main focus would be to insist on the difficulties and the limitations of this very powerful approach of cell biology in the case of Ca^{++} monitoring in Plant protoplasts. We think that it would be a gain of time for other users to be aware of these informations.

I - DIGITALIZED VIDEO FLUORESCENCE MICROSCOPY

The complete system we are using is described on Fig.1. Other similar set-ups are reported in other reviews (Inoue, 1986; Weiss et al, 1989). The protoplast preparation is observed under an inverted fluorescence microscope by means of a very sensitive video camera. The video signal, which is a scanning of the fluorescence emission, is digitized on-line and stored in a computer.

I - A - MICROSCOPE

As in many studies, one wants to be able to manipulate the protoplasts, we consider that an inverted microscope is more user friendly (Ploem, 1989). But the use of high power objectives is then not very easy. The design of special chambers where the bottom is a coverslip is then requested and has been proved to be very easy to use.

As fluorescence emission is, in most cases, very temperature sensitive, a control of this parameter is needed. A heated microscope stage should be provided.

The commercial light source is suitable for most studies. Nevertheless, due to some overlapping between the excitation and emission filters, a laser would solve many troubles when the fluorescence emission is very low and difficult to extract from the background level.

As with many fluorescent probes, two different excitation (or emission) wavelengths are requested (ratio imaging), the switch between the filter blocks must be very fast.

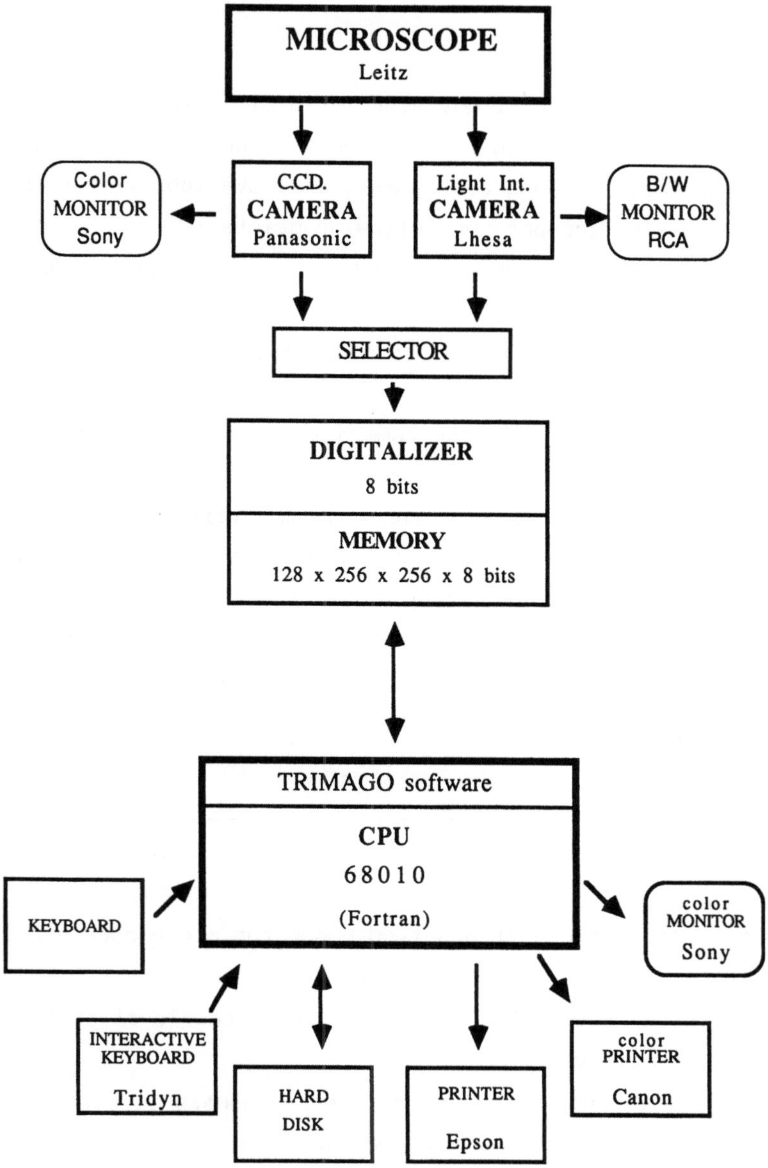
MICROSCOPE
Leitz
Color
MONITOR
Sony
C.C.D.
CAMERA
Panasonic
Light Int.
CAMERA
Lhesa
B/W
MONITOR
RCA
SELECTOR
DIGITALIZER
8 bits
MEMORY
128 x 256 x 256 x 8 bits
TRIMAGO software
CPU
68010
(Fortran)
KEYBOARD
color
MONITOR
Sony
INTERACTIVE
KEYBOARD
Tridyn
HARD
DISK
PRINTER
Epson
color
PRINTER
Canon

- Fig. 1 -

I - B - VIDEO CAMERA

As the level of the light associated with fluorescence emission is low, the use of light intensifying camera is needed (Spring and Lowy, 1989). Many different brands are commercially available. They should be choosen by taking into account the following points :

- The gain must be manual in order to compare quantitatively the level of emission.
- The linearity of the response must be checked. This can be corrected by the computer if needed.
- The sensitivity of the camera must be as uniform as possible. Here again, correction by the computer is a great advantage because the illumination of the sample is not uniform. Correction by light fibers (Inoue, 1986) is not operative for fluorescence detection due to the very weak transmission of the set-up.

As described in Fig.2, the correction needed to obtain a uniformity of detection is easily obtained by use of the associated computer.

---

1 - A reference slide is built by a very thin layer of a fluorescent solution.

2 - This reference sample is observed under the microscope and stored as reference matrix in the computer (I1).

3 - The matrix associated to the background level of the camera is stored in the computer (I0).

4 - The fluorescence of the reference sample is corrected from the background by substraction of the two matrix : $I2 = I1 - I0$

5 - The cell preparation is observed and the sample matrix is stored in the computer (I3).

6 - The background level is substracted : $I4 = I3 - I0$

7 - The definitive correction is obtained by dividing I4 by I2 : I5

- Fig. 2 -

PROTOCOL FOR CORRECTING THE UNHOMOGENEITY IN FLUORESCENCE MICROSCOPY VIDEO MONITORING

I - C - DIGITALIZATION

The video signal is digitalized and stored as a matrix in the computer. In most cases, the conversion is operated on 8 bits i.e 256 different light levels. The size of the matrix can be selected as a function of the optical resolution. 256 x 256 (64 K data points) or 512 x 512 (250 K data points) are routinely used. Limitations are computer time (doubling the size of the matrix increases it by a factor 4) and size of the image memory. This last component should be as large as possible (in our system, we can store up to 128 images). This gives direct access to many informations and saves time during data processing (Castelman, 1979).

Several optional features would improve the working conditions. The video signal associated to fluorescence emission is always affected by a high level of electronic noise. As noise is a random process, the signal to noise ratio can be improved by averaging several pictures. This can be operated by the computer, but, as such, is time consuming, or by feeding the video signal first in a real time averager before storing it. A very fast video/digital converter is then needed in order to work on line. The video picture is converted in the matrix in 40 mS (in Europe).

I - D - DATA HANDLING

In the computer, the picture is stored as a mathematical matrix where each element is the light level at a particular position, $I(x,y)$. In the case of the fluorescence emitted by an ion indicator, this light level is directly related to the ionic concentration (see below). Dedicated softwares handle these mathematical data in order to extract the biological informations.

As described in I-A and I-B, data must first be corrected from artefacts associated to the detection system (microscope and video camera). Several reference pictures should be stored for mimizing the problems of linearity and spatial uniformity. One should emphasize that most of these corrections need a very localized definition almost at the level of every data point. This explains why the image memory space should be large to allow the storage of these reference data.

Another advantage of digitizing the picture is the opportunity to improve their quality. After data handling, the matrix can be converted back in'an image which is observed on a monitor. Contrast can be improved, background can be substracted, all these operations allow the detection of details buried in the initial picture.

One should insist on the advantage that with most systems, the dedicated software is provided with the hardware. In some cases, the computational language is Fortran allowing the development of programs tailored for specific applications. But except for this improvement, the use of a computerized video system is in most cases user friendly and does not require special gifts in computer sciences. Nevertheless, a core of basic subroutines must be provided as summarized in Fig. 3 . The arithmetic subroutines are essential for the quantitative fluorescent observation of single cells.

---

1 - Statistics :
- Histogram (gives the range of light levels)
- Pseudo color classifications.

2 - Arithmetics :
- Addition (averaging)
- Substraction (correction)
- Division (correction).

3 - Geometry :
- Homothety (for zooming).

4 - Logic function :
- Thresholding (for the selection of areas of interest).

5 - Graphics :
- Level of light along a given line.

6 - Transfer :
- Modification of the light levels (contrast improvement).

7 - Filtering :
- Improvement of the quality of the image.

- Fig. 3 -

SUBROUTINES NEEDED FOR IMAGE PROCESSING

I - E - PERIPHERALS

Matrix can be stored as files on a hard disk but another mass storage device is needed for the volume of data is rapidly enormous.

Rebuilt images and graphs are observed on a video monitor and stored as photographic pictures. Graphs with pseudo color can be printed on a color printer.

II - FLUORESCENT CALCIUM INDICATORS

II - A - NATURE

Since 1985, new probes have developped such as Fura 2, Indo 1 or Fluo 3 thanks to the work of the group in Berkeley (Tsien, 1989). They are now commercially available from Molecular Probes. The basic idea is to chemically modify the well known Ca++ chelator in order to obtain a compound with a high affinity for Ca++ and which is highly fluorescent only when Ca++ is bound:

$$Ca^{++} + \text{Indicator} \rightleftharpoons \text{Complex} \qquad (K_d)$$

The emission is the sum of the emission by free dyes (Indicator) and bound forms (Complex).

Several criterion are requested :

a) high optical absorption

b) high quantum yield

c) good water solubility

d) suitable wavelengths for excitation

e) Excitation above 360 nm for their use with a microscope without quartz optics

f) a good affinity for Ca^{++}

As far as this last point is concerned, too high an affinity would buffer out Ca^{++} transient changes and is not a technical advantage.

II - B - PROBLEMS

One of the problems associated with fluorescence microscopy is fading, i.e photobleaching of the fluorescent molecule due to the very strong illumination (Lansing-Taylor and Salmon, 1989). This photochemical destruction of the Ca^{++} indicator alters the quantitative evaluation and should be reduced by decreasing the intensity of illumination (either by using neutral density filters or by defocusing the lamp). Of cource, the signal to noise ratio will be decreased and a compromise must be found.

Toxicity of these compounds can be reduced by using a low level of loading. Of cource the emission may be reduced but the new compounds can be used with cytoplasmic concentration around 20 uM.

The protoplasts autofluorescence is one of the major troubles. In commercial microscope filters, the emission filter is a high pass filter which allows the detection of the endogeneous emission. This can be solved by adding a band pass filter which rejects the red light.

Specificity for Ca^{++} is altered by the binding of magnesium or other metals. This point has been improved with the new compounds and can be partly corrected by using chelators for metals.

II - C - CALCIUM QUANTIFICATION

Ca^{++} binding induces an increase in the quantum yield of the probe and a shift in the emission or excitation wavelengths (Tsien, 1989). The ratio of the light emitted by the free and the bound form of the dye depends on the wavelength. Advantages can be taken from the excitation shift to obtain an experimental parameter which is a function of only the Ca^{++} concentration but does not depend on the amount of dye (the so called 340/380 ratio)(Tsien, 1989). The calcium concentration at each position in the protoplast is then obtained by dividing 2 pictures. A limit of the technique is the level of the noise associated to the signal. When the signal is low, i.e. when the Calcium level is weak, the fiability of the division is very poor and this method is then prone to artifacts. When working at fixed wavelengths, binding of Ca++ will change the intensity of emission of the dye and local changes will induce localized modifications in the emission. These local changes would be detected and localized by the video system even if they are transient (the time resolution of the system being less than 1 S). But this requires that during the measurement, the local concentration in dyes remains constant. Furthermore, the quantitative evaluation of the change in Ca^{++} concentration would be difficult but the temporal aspect of the changes is directly accessible.

Quantification is only obtained by comparison with a reference curve obtained under controlled Ca^{++} concentration by use of ionophores. This approach is nevertheless prone to artefacts by altering the ionic cytoplasmic content because K_d, the dissociation constant of Ca^{++} to the dye, is sensitive to pH and ionic strength.

III - INDICATOR LOADING

One of the major problems is the introduction of the exogeneous dye in the cytoplasm under conditions which do not affect the protoplast viability.

---

1 - Preparation of protoplasts.
2 - Loading of the fluorescent indicator.
3 - Observation under the microscope.
4 - Digitization and storage of informations.
5 - Correction :
- Background substraction.
- Illumination.

- Fig. 4 -

MICROLOCALIZED IONIC QUANTIFICATION BY VIDEO IMAGING

III - A - USE OF ACETOXYMETHYL ESTER DERIVATIVES

These products are less hydrophilic than the acid form and can cross the plasma membrane (Tsien, 1989). Cytoplasmic esterases will then free the acid form which is then trapped in the cytoplasm. A major advantage of this methodology is that only the acid form is Ca^{++} sensitive. A drawback is associated to the release in the cytoplasm of the acetoxymethyl ester and of its degradation products which may be toxic for the cell and in any case will alter its properties. This difficulty can be limited if a low level of loading is proved to be workable.

Another technical problem is that this methodology is very dependent on the cell species. It has indeed been reported that it was not succesful with plant protoplasts (Bush and Jones, 1989).

III - B - ACIDIC INCUBATION

The acidic form of one indicator (Indo 1) was observed to accumulate in plant protoplasts when incubated at a low pH (4.5)(Bush et al, 1988; Bush and Jones, 1989). A high level of loading can be obtained which is a function of the external concentration. When entrapped in the cytoplasm, the dye does not leak out and uniformly dispersed in the cytoplasm.

III - C - MICROINJECTION

A given amount of the dye solution can be injected in the cytoplasm by use of micromanipulators. This technology requires a rather expensive equipment and its use is limited to protoplasts where the cytoplasmic volume is large when compared to the one of the vacuole. The mechanical damage affecting the plasmalemma is apparently rapidly repaired and does not affect the viability.

III - D - ELECTROPERMEABILIZATION

Short duration electric field pulses are known to reversibly alter plant protoplast membrane organization giving a transient permeabilization (Jones et al, 1987). When used under strictly controlled conditions, this treatment does not affect the cell viability and allows the loading of exogeneous molecules (Montané et al, 1988; Tempelaar and Jones, 1988; Jones, 1988).

This methodology has been used for loading plant protoplasts with Quin 2, a Ca^{++} indicating fluorescent dye (Gibroy et al, 1986). The amount of loaded indicator is directly controlled by the external concentration. This approach appears of general use and seems the method of choice for loading indicators in cells and protoplasts.

IV - LOCALIZED DETECTION OF Ca^{++} CHANGES

Quantification of Ca^{++} concentration at the single cell level is then possible but difficult. The use of computer assisted video detection allows the detection of the light level with a very good spatial resolution (say 1 um). But nevertheless one should take into account the working distance of a microscope and the fact that a layer is indeed observed. In other words, the lateral resolution is appaently good but may be affected by the vertical resolution. High power objectives with a very short working distances would partiallly solve this problem. The temporal aspect of the local Calcium changes is more directly accessible by this approach and would give information on the compartimentalization of the cytoplasmic fluctuations.

Following the changes in light emission at every point in the cell can after correction indicate the relative variation in Ca^{++} in different places inside the protoplast. Ionic flows would then be monitored if present.

ACKNOWLEDGEMENTS : Financial supports were provided by the CNRS, the MRES and the INSERM (grant 868039).

References

D.S. BUSH, A.K. BISWAS, R.L. JONES, (1988) Measurement of cytoplasmic Ca++ and H+ in Baley Aleurone Protoplasts, p 139, in "Progress in plant protoplast research", Kluwer Academic Pub., Dordrecht.

D.S. BUSH, R.L. JONES, (1989) Measurement of cytoplasmic calcium in aleurone protoplasts, Cell calcium, 8, p 455.

K.R. CASTELMAN, (1979), Digital Image Processing, Prentice Hall, Englewood Cliffs.

S. GIBROY, W.A. HUGHES, A.J. TREWAVAS, (1986) The measurement of intracellular calcium levels in protoplasts from higher plant cells using INDO-1 and FURA-2, FEBS lett., 199, p. 217.

S. INOUE, (1986), Video Microscopy, Plenum, New York.

H. JONES, M.J. TEMPELAUR and M.G.K. JONES, (1987), Recent Advances of Plant Electroporation, p. 347, Oxford Survey of plant Molecular and Cell Biology.

H. JONES, (1988), Fusing plant protoplasts, Trends in biotechnology, 6, p. 153.

D. LANSING TAYLOR and E.D. SALMON, (1989), Basic fluorescence microscopy, p 208 in "Methods in Cell Biology", Vol.29., Academic Press, San Diego

M.M. MONTANE, G. ALIBERT, J. TEISSIE, (1988), Plant protoplast electropermeabilisation : role of electrical and biological parameters, p 315, in "Progress in plant protoplast research", Kluwer Academic Pub., Dordrecht.

J.S. PLOEM, (1989), Fluorescence microscopy, p. 163 , in "Light microscopy in Biology", IRL press, Oxford.

K.R. SPRING and R.J. LOWY, (1989), Characteristics of low light level television cameras, p 270, in "Methods in Cell Biology", Vol.29, Academic Press, San Diego.

M.J. TEMPELAR, M.G.K. JONES, (1988), Electromanipulation of protoplasts, p 311, in "Progress in plant protoplast research", Kluwer Academic Pub., Dordrecht.

R.Y. TSIEN, (1989), Fluorescent indicators of ion concentrations, p 127, in "Methods in cell biology", Vol.30, D. Lansine Taylor and Yu-Li Wang Eds., Academic Press, San Diego.

D.G. WEISS, W. MARLE and R.A. WICK (1989) Video microscopy, p 221, in "Light Microscopy in Biology", IRL Press, Oxford

PHOSPHOLIPID DERIVED MESSENGERS

Wendy F. Boss
Abdul R. Memon[1]
Qiuyun Chen
Department of Botany
Box 7612
North Carolina State University
Raleigh, NC 27695
USA

Since the discovery of the regulatory role of inositol phospholipids in animal cell signalling, there has been considerable interest in these inositol phospholipids as potential sources of second messengers in plants (Boss, 1989). In animal cells, the negatively charged phospholipid, phosphatidylinositol bisphosphate (PIP_2), is present in the plasma membrane and in response to external stimuli is cleaved by phospholipase C to produce the second messengers, inositol-1,4,5-trisphosphate (IP_3) and diacylglycerol (DAG) (Billah and Lapetina, 1983; Berridge and Irvine, 1984; Downes and Michell, 1985; Michell, 1986; Majerus et al., 1986; Berridge, 1987). DAG activates protein kinase C (Nishizuka, 1984) and IP_3 releases calcium from non-mitochondrial intracellular stores (Streb et al., 1983) thus activating calcium-dependent enzymes.

While a similar polyphosphoinositide pathway may occur in selected plant tissue (e.g. Samanea saman, Morse et al., 1987, 1989b) or at selected times during the cell cycle (e.g. Ettlinger and Lehle, 1988), it does not appear to be universally involved in cell signalling in higher plants. This symposium provides us with the opportunity to reassess the role(s) of the polyphosphoinositides specifically and phospholipids in general in plant growth and development. The goal is to foster new ideas for future research. For this reason, we will focus on alternative mechanisms by which the

---

[1]Department of Biology, Middle East Technical University, Ankara, Turkey

NATO ASI Series, Vol. H 47
Signal Perception and Transduction in Higher Plants
Edited by R. Ranjeva and A. M. Boudet

inositol phospholipids can regulate plant growth and development. First consider where the inositol phospholipids are found in the cell.

Distribution of the Phosphoinositides Within Cells

The plasma membranes of higher plants (Wheeler and Boss, 1987) and algae (Einspahr et al., 1988) are enriched in the phosphoinositides, phosphatidylinositol monophosphate (PIP) and PIP_2. Although the plasma membrane contains the largest pool of the polyphosphoinositides based on membrane protein or lipid, other membranes also contain these negatively charged lipids (Hendrix et al., 1989). Using a wild carrot culture cell line as a model, we have characterized the inositol lipid distribution in the various membrane fractions after 12 to 18 h of labeling the cells with myo[2-^{3}H]inositol. These data are summarized in Table 1.

Table 1. Distribution of [^{3}H]Inositol-Labeled Lipids of Membranes from [^{3}H]Inositol-Labeled Fusogenic Carrot Cells*

Cell type	PIP_2	LPIP	PIP	LPI	PI
Whole cells	0.6 ± 0.3	4.7 ± 1.3	7.0 ± 3.4	7.4 ± 1.6	76.0 ± 6.0
Plasma membrane (upper phase)	1.3 ± 0.1	18.9 ± 3.3	14.2 ± 1.5	4.5 ± 0.5	54.0 ± 4.5
Microsomes (lower phase)	0.3 ± 0.1	2.0 ± 0.6	2.7 ± 0.6	18.6 ± 1.4	73.5 ± 1.6
Nuclei isolated from protoplasts	3.6 ± 1.1	11.5 ± 0.9	5.0 ± 0.3	12.3 ± 0.1	55.7 ± 2.8
Vacuoles isolated from protoplasts	--	18.4 ± 6.7	0.9 ± 0.1	1.7 ± 0.7	79.1 ± 5.9

*Percent of total [^{3}H]inositol lipid recovered ± SD from at least three separate experiments. PIP_2, phosphatidylinositol-4,5-bisphosphate; LPIP, lysoPIP; PIP, phosphatidylinositol-4-monophosphate; LPI, lysoPI; PI, phosphatidylinositol.

Interestingly, unlike most animal cells, lysophosphoinositides also are very prevalent in the tissue culture cells. These lipids were shown not to be generated during extraction and like the parent lipids, they partitioned into distinct membrane fractions (Wheeler and Boss, 1987). For example, over 96% of the recovered lysophosphatidylinositol (LPI) was found in the lower phase fraction (intracellular membranes) and about 4 % in the plasma membrane fraction. In contrast, lysophosphatidylinositol monophosphate (LPIP), which can be synthesized by the phosphorylation of LPI (Wheeler and Boss, 1989), was more prevalent in the plasma membrane fraction. The tonoplast (vacuolar membrane) and isolated vacuoles also were found to contain LPIP as the predominant inositol lipid other than PI. The tonoplast, however, contained little or no PIP or LPI (Table 1, Cho et al., unpublished results). Isolated nuclei, on the other hand, had detectable levels of PIP_2, PIP and LPI as well as LPIP and PI. In spite of the fact that $[^3H]PIP_2$ was 5.3% of the total inositol lipid recovered from the nuclei, the total inositol lipid per mg protein or per $[^{14}C]$ myristate-labeled lipid was much less than that found in other membranes or organelles studied (Hendrix et al., 1989). However, the fact that the inositol phospholipids and lysolipids were found in the nuclear membrane is of interest because of their purported role in regulating cell proliferation (Berridge, 1987).

The lipid distribution experiments were done after 12 to 18 h of labeling, short term labeling of 1/2 to 1 h indicated that the $[^3H]$inositol-labeled PIP, LPIP and PIP_2 were present initially in the lower phase membrane fraction. The data are consistent with the studies of Sandelius and Sommarin (1986) who showed that PI and PIP kinase activities were found in the lower phase fraction as well as the plasma membrane fraction and with those of Wheeler and Boss (1989) who showed that LPI kinase activity was found in both membrane fractions. We also have found PI kinase activity associated with isolated nuclei (Hendrix et al., 1989). In addition, in animal cells, PI kinase activity has been reported to be associated with the Golgi apparatus (Jergil and Sundler, 1983), lysosomes (Collins and Wells, 1983), coated vesicles (Campbell et al.,1985) and the nuclear envelope (Smith and Wells, 1983). Thus, the polyphosphoinositides can be synthesized from PI on intracellular membranes and transported to the plasma membrane as well as being synthesized from PI directly on the plasma membrane.

IP_3 as a Second Messenger

If IP_3 is a second messenger, then according to the animal paradigm it should increase in response to external stimuli. Quantitating IP_3 in plants, however, is thwarted by the fact that inositol is metabolized to glucuronic acid and other cell wall metabolites (Loewus and Loewus, 1983) which will coelute with the inositol phosphates on anion exchange columns (Coté et al, 1987; Rincón et al., 1989). Therefore, one must use HPLC or paper electrophoresis to obtain adequate separation (Coté et al., 1989). Furthermore, the levels of PIP_2 and thus IP_3 have been shown to be at or near the limits of detection in most plants studied. For example, in cultured carrot cells, PIP_2 was estimated to be 0.05 % or less of the total recovered phospholipids (Drøbak et al., 1988) and 0.1% to 0.8 % of the total recovered inositol lipid (Boss and Massel, 1985; Wheeler and Boss, 1987; Rincón et al., 1989). Heim and Wagner (1989) reported that $[^3H]PIP_2$ and IP_3 were undetectable in stationary phase Catharanthus roseus cells after 24 h of labeling with $[^3H]$inositol, and that PIP_2 was less than 1 % of the total $[^3H]$inositol-labeled lipid in rapidly growing cells. We also have not been able to detect IP_3 in tissue culture cells.

If IP_3 is produced by phospholipase C in the plasma membrane of plant cells, it still must reach the vacuole which has been shown to be the primary IP_3-sensitive calcium store (Schumaker and Sze, 1987; Ranjeva et al., 1988). Membrane and soluble protein fractions (40,000g pellet and supernatant, respectively) from the carrot cells metabolize IP_3 rapidly (Memon et al., 1989). With the membrane fraction 90 % of the IP_3 was metabolized in 15 min at a rate of 55 pmoles mg^{-1}protein min^{-1} and IP_2 was the major product. With the soluble fraction, the rate of metabolism was slightly faster (70 pmoles mg^{-1}protein min^{-1}); 99 % of the IP_3 was metabolized and IP_2 and IP were the major products at 60 and 40 % of the recovered ^3H, respectively. The metabolism of IP_3 in both fractions was not sensitive to 10 mM lithium, a commonly used inhibitor of I-1,4-P_2 phosphatase in animal cells. However, we did observe inhibition of IP_3 metabolism when 1 mM molybdate was added to the membrane fraction.

While IP_3 may be produced in selected tissue of higher plants (e.g. pulvini, Morse et al., 1987) or in lower plants (e.g. mosses, Hartmann and Pfaffmann, 1989), no one has shown that IP_3 produced in vivo causes an increase in cytosolic calcium. There has been only one report of exogenously added IP_3 inducing a calcium efflux in plant cells. This was with protoplasts derived from wild carrot culture cells that had only a few small vacuoles (Rincón and Boss, 1987). When protoplasts from cells with a large central vacuole were used, IP_3 had no significant effect (Reddy and Poovaiah, 1987; Boss, 1989). IP_3 was metabolized rapidly by cytosolic as well as membrane phosphatases to IP_2 and IP from carrot cells which had a large central vacuole and from those which did not (Rincón and Boss, 1989; Memon et al., 1989) and IP_3 could not be detected in either. Therefore, the differences in the responses to IP_3 may be due to differences in calcium flux as previously observed (Grimes and Boss, 1985) or to the reuptake of calcium into the larger vacuole so that a net efflux was not detected.

Studies of phospholipase C activity in isolated membranes have shown no evidence for activation by growth regulators. Several laboratories have shown that exogenously added PI, PIP, and PIP_2 are metabolized by plant plasma membrane to IP, IP_2, and IP_3 (Melin et al., 1987; Pfaffmann et al., 1987; Murthy and Irvine, 1988); however, to date there is no reproducible evidence for significant auxin stimulation of membrane-associated phospholipase C (Pfaffmann et al., 1987, Sandelius and Morré, 1987; Morré, et al., 1989; Morré, 1989; Sandelius and Sommarin, 1989).

DAG as a Second Messenger

There is no evidence for a role for DAG plus calcium in activating a specific protein kinase in plants. Several have reported that phospholipids (specifically phosphatidylserine) plus DAG or phospholipids plus calcium or tumor promoter enhance protein kinase activity (Elliott and Skinner, 1986; Olah and Kiss, 1986; Schäfer et al., 1985; Favre and Turian, 1987). Morré et al. (1984b) reported a nonspecific increase in phosphorylation of membrane proteins in the presence of DAG and calcium. However, although the plant protein kinases have been shown to be immunologically related to protein kinase C in animal cells (Elliot and Koke, 1987), they do not bind the phorbol ester tumor promoters and do not show both a calcium and DAG dependence (for review see Harmon, 1989; Budde and Randall, 1989). In addition, there have been no reports of activation of protein phosphorylation by added DAG or DAG analogues in vivo.

Alternative Roles for Phospholipids in Signal Transduction

If the primary role of PIP_2 is not to produce IP_3 and DAG, then what is it? In sarcoplasmic reticulum where PIP_2 levels, like those of higher plants, are relatively low, PIP has been shown to activate the Ca^{2+} ATPase (Schäfer et al., 1987). Schäfer proposed that the primary role of the phosphoinositides in this system was to act as membrane effectors rather than as a source of second messengers. If this is true then regulating the levels of PIP and PIP_2 by regulating the PI and PIP kinase or PIP and PIP_2 phosphatase activity would be a critical step in cell signalling.

We have observed rapid (within seconds) changes in PIP kinase activity and corresponding changes in the plasma membrane ATPase activity in two systems with two separate stimuli. In both instances there was no evidence for activation of phospholipase C. In addition, while we have not shown a direct relationship between the activation of the kinase and the ATPase, we have shown that adding PIP or PIP_2 to isolated membranes almost doubles the ATPase activity. These data are briefly described in the paragraphs that follow.

<u>Suspension culture cells respond rapidly to cell wall digestion enzymes.</u> There is a dramatic increase in PIP_2 and loss in the lower R_f analogue of LPIP (<u>sn</u>-1-palmitoyl lysophosphatidylinositol monophosphate) after treating [^{3}H]inositol labeled carrot culture cells for 10 min with Driselase (Rincón et al., 1989; Chen and Boss, unpublished results). Driselase is a crude mixture of fungal cellulases used for plant cell wall digestion. The precise stimulus in the Driselase is as yet unidentified; however, the active component is heat sensitive and the response cannot be mimicked by protease or pectinase.

One possibility is that the response to Driselase is due to elicitors released from the cell wall during treatment; however, this is not likely since Driselase which had cell wall fragments but was heat treated (i.e. Driselase which first had been added to cells for 10 min, removed from the cells, and then boiled) did not increase the PIP_2. These data are in agreement with the work of Strasser et al. (1986) who found no effect of elicitors on inositol lipid metabolism and IP_3 production.

Driselase treatment caused an increase in PIP kinase activity that was detected in an in vitro phosphorylation assay after as little as 5 sec treatment with Driselase in vivo. As with the whole cell experiments, the response was not seen with boiled Driselase. The effect of the in vivo Driselase treatment was not due to residual Driselase in the in vitro assay since when Driselase was added directly to the membranes, all lipid phosphorylation decreased. The same effect had been observed for plasma membrane protein phosphorylation (Blowers et al., 1988).

In addition to activating PIP kinase, treatment of cells with 2 % (w/v) Driselase resulted in activation of the plasma membrane ATPase which is vanadate-sensitive (Table 2). The increase in vanadate-sensitive ATPase activity also was seen after 5 sec of in vivo treatment; however, it was not detectable at lower concentrations (0.04 %, w/v) of Driselase where the PIP kinase activity was enhanced (Table 2). These data suggest that kinase activation precedes ATPase, but more time course experiments at the lower concentrations are necessary before the sequence of events can be determined.

Table 2. PIP kinase and vanadate-sensitive ATPase activities at 0.04 % and 2 % (w/v) Driselase in 0.4 molal sorbitol. Numbers are the average of duplicate values from one experiment. The experiment has been repeated twice and the trends are consistent

Treatment	Percent ^{32}P-labeled PIP_2 as percent sorbitol control	Vanadate-sensitive ATPase activity as percent sorbitol control
1 min Driselase 0.04 %	142 %	106 %
5 sec Driselase 2 %	203 %	554 %
10 sec Driselase 2 %	151 %	176 %
1 min Driselase 2 %	242 %	305 %

<u>PI and PIP kinase activities in etiolated sunflower hypocotyls are sensitive to light.</u> When the etiolated hypocotyls were exposed to 1 min of white light (140 μEinsteins m^{-2} sec^{-1}), the [^{32}P]PIP_2 produced by phosphorylation of endogenous plasma membrane lipids with [γ^{32}P]ATP was 58 % less than that produced in the dark controls (Fig. 1, Table 3). The[^{32}P]PIP was slightly less in the light-irradiated plants. These data suggested that the in vivo light treatment had in some manner decreased the plasma membrane PI and PIP kinase activities.

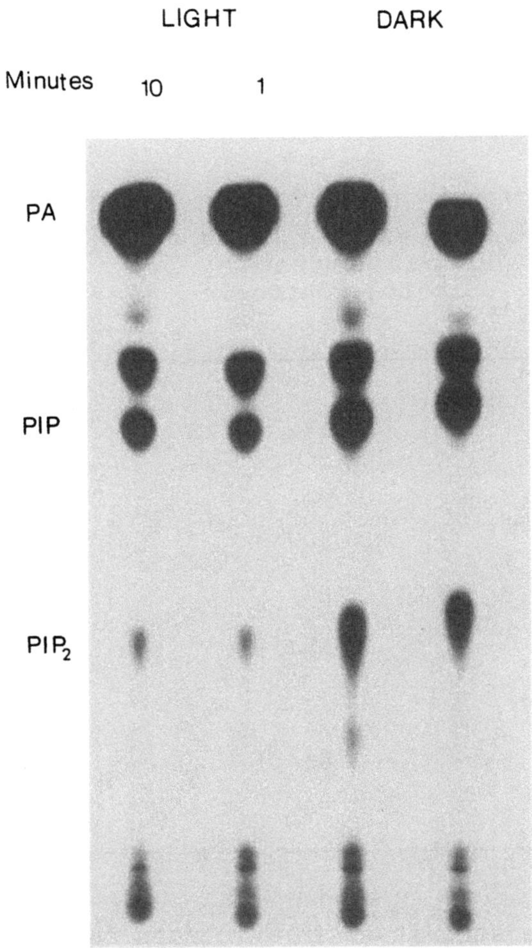

Fig. 1 An autoradiogram showing the effects of light on the activity of the plasma membrane lipid kinases from sunflower hypocotyls. Plants were grown in the dark for 7 days at 25° C. The hypocotyls were excised and kept in the dark or exposed to light for 1 or 10 min; homogenized; the plasma membrane isolated by aqueous two-phase partitioning, and the in vitro phosphorylation was done with [γ^{32}P]ATP. The experiment has been repeated three times and the trends are consistent

Table 3. Vanadate-sensitive ATPase and lipid kinases from the plasma membrane of dark grown and 10 min light irradiated sunflower hypocotyls. The ATPase activity is expressed as percent of control and the distribution of the ^{32}P-labeled lipids is expressed as percent of total ^{32}P incorporated. Phosphorylation was done with endogenous substrate

	Dark	Light
ATPase Activity (% of Control)	100	68 ± 2
PIP_2 (% of total ^{32}P)	12 ± 1	5 ± 0.4
PIP	18 ± 2	16 ± 0.5
PA	32 ± 1	41 ± 0.4

Another explanation for the decrease in PIP and PIP_2 in the light-irradiated plants was that light activated phospholipase C as proposed by Morse et al. (1987; 1989a,b) for Samanea saman. However, analysis of the water soluble products of the in vitro phosphorylation reaction proved that this was not true for sunflower hypocotyls. When the water soluble products were analyzed by paper electrophoresis, $[^{32}P]IP_3$ recovered from the reaction was just above background levels in both light and dark treated plants (20 to 30 cpm μg^{-1} protein min^{-1}). In two separate experiments there was slightly more $[^{32}P]IP_2$ recovered from the dark grown plants than light-irradiated plants (e.g., 400 cpm μg^{-1} protein min^{-1} for the dark grown plants and 300 cpm μg^{-1} protein min^{-1} for those irradiated with light). The IP_2 recovered was probably from the hydrolysis of $[^{32}P]PIP$ and not the dephosphorylation of IP_3 since the reaction was run in the presence of molybdate, a nonspecific phosphatase inhibitor, which inhibits membrane associated IP_3 phosphatase (Memon et al., 1989). In either event, the decrease in PIP and PIP_2 in membranes from light-irradiated hypocotyls was not due to light activation of phospholipase C.

In addition to inhibiting the PI and PIP kinase activities, the light treatment also decreased the plasma membrane ATPase activity of the sunflower hypocotyls (Table 3). With both the carrot and sunflower plasma membranes, exogenously added PIP and PIP_2 activate the vanadate-sensitive membrane ATPase.

Table 4. The effect of phospholipids and inositol phosphates on the vanadate-sensitive ATPase activity of plasma membrane isolated from dark grown and 10 min light irradiated sunflower hypocotyls. The ATPase activities in the dark grown and light-irradiated controls were 157 $\pm$ 4 and 109 $\pm$ 8 nmoles Pi mg^{-1} protein min^{-1}, respectively. The ATPase activity in the treatments are expressed as percent of their respective controls

Compound Added	Dark Grown	Light Irradiated
--	100	100
PI	105	110
PIP	202	293
PIP_2	178	255
IP_2	96	95
IP_3	101	104
LPI	112	125

The percentage increase in the ATPase specific activity is greater in the light-irradiated plants compared to the dark controls; however, the maximal specific activity achieved in the presence of PIP or PIP_2 is the same in both instances. The response was not due to the metabolism of PIP or PIP_2 since IP_3, IP_2, or DAG did not affect enzyme activity (Table 4, Memon et al., unpublished results).

Curiously, light inhibition of the PI and PIP kinase activities was overcome by adding vanadate (0.2 mM) to the in vitro phosphorylation reaction mixture. The effect of vanadate is probably not due to a decreased in demand for ATP by the vanadate-inhibited ATPase since the dark grown plants, which have higher ATPase activity, are not affected by the vanadate.

The fact that vanadate overcomes the light inhibition of PI and PIP phosphorylation suggests that a phosphorylated protein regulates kinases. Vanadate is a phosphatase inhibitor (Macara, 1980). If the E-P form of a protein activates the sunflower inositol phospholipid kinases, then light could have either decreased the phosphorylation of E or activated a specific vanadate-sensitive phosphatase that dephosphorylated E-P. Protein phosphorylation and dephosphorylation has been shown to be important in regulating several plant enzymes (Ranjeva et al., 1983 and see Ranjeva and Boudet, 1987 and Budde and Chollet, 1988 for review).

Whitman et al. (1987) found that adding vanadate enhanced the activity of the type-I PI kinase from transformed fibroblasts and proposed that a phosphorylated protein was regulating enzyme activity. A phosphotyrosine phosphorylated protein was associated with the partially purified PI kinase and it was presumed that the effect of vanadate was to inhibit dephosphorylation during isolation. Kiehl et al. (1987) reported increased PI phosphorylation in the presence of 0.3 mM vanadate using detergent solubilized nicotinic acid acetylcholine receptor. Like the fibroblast kinase, the receptor contained a tyrosine phosphorylated subunit. Since there is only one report of tyrosine phosphorylation in plants and this was associated with the nuclear fraction (Torruella et al., 1986) and since no tyrosine kinase activity has been found in the microsomal membranes isolated from plants (Blowers and Trewavas, 1989), it is unlikely that a phosphotyrosine intermediate is involved. However, a vanadate-sensitive step appears to be involved in the light inhibition of PIP kinase activity.

The decrease in PI and PIP kinase activities in response to light is consistent with a preliminary report by Peeler and Thompson (1989) suggesting that incorporation of $^{32}P_i$ into the inositol phospholipids of <u>Dunaliella</u> <u>salina</u> is greater in the dark than in the light. In addition, rapid effects of light on the phosphorylation of plasma membrane proteins have been observed in the laboratory of Dr. Winslow Briggs. Gallagher et al. (1988) exposed etiolated pea stems to 10 sec of blue light; isolated plasma membranes, and found decreased phosphorylation of a 120 kDa polypeptide after in vitro phosphorylation of membrane proteins with [γ^{32}P]ATP. They have suggested that the 120 kDa protein may be a photoreceptor.

We are not proposing that the phospholipids or the phospholipid kinases are photoreceptors in our system. The PI and PIP kinases, however, do respond rapidly to light given in vivo and may be affected by the phosphorylation of membrane proteins not unlike the small molecular weight autophosphorylating protein (pp18) reported by Blowers and Trewavas (1988) also to contain PI kinase activity (1988).

<u>PIP and PIP_2 may act as direct effectors of membrane function.</u> PIP and PIP_2 have been shown to increase ATPase activity of several E_1-E_2 type or P-ATPases which operate through a phosphorylated intermediate such as the canine renal Na^+/K^+ ATPase (Lipsky and Lietman (1980), the Ca^{2+}-transport ATPase of sarcoplasmic reticulum (Varsanyi et al., 1983), the plasma

membrane Ca^{2+}-transport ATPase of erythrocytes (Choquette et al., 1984) and hepatocytes (Lin and Fain, 1984) and the nuclear envelope ATPase of hepatocytes (Smith and Wells, 1983). As with our results, the effect of the lipids on the animal cell ATPase was not due to metabolites but rather the direct effect of the negatively charge phospholipids (Schäfer et al., 1987). While the exact mechanism by which PIP and PIP_2 affected the ATPases is unknown, PIP has been shown to be tightly associated with the Ca^{2+}-ATPase of sarcoplasmic reticulum (Schäfer et al., 1987).

In addition, Chauhan and Brockerhoff (1988) have shown that PIP and PIP_2 can regulate protein phosphorylation. They proposed that in some cells, a PI $\rightleftharpoons$ PIP $\rightleftharpoons$ PIP_2 shuttle occurs which is maintained by phospholipid kinases and phosphatases and that it is the ratio of PI:PIP:PIP_2 rather than the production of DAG that is important in regulating protein phosphorylation. To our knowledge, the effect of PIP and PIP_2 on protein phosphorylation in plants has not been tested.
Other structural roles of the inositol phospholipids include binding to cytoskeletal proteins (Anderson and Marchesi, 1985; Lassing and Lindberg, 1985; 1988) and serving as anchors for membrane proteins (Low and Kincade, 1985; Low and Saltiel, 1988; Ferguson and Williams, 1988). PI-phospholipase C sensitive proteolipids were not detected in Dunaliella salina cells (Stephenson et al., 1989) or in carrot suspension culture cell plasma membranes (Okpodu and Boss, unpublished results). There are, however, in both systems many proteins which label when cells are grown in [^{14}C] myristate and these fatty acid acylated proteins may be involved in cell signalling.

A-type phospholipases may be involved in cell signalling.
Unlike the C-type phospholipases that cleave the phospholipid head group from the glycerol backbone, the A-type phospholipases cleave the fatty acids either in the sn-1

(phospholipase A_1) or sn-2 (phospholipase A_2) position. Activation of phospholipase A_2 by increased cytosolic calcium is an important step in the signalling pathway and leads to the production of eicosanoids such as prostaglandins in many animals cells (Billah and Lapetina, 1982; Berridge, 1984; Majerus et. al., 1986; Margolis, 1988). Even though arachidonic acid is not found in most higher plants, the metabolites of linolenic and linoleic acids may serve a similar role to eicosanoids in plant tissue (for review see Anderson, 1989) or the fatty acids and lysolipids themselves, may play important roles as second messengers in plants.

Little is known about A-type phospholipases in plants. They have been proposed to play a role in phospholipid metabolism during senescence (Lesham, 1987). We have shown that Driselase activate the plasma membrane phospholipase A_2 measured with NBD-phosphatidylsholine as a substrate (Dengler et al., 1988). Phosphatidylcholine has been proposed to be an important alternative source of lipid second messengers in animal cells (Pelech and Vance, 1989). Since the phospholipase C activity specific for phosphatidylinositols has not been shown to be auxin responsive (Pfaffmann et al., 1987; Sandelius and Morré, 1987; Morré et al. 1989; Morré, 1989; Sommarin and Sandelius, 1989), phospholipase A or D activity may be responsible for the marked auxin-induced lipid breakdown observed with isolated plasma membranes (Morré et al. 1984a) or the loss of ^{32}P-labeled phospholipid observed by Zbell and Walter (1987).

Importantly, unlike DAG, both lysolipids and fatty acids have been shown to activate several plasma membrane enzymes in plants. These include the vanadate-sensitive ATPase (Palmgren et al., 1988), the auxin-stimulated NADH oxidase (Brightman et al., 1988), the 1,3-β-D-glucan synthetase (Kauss and Jeblick, 1986), phytase (Lim and Tate, 1971), and protein kinases from plants (Scherer et al., 1988; Martiny-Baron and Scherer, 1989;

Lucantoni and Polya, 1987; Klucis and Polya, 1987; Harmon, 1989). In addition, lysolipids inhibit the PI and PIP and DAG kinases and thus may be important in regulating the biosynthesis of the inositol phospholipids (Wheeler and Boss, 1989).

In summary, we propose that in plant cells, the inositol phospholipids may play roles other than as the source of the second messengers, IP_3 and DAG. In the two systems studied, carrot suspension culture cells and sunflower hypocotyls, the activity of PIP kinase but not PIP_2-specific phospholipase C changed rapidly in response to external stimuli. The negatively charged polyphosphoinositides in and of themselves, were shown to regulate the plasma membrane ATPase. Therefore, the ratio of PI:PIP:PIP_2 may be a critical factor in regulating plant growth and development. In addition, other phospholipases, A and D, may play important roles in generating alternative second messengers such as fatty acids and lysolipids. A diagram outlining potential pathways for signal transduction in response to fungal cellulases, light, and auxin is given in Figure 2.

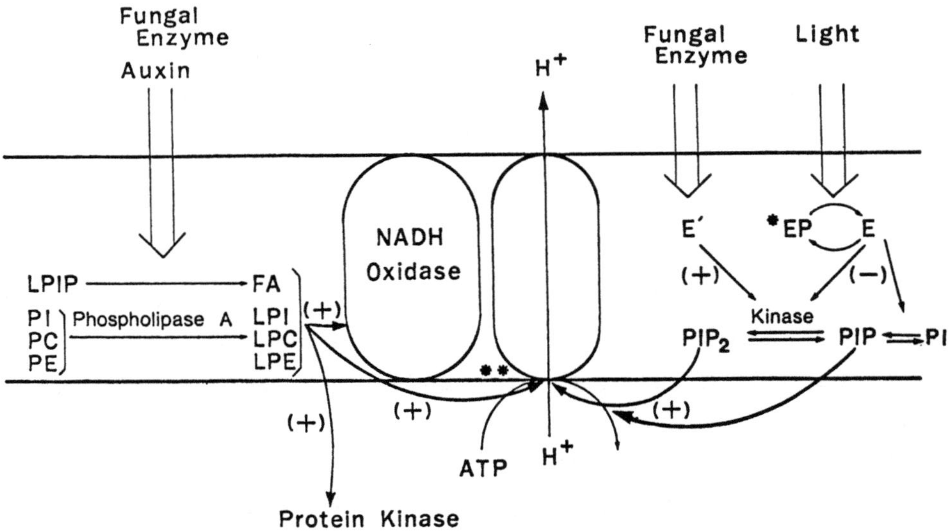

Fig. 2 Possible signal transduction pathways involving phospholipids. External stimuli activate (+) or inhibit (-) the PI and PIP kinases possibly via a phosphorylated intermediate (EP). Increasing PIP and PIP_2 increases the plasma membrane ATPase activity. Stimuli also activate phospholipases which produce fatty acids (FA) and lysolipids (LPI, LPC, LPE). These in turn activate the plasma membrane ATPase, the auxin-sensitive NADH oxidase, and protein kinases

References

Anderson JM (1989) Membrane-derived fatty acids as precursors to second messengers. In: Boss WF, Morré DJ (eds) Second Messengers in Plant Growth and Development. Alan R. Liss, New York, pp 181-212

Anderson RA, VT Marchesi (1985) Regulation of the association of membrane skeletal protein 4.1 with glycophorin by a polyphosphoinositide. Nature 318:295-298

Berridge MJ, RF Irvine (1984) Inositol trisphosphate, a novel second messenger in cellular signal transduction. Nature 312:315-321

Berridge MJ (1987) Inositol trisphosphate and diacylglycerol: Two interacting second messengers. Ann Rev Biochem 56:159-193

Billah MM, EG Lapetina (1983) Platelet-activating factor stimulates metabolism of phosphoinositides in horse platelets: Possible relationship to Ca^{2+} mobilization during stimulation. Proc Natl Acad Sci USA 80:965-968

Billah MM, EG Lapetina (1982) Formation of lysophosphatidylinositol in platelets stimulated with thrombin or ionophore A23187. J Biol Chem 257:5190-5200

Blowers DP, AJ Trewavas (1989) Second Messenger: Their existence and relationship to protein kinases. In: WF Boss and DJ Morré (eds) Second Messengers in Plant Growth and Development. Alan R. Liss New York, pp 1-28

Blowers DP, WF Boss, AJ Trewavas (1988) Rapid changes in plasma membrane protein phosphorylation during cell wall digestion. Plant Physiol 86:505-509

Blowers DP, AJ Trewavas (1988) Phosphatidylinositol kinase activity of a plasma membrane-associated calcium-activated protein kinase from peas. FEBS Lett 238:87-89

Boss WF, MO Massel (1985) Polyphosphoinositides are present in plant tissue culture cells. Biochem Biophysics Res Commun 132:1018-1023

Boss WF (1989) Phosphoinositide metabolism: Its relation to signal transduction in plants. In: Boss WF and DJ Morré (eds) Second Messengers in Plant Growth and Development. Alan R. Liss, New York, pp 29-56

Brightman, AO, R Barr, FL Crane, DJ Morré (1988) Auxin-stimulated NADH oxidase purified from plasma membrane of soybean. Plant Physiol 86:1264-1269

Budde RJA, R Chollet (1988) Regulation of enzyme activity in plants by reversible phosphorylation. Physiol Plant 72:435-439

Budde RJA, Randall DD (1988) Protein kinases and future prospects. In: Morré DJ, Boss WF, Loewus FA, (ed) Inositol metabolism in plants. Allan R. Liss, New York

Campbell CR, JB Fishman, RE Fine (1985) Coated vesicles contain a phosphatidylinositol kinase. J Biol Chem 260:10948-10951

Chauhan VPS, H Brocherholt (1988) Phosphatidylinositol-4,5-bisphosphate may antecede diacylglycerol as activation of protein kinase C. Biochem Biophys Res Comm 155:18-23

Choquette D, E Hakim, AE Filoteo, EA Plishker, JR Bostwick, JT Penniston (1984) Regulation of plasma membrane Ca^{2+} ATPases by lipids of the phosphatidylinositol cycle. Biochem Biophys Res Comm 125:908-915

Collins CA, WW Wells (1983) Identification of phosphatidylinositol kinase in rat liver lysosomal membranes. J Biol Chem 258: 2130-2134

Coté GG, MJ Morse, RC Crain, RL Satter (1987) Isolation of soluble metabolites of the phosphatidylinositol cycle from _Samanea saman_. Plant Cell Reports 6:352-355

Coté GG, LM Quarmby, RL Satter, MJ Morse, RC Crain (1988) Extraction, separation and characterization of the inositol phospholipid cycle. In: Morré DJ, Boss WF, Loewus FA, (eds) Inositol Metabolism in Plants. New York, Alan R Liss

Dengler LA, M Rincón, WF Boss (1988) NBD-PC: A tool to study endocytosis in plant protoplasts. In: Morré DJ, KE Howell, GMW Cook, WH Evans (eds) Cell-Free Analysis of Membrane Traffic. New York, Alan R Liss

Downes, CP, RH Michell (1985) Inositol phospholipid breakdown as a receptor-controlled generator of second messengers. In: Cohen P, MD Houslay (eds) Molecular Mechanisms of Transmembrane Signalling. Elsevier, Amsterdam, pp 3-56

Drøbak BK, IB Ferguson, AP Dawson, RF Irvine (1988) Inositol-containing lipids in suspension cultured plant cells: An isotopic study. Plant Physiol 87:217-222

Einspahr KJ, TC Peeler, GA Thompson, Jr (1988) Rapid changes in polyphosphoinositide metabolism associated with the response of Dunaliella salina to hyposmotic shock. J Biol Chem 263:5775-5779

Elliott DC, JD Skinner (1986) Calcium-dependent, phospholipid-activated protein kinase in plants. Phytochem 25:39-44

Elliott DC, YS Kokke (1987) Cross-reaction of a plant protein kinase with antiserum raised against a sequence from bovine brain protein kinase C regulatory sub-unit. Biochem Biophys Res Commun 145:1043-1047

Ettlinger C, L Lehle (1988) Auxin induces rapid changes in phosphatidylinositol metabolites. Nature 331:176-178

Favre B, G Turian (1987) Identification of a calcium- and phospholipid-dependent protein kinase (protein kinase C) in Neurospora crassa. Plant Sci 49:15-21

Ferguson MAJ, AF Williams (1988) Cell-surface anchoring of proteins via glycosyl-phosphatidylinositol structures. Ann Rev Biochem 57:285-320

Gallagher S, TW Short, PM Ray, LH Pratt, WR Briggs (1988) Light-mediated changes in two proteins associated with plasma membrane fractions from pea stem sections. Proc Natl Acad Sci USA 85:8003-8007

Grimes HD, WF Boss (1985) Intracellular calcium and calmodulin involvement in protoplast fusion. Plant Physiol 79:253-258

Harmon AC (1989) Lipid activated Protein Kinases. In: Morré DJ, Boss WF, Loewus FA, (eds) Inositol metabolism in plants. Alan R. Liss, New York

Hartmann E, H Pfaffmann (1989) Mosses as a model system for involvement of phosphatidylinositol metabolism in signal transduction. In: DJ Morré, WF Boss, F Loewus (eds) Inositol Metabolism in Plants. Alan R. Liss, New York

Heim S, KG Wagner (1989) Inositol phosphates in the growth cycle of suspension cultured plant cells. Plant Sci in press

Hendrix KW, HA Assefa, WF Boss (1989) Polyphosphoinositides are Present in Nuclei Isolated from Carrot Protoplasts. Protoplasma in press

Jergil B, R Sundler (1983) Phosphorylation of phosphatidylinositol in rat liver Golgi. J Biol Chem 258:7968-7973

Kiehl R, M Varsanyi, E Neumann (1987) Phosphorylation of

phosphatidylinositol associated with nicotinic acetylcholine receptor of Torpedo californica. Biochem Biophys Res Comm 147:1251-1258
Klucis E, GM Polya (1987) Calcium-dependent activation of two plant leaf calcium-regulated protein kinases by unsaturated fatty acids. Biochem Biophys Res Comm 147:1041-1047
Lassing I, U Lindberg (1988) Evidence that phosphatidylinositol cycle is linked to cell motility. Exp Cell Res 174:1-15
Lassing I, Lindberg U (1985) Specific binding between phosphatidylinositol 4,5-bisphosphate and profilactin. Nature 314: 472-474
Leshem YY (1987) Membrane phospholipid catabolism and Ca^{2+} activity in control of senescence. Physiol Plant 69:551-559
Lin SH, JN Fain (1985) Calcium-magnesium ATPase in rat hepatocyte plasma membranes: inhibition by vasopressin and purification of the enzyme. Prog Clin Biol Res 168:25-30
Lipsky JJ, PS Lietman (1980) Neomycin Inhibition of Adenosine triphosphatase: Evidence for a neomycin phospholipid interaction. Antimicrobial Agents and chemotherapy 18:532-535
Loewus FA, MW Loewus (1983) *myo*-Inositol: Its biosynthesis and metabolism. Annu Rev Plant Physiol 34:137-161
Low MG, PW Kincade (1985) Phosphatidylinositol is the membrane-anchoring domain of the thy-1 glycoprotein. Nature 318:62-64
Low MG, AR Saltiel (1988) Structural and functional roles of glycosyl-phosphatidylinositol in membranes. Science 239:268-275
Lucantoni A, GM Polya (1987) Activation of wheat embryo calcium-regulated protein kinase by unsaturated fatty acids in the presence and absence of calcium. FEBS Lett 221:33-36
Macara, IG (1980) Vanadium-an element in search of a role. Trends in Biochem Sci 5:92-94
Majerus PW, TM Connolly, H Deckmyn, TS Ross, TE Bross, H Ishii, VS Bansal, DB Wilson (1986) The metabolism of phosphoinositide-derived messenger molecules. Science 234:1519-1526
Margolis BL, B Holub, DA Troyer, KL Skorecki (1988) Epidermal growth factor stimulates phospholipase A_2 in vasopressin-treated rat glomerular mesangial cells. Biochem J 256:469-474
Marthiny-Baron G, GFE Scherer (1988) A plant protein kinase and plant microsomal H^+ transport are stimulated by the ether phospholipid. platelet-activating factor. Plant Cell Rep 7:579-582
Melin PM, M Sommarin, AS Sandelius, B Jergil (1987) Identification of Ca^{2+}-stimulated polyphosphoinositide phospholipase C in isolated plant plasma membranes. FEBS Lett 223:87-91
Memon, A., M. Rincon and W. F. Boss (1989) Inositol trisphosphate metabolism in carrot (Daucus carota L.) cells. Plant Physiol in press
Michell RH (1986) Inositol lipids and their role in receptor

function: History and general principles. In: Putney JW, Jr. (ed) Receptor Biochemistry and Methodology, Phosphoinositide and Receptor Mechanisms. Vol 7, Alan R Liss Inc, New York, pp 1-24

Morré DJ (1989) Stimulus-response coupling in auxin regulation of plant cell elongation. In: Boss WF and DJ Morré (eds) Second Messengers in Plant Growth and Development. Alan R Liss, New York pp 29-56

Morré DJ, B Gripshover, A Monroe, JT Morré (1984a) Phosphatidylinositol turnover in isolated soybean membranes stimulated by the synthetic growth hormone 2,4-dichlorophenoxyacetic acid. J Biol Chem 259:15364-15368

Morré DJ, JT Morré RL Varnold (1984b) Phosphorylation of membrane located proteins of soybean in vitro and response to auxin. Plant Physiol 75:265-268

Morré DJ, B Drobes, H Pfaffmann, FE Wilkinson, E Hartmann (1989) Diacylglycerol levels unchanged during auxin-stimulated growth of excised hypocotyl segments of soybean. Plant Physiol 90:275-279

Morse MJ, RC Crain, GG Coté, RL Satter (1989a) Light-stimulated inositol phospholipid turnover in Samanea saman pulvini. Increased levels of diacylglycerol. Plant Physiol 89:724-727

Morse MJ, RC Crain, GG Coté, RL Satter (1989b) Light-signal transduction via accelerated inositol phospholipid turnover in Samanea saman pulvini. In: Morré DJ, WF Boss, FA Loewus (eds) Inositol Metabolism in Plants. Alan R Liss, New York

Morse MJ, RC Crain, RL Satter (1987) Light-stimulated inositol phospholipid turnover in Samanea saman leaf pulvini. Proc Natl Acad Sci USA 84:7075-7078

Murthy WC, RF Irvine (1988) Phosphatidylinositol 4,5-bisphosphate phosphodiesterase in higher plants. Biochem J. 249:877-881

Nishizuka Y (1984) Turnover of inositol phospholipids and signal transduction. Science 225:1365-1370

Olah Z, Z Kiss (1986) Occurrence of lipid and phorbol ester activated protein kinase in wheat cells. FEBS Lett 195:33-37

Palmgren MG, M Sommarin, P Ulvskov, PL Jorgenson (1988) Modulation of plasma membrane H^+ ATPase from oat roots by lysophosphatidylcholine, free fatty acids and phospholipase A_2. Physiologia Plantarum 74:11-19

Peeler TC and GA Thompson, Jr (1989) Effects of light on inositol phospholipid metabolism in Dunaliella salina. Plant Physiol 89S:895

Pelech SL, DE Vance (1989) Signal transduction via phosphatidylcholine cycles. Trends in Biochem Sci 14:29-30

Pfaffmann H, E Hartmann, AO Brightman, DJ Morré (1987) Phosphatidylinositol specific phospholipase C of plant stems: Membrane associated activity concentrated in plasma membranes. Plant Physiol 85:1151-1155

Ranjeva R, G Refeno, AM Boudet, D Marmé (1983) Activation of plant quinate:NAD 3-oxidoreductase by Ca^{2+} and calmodulin.

Proc Natl Acad Sci 80:5222-5224

Ranjeva R, AM Boudet (1987) Phosphorylation of proteins in plants: regulatory effects and potential involvement in stimulus/response coupling. Annu Rev Plant Physiol 38:73-93

Ranjeva R, A Carrasco, AM Boudet (1988) Inositol trisphosphate stimulates the release of calcium from intact vacuoles isolated from Acer cells. FEBS Lett 230:137-141

Reddy ASN, BW Poovaiah (1987) Inositol 1,4,5-trisphosphate induced calcium release from corn coleoptile microsomes. J Biochem 101:569-573

Rincón M, Q Chen, WF Boss (1989) Characterization of inositol phosphates in carrot (Daucus carota L.) cells. Plant Physiol. 89:126-132

Rincón M, Boss WF (1989) The second messenger role of Phosphoinositides. In: Morré DJ, Boss WF, Loewus FA, (eds) Inositol Metabolism in Plants. Alan R. Liss, New York

Rincón M, WF Boss (1987) myo-Inositol trisphosphate mobilizes calcium from fusogenic carrot (Daucus carota L.) protoplasts. Plant Physiol 83:395-398

Saltiel, AR. JA Fox, P Sherline, N Sahyoun, P Cuatrecasas (1987) Purification of phosphatidylinositol kinase from rat brain myelin. Biochem J 241:759-763

Sandelius AS, DJ Morré (1987) Characteristics of phosphatidylinositol exchange activity of soybean microsomes. Plant Physiol 84:1022-1027

Sandelius AS, M Sommarin (1986) Phosphorylation of phosphatidylinositols in isolated plant membranes FEBS Lett 201:282-286

Sandelius AS, M Sommarin (1989) Membrane-localized reactions involved in polyphosphoinositide turnover. In: Morré DJ, Boss WF, Loewus FA, (eds) Inositol Metabolism in Plants. Alan R Liss, New York

Schäfer A, F Bygrave, S Matzenauer, D Marmé (1985) Identification of a calcium and phospholipid-dependent protein kinase in plant tissue. FEBS Lett 187:25-28

Schäfer M, G Behle, M Varsanyi, LMG Heilmeyer, Jr (1987) Ca^{2+} regulation of 1-(3-sn-phosphatidyl)-1D-myo-inositol 4-phosphate formation and hydrolysis on sarcoplasmic-reticular Ca^{2+}-transport ATPase: A new principle of phospholipid turnover regulation. Biochem J 247:579-587

Scherer GFE, G Martiny-Baron, B Stoffel (1988) A new set of regulatory molecules in plants: A plant phospholipid similar to platelet activating factor stimulates protein kinase and proton-translocating ATPase in membrane vesicles. Planta 175:241-253

Schumaker KS, H Sze (1987) Inositol 1,4,5-trisphosphate releases Ca^{2+} from vacuolar membrane vesicles of oat roots. J Biol Chem 262:3944-3946

Smith CD, WW Wells (1983) Phosphorylation of rat liver nuclear envelopes II. Characterization of in vitro lipid phosphorylation. J Biol Chem 258:9368-9373

Sommarin M, AS Sandelius (1987) Phosphatidylinositol and

phosphatidylinositol phosphate kinases in plant plasma membranes. Biochim Biophys Acta 958:268-278
Stephenson M, PE Ryals, GA Thompson, Jr (1989) Fatty acid acylated proteins of the halotolerant alga Dunaliella salina. Plant Physiol 90:549-552
Strasser H, C Hoffman, H Grisebach, U Matern (1986) Are polyphosphoinositides involved in signal transduction of elicitor-induced phytoalexin synthesis in cultured plant cells? Z. Naturforsch 41c:717-724
Streb H, RF Irvine, MJ Berridge, I Schulz (1983) Release of Ca^{2+} from a nonmitochondrial intracellular store in pancreatic acinar cells by inositol-1,4,5-trisphosphate. Nature 306:67-68
Torruella M, LM Casano, RH Vallejos (1986) Evidence of activity of tyrosine kinase(s) and of the presence of phosphotyrosine in pea plantlets. J Biol Chem 261:6651-6653
Varsanyi M, HG Tolle, LMG Heilmeyer, Jr, RMC Dawson, RF Irvine (1983) Activation of sarcoplasmic reticular Ca^{2+} transport ATPase by phosphorylation of an associated phosphatidylinositol. EMBO J 2:1543-1548
Wheeler JJ, Boss WF (1989) Inositol lysophospholipids. In: Morré DJ,Boss WF, Loewus FA, (eds) Inositol Metabolism in Plants. Alan R Liss, New York
Wheeler JJ, WF Boss (1987) Polyphosphoinositides are present in plasma membrane from fusogenic carrot cells. Plant Physiol 85:389-392
Whitman M, D Kaplan, T Roberts, L Cantley (1987) Evidence for two distinct phosphatidylinositol kinases in fibroblasts. Biochem J. 247:165-174
Zbell B, G Walter (1987) About the search for the molecular action of high affinity auxin-binding sites on membrane-localized rapid phosphoinositide metabolism in plant cells. In: KlÀmbt D (ed) Plant Hormone Receptors. Springer-Verlag, Berlin pp 141-153

REGULATION OF FREE CYTOPLASMIC CALCIUM : A ROLE FOR PHOSPHATIDYLINOSITOL METABOLISM AND G-PROTEINS

Dieter Marmé
Molecular Cell Biology Goedecke AG
Mooswaldallee 1-9
D-7800 Freiburg
FRG

INTRODUCTION

The molecular mechanisms by which plant convey environmental and hormonal information into growth and developmental responses are largely unknown. Over the last years considerable evidence has been accumulated that free calcium ions in the cytoplasm play an important role in signal transduction (for refs. see Marmé, 1988). Cyclic nucleotides such as cAMP and cGMP do not seem to serve any regulatory functions in plants, at least not by activating specific protein kinases (Salimath and Marmé, 1983). The important question is then : how does the plant cell regulate its cytoplasmic free Ca^{2+}. The existence of a Ca^{2+}, calmodulin-activated (Ca^{2+} + Mg^{2+}) ATPase has been demonstrated in higher plants (Dieter and Marmé, 1981a). It has also been shown that the transport activity is under the control of light (Dieter and Marmé, 1981a). This regulatory mechanism is clearly specific for plants. The animal (Ca^{2+} + Mg^{2+}) ATPase does not seem to be under the control of a hormonal signal. The existence of Ca^{2+} channels which are important components for the regulation of cytoplasmic Ca^{2+} in animals is still a matter of intense investigation in the plants field. The occurrence of binding sites for Ca^{2+} -antagonists in plants (Andrejauskas et al., 1985) and the control of Ca^{2+} flux into protoplasts by Ca^{2+} -antagonists (Graziana et al., 1988) indicate that a similar mechanism to control free cytoplasmic Ca^{2+} might also exist in plants. However patch clamp analysis of plant Ca^{2+} channels has to be awaited before a clear answer to this problem can be given. The recent discovery that the degradation product of phosphatidylinositol 4,5-bisphosphate, inositol 1,4, 5-trisphosphate is able to release Ca^{2+} from intracellular stores in animal cells (for refs. see Berridge and Irvine, 1984) has

NATO ASI Series, Vol. H 47
Signal Perception and Transduction in Higher Plants
Edited by R. Ranjeva and A. M. Boudet

prompted many researchers to postulate a similar mechanism for plant cells too. This review will put together evidence for the existence of such a control system in plants.

ENZYMES OF THE POLYPHOSPHOINOSITIDE METABOLISM

The essential components of the polyphosphoinositides in the plasma membrane are phosphatidylinositol (PI), phosphatidylinositol 4-phosphate (PIP) and phosphatidylinositol 4,5-bisphophate (PIP_2). PIP_2 is formed from PI by two sequential phosphorylation reactions which are catalysed by a PI kinase and a PIP kinase. The presence of PI, PIP and PIP_2 was first demonstrated by Boss and Massel (1985). These authors could show that wild carrot cells grown in suspension culture after labeling with myo-[2-^{3}H] inositol contained 93% PI, 1.7% PIP and 0.8% PIP_2. The existence of these polyphosphoinositides was further reported for pulvini of Samanea saman (Morse et al., 1986) and in plasma membranes of fusogenic carrot cells (Wheeler and Boss, 1987). Irvine and colleagues (1989) had deacylated myo-[2-^{3}H] inositol labeled phosphatidylinositol and clearly shown that the resulting inositolphosphates were D-myo-inositol (1,4,5) trisphosphate and D-myo-inositol (1,4) bisphosphate. They concluded that plants contain phophatidylinositol 4-phosphate and phosphatidylinositol 4,5-bisphosphate as the only polyphosphoinositides. Recently the activities of PI and PIP kinases have been detected and characterized in plasma membranes from Triticum aestivum (Sommarin and Sandelius, 1986 ; Sommarin and Sandelius, 1988). In animal cells PIP_2 is specifically degraded by a PIP_2-specific phospholipase C (for refs. see Berridge and Irvine, 1984). This enzyme has also been shown to be present in purified plasma membranes from wheat seedlings (Melin et al., 1987). These authors also demonstrated that the enzyme could be stimulated by Ca^{2+} in the µM range. Furthermore the enzyme catalysed the hydrolysis of both PIP and PIP_2. Another group (Mc Murray and Irvine, 1988) found that a particulate fraction from celery contained a polyphosphoinositide phospholipase C which could be activated by desoxycholate and which was also Ca^{2+}-stimulatable in the µM range. The activity to hydrolyse PIP was always greater than that observed with PIP_2 as substrate.

In a plasma membrane fraction from Dunaliella salina Einsphar et al., (1989) were able to measure a phospholipase C activity when radioactive PIP_2 was used as a substrate. Again this enzyme exhibited a considerable Ca^{2+} dependence in the µM range. In addition and in contrast to the enzyme activity described by Melin et al. (1987) and Mc Murray and Irvine (1988) this activity could be stimulated by 100 µM GTPτS over a broad range of different Ca^{2+} concentrations. This indicates a possible involvement of G-proteins in the stimulation of phospholipase C. Evidence for such a mechanism has previously been published by Dillenschneider et al. (1986) for membranes of Acer pseudoplanus.

INOSITOL 1, 4, 5 - TRISPHOSPHATE - INDUCED CALCIUM RELEASE

Inositol 1, 4, 5 -trisphosphate (IP_3) binds to specific receptors most probably located at the endoplasmic reticulum of animal cells and releases Ca^{2+} from this storage compartment into the cytoplasm (For refs. see Berridge and Irvine, 1984). Because of the presence of the polyphosphoinositol metabolism in plants it is an intriguing idea to postulate a similar role for IP_3 also for plant cells. The first report indicating that such an IP_3-induced Ca^{2+} release mechanism indeed exists in plants was published by Drobak and Ferguson in 1985. These authors could show that about 30% of the Ca^{2+} taken up by microsomes in the presence of ATP was released by µM concentrations of IP_3. Schumaker and Sze (1987) provided evidence that IP_3 but not IP_2 or IP caused a transient decrease of Ca^{2+} in tonoplast vesicles from oat roots. This process was concentration dependent and had an apparent K_m of 0.6 µM. Furthermore Cornelius et al. (1989) reported on similar findings with vacuoles from Neurospora crassa. They were able to stimulate Ca^{2+} release from vacuoles with an apparent K_m of 5.28 µM. Similar results with vacuoles from Acer pseudoplanus were published by Ranjeva et al. (1988). Using protoplasts from carrot cells Rincon and Boss (1987) could demonstrate a 17% net loss of accumulated $^{45}Ca^{2+}$ after addition of IP_3.
Phytic acid (IP_6) was also capable to induce the same response. From all these data one might conclude that IP_3 mediated Ca^{2+} release as observed for animal cells occurs also in plants. However it is still too early to make any statement about the specificity of this process with respect to

inositolphosphate isomers, receptors and identity of organelles. Clearly much more work is needed too substantiate the hypothesis of IP_3 being a compound which controls cytoplasmic free Ca^{2+} by releasing this cellular mėssenger from specific sites in the cell.

CHANGES OF POLYPHOSPHOINOSITIDE METABOLISM BY HORMONES AND LIGHT

The most crucial point in identifying the regulatory role of the PI metabolism is the demonstration that components of this metabolism can be changed by environmental and/or internal stimuli such as hormones. Even in animal cells the situation has been complicated by the fact that growth factor receptors upon activation can not only stimulate the generation of IP_3 but can also interfere with the phosphoinositol kinase(s). However it still remains to be shown that this is implicated in cellular signaling. Nevertheless it demonstrates that the regulatory effects of these components have to be considered also in plants. Several reports indicate that regulatory signals such as light, hormones and elicitors are capable to generate inositolphosphates.

The first demonstration was published by Morse et al., (1987). These authors could show that a brief irradiation of Samanea saman pulvini with light caused a decrease of PIP and PIP_2 accompagnied with a concomittent increase of IP_3 and IP_2. From the same group it was also reported the diacylglycerol, the second degradation product of PIP_2, increased after 30 seconds of white light and returned to below the control level after 10 minutes of white light (Morse et al., 1988). The plant growth hormone auxin has also been implicated in the control of PIP_2 degradation.

Ettlinger and Lehle (1988) have demonstrated that auxin can generate transient changes in IP_2 and IP_3 within minutes in Catharanthus roseus cells. They also reported that these changes were accompagnied by a redistribution within the polyphosphoinositide fraction. Kurosaki et al., (1987) gave evidence for the generation of IP_3 by the stimulation of carrot cells with an elicitor of the phytoalexin 6-methoxymellein. Thus it seems that a wide variety of stimuli can interfere with the generation of IP_3. However results obtained by Melin et al., (1987) show that the phospholipase C activity in the plasma membrane isolated from light-grown wheat shoots was lower than from dark grown ones. These data might not be in contrast to the ones summarized above because these plants were grown under a different light-dark regime.

EVIDENCE FOR G-PROTEIN LIKE PEPTIDES

G-proteins have been implicated in the transduction of hormonal and light signals from the receptors to the cytoplasmic effector molecules such as adenylate cyclase, phospolipase C and cGMP phosphodiesterases for refs. see Casey and Gilman, (1988). Cholera toxin and pertussis toxin have been shown to interfere with the activation of adenylate cyclase, mediated by G_s, and the inhibition of adenylate cyclase, mediated by G_i, respectively. These toxins are also valuable tools for radiolabeling G-proteins because of their NAD-dependent ADP-ribosylation activity. G-proteins can also be detected by their ability to bind GTP to their a-subunit and their GTPase activity.

Recently several papers have been published with the aim to demonstrate the existence of G-protein like peptides also in plants. Hasunuma and Funadera (1987) have demonstrated the occurence of proteins binding [^{35}S] GTPτS in Lemna. They found a K_m for GTP-binding of 8 nM which compares well with GTP-binding to transducin and G_i. Seven fractions of GTP-binding proteins were detected by gel filtration of an extract of epicotyls of Pisum sativum seedlings.

These fractions were also ADP-ribosylated by pertussis toxin. K_m's for GTP-binding were in the range of 1 to 50 nM. Drobak et al. (1988) were able tho show the existence of specific GTP-binding proteins in microsomal fractions obtained from zucchini. These authors blotted the microsomal peptides, separated by SDS-PAGE, on nitrocellulose and detected the binding activity by incubation with [α-^{32}P] GTP or [τ-^{32}P] GTP. Four peptides were identified to pick up the label specifically. The relative molecular weights ranging from roughly 23 to 28 kDa were comparable to a novel class of low molecular weight GTP-binding proteins (Bhullar and Haslam, 1988). Similarly GTP-binding using [^{35}S] GTPτS was obtained with zucchini hypocotyl membranes (Jacobs et al., 1988). In the same paper it was also reported that antiserum against a synthetic peptide, deduced from the a-submit of G_s, was able to label on western blots two peptides of 50 and 33 kDa. Blum and colleagues (1988) also using the western blotting technique could show that various peptides from different plants were labelled with antibodies against a sequence occuring in the a-submit of G_s, G_i, G_o and G_t. The relative molecular weights ranged from 31 to 38 kDa.

Recently we were able to demonstrate ADP-ribosylation with cholera toxin of a peptide from corn coleoptiles slightly above 30 kDa (Aducci, deBoer, Gaul, Marmé, unpublished). All these results do not conclusively demonstrate the existence of G-proteins in plants. They do not at all give any indication of a putative physiological role of these proteins. Thus clearly much more work is needed to further clarify the significance of these findings.

CONCLUSION

Assuming that G-protein-mediated activation of polyphosphoinositide metabolism plays a significant role in plant signal transduction has prompted many investigators to identify the components involved in this mechanism in plants.

As it looks now there is convincing evidence that most of the components exist and there is also clear indication the external signals interfere with the PI metabolism. As we know from the animal field protein kinase C is one of the major signal carriers involved in the control of differention and proliferation (Nishizuka, 1988). Morse et al. (1989) has shown for the first time that diacylglycerol is formed upon irradiation of Samanea saman pulvini. Diacylglycerol is the physiological activator of protein kinase C in animals. There are some indications that a Ca^{2+} and phospholipid-activated protein kinase exists also in plants (Schäfer et al., 1985 ; Elliot and Skinner, 1986 ; Olah and Kiss, 1986). However phorbol esters, tumor promotors binding to the diacylglycerol site, and diacylglycerol failed to show convincingly a stimulation of a plant Ca^{2+} and phospholipid-dependent protein kinase. Thus further speculations about the existence of a protein kinase C - like enzyme and its putative function have to be handled with extreme caution.

Despite the lack of definite and conclusive evidence for the complete signal transducing system in plants the available results are very promissing and justify further investigations.

REFERENCES

Andrejauskas E, Hertel R, Marmé D (1985) Specific binding of the calcium antagonist [^{3}H] verapamil to membrane factions from plants. **J. Biol. Chem. 260: 5411-5414**

Berridge MJ, Irvine RF (1984) Inositol triphosphate, a novel second messenger in cellular signal transduction. **Nature 312: 315-321**

Bhullar RP, Haslam RJ (1987) Detection of 23-27 kDa GTP-binding proteins in platelets and other cells. **Biochem. J. 245: 617-620**

Blum W, Hinsch KD, Schulz G, Weiler EW (1988) Identification of GTP-binding proteins in the plasma membrane of higher plants. **Biochem. Biophys. Res. Commun. 156: 954-959**

Boss WF, Massel MO (1985) Polyphosphoinositides are present in plant tissue culture cells. **Biochem. Biophys. Res. Commun. 132: 1018-1023**

Casey PJ, Gilman AG (1988) G protein involvement in receptoreffector coupling. **J. Biol. Chem. 263: 2577-2580**

Cornelius G, Gebauer G, Techel D (1989) Inositol trisphosphate induces calcium release from Neurospora crassa vacuoles. **Biochem. Biophys. Res. Comm. 162: 852-856**

Dillenschneider M, Hetherington A, Graziana A, Alibert G, Berta P, Haiech J (1986) The formation of inositol phosphate derivatives by isolated membranes from Acer pseudoplantanum is stimulated by guamine nucleotides. **FEBS Letters 208: 413-417**

Drobak BK, Ferguson IB (1985) Release of Ca^{2+} from plant hypocotyl microsomes by inositol 1, 4, 5-trisphosphate. **Biochem. Biophys. Res. Commun. 130: 1241-1246**

Drobak BK, Allan EF, Comerford JG, Roberts K, Dawson AP (1988) Presence of guamine nucleotide-binding proteins in a plant hypocotyl microsomal fraction. **Biochem. Biophys. Res. Comm. 150: 899-903**

Einsphar KJ, Peeler TC, Thompson GA (1989) Phosphatidylinsoitol 4, 5-bisphosphate phospholipase C and phosphomonoesterase in Dunaliella salina membranes. **Plant Physiol. 90: 115-1120**

Elliot DC, Skinner JD (1986) Calcium-dependent, phospholipidactivated protein kinase in plants. **Phytochemistry 25: 39-44**

Ettlinger C, Lehle L (1988) Auxin induces rapid changes in phosphatidylinositol metabolites. **Nature 331: 176-178**

Graziana A, Fosset M, Ranjeva R, Hetherington AM, Lazdunski M (1977) Ca^{2+} channel inhibitors that bind to plant cell membranes block Ca^{2+} entry into protoplasts. **Biochemistry 24: 764-768**

Hasunuma K, Furukawa K, Tomita K, Mukai C, Nakamurot (1987) GTP-binding proteins in etiolated epicotyls from Pisum sativum seedlings. **Biochem. Biophys. Res. Commun. 148: 133-139**

Hasunuma K, Funadera K (1987) GTP-binding proteins in green plant, Lemna paucicostata. **Biochem. Biophys. Res. Commun. 143: 903-912**

Irvine RF, Letcher AJ, Lander DJ, Drobak BK, Dawson AP, Musgrave A (1989) Phosphatidyl 4, 5-bisphosphate and phosphatidylinositol 4-phosphate in plant tissue. **Plant Physiol. 89: 888-892**

Jacobs M, Thelen MP, Farndale RW, Astle MC, Rubery PH (1988)Specific guanine nucleotide binding by membranes form Cucurbita pepo **Biochem. Biophys. Res. Commun. 155: 1478-1484**

Kurosaki F, Tsurusawa Y, Nishi A (1987) Breakdown of phosphatidylinositol during the elicitation of phytoalexin production in cultured carrot cells. **Plant Physiol. 85: 601-604**

Marmé D (1988) The role of calmodulin in plants. In : **Calmodulin (Cohen P and Klee C, eds.) Elsevier, Amsterdam, pp 313-328**

Mc Murray WC, Irvine RF (1988) Phosphatidylinositol 4,5-bis-phosphate phosphodiesterase in higher plants. **Biochem. J. 249: 877-881**

Melin PM, Sommarin M, Sandelius AS, Jergil B (1987) Identification of Ca^{2+} -stimulated polyphosphoinositide phospholipase C in isolated plant plasma membranes. **FEBS Letters 223: 87-91**

Morse MJ, Crain RC, Satter RL (1986) Phosphatidylinositol cycle metabolites in Samanea saman pulvini. **Plant Physiol. 83: 640-644**

Morse MJ, Crain RC, Satter RL (1987) Light stimulated inositolphospholipid turnover in Samanea saman leaf pulvini. **Proc. Natl. Acad. Sci USA 84: 7075-7085**

Morse MJ, Crain RC, Coté GG, Satter RL (1989) Light-stimulated inositol phospholipid turnover in Samanea saman pulvini: Increased levels of diacyglycerol. **Plant Physiol. 89: 724-727**

Nishizuka Y (1988) The molecular heterogenecity of protein kinase C and its implications for cellular regulation. **Nature 334: 661-665**

Olah Z, Kiss Z (1986) Occurence of lipid and phorbol ester activated protein kinase in wheat cells. **FEBS Letters 195: 33-37**

Ranjeva R, Carrasco A, Boudet AM (1988) Inositol trisphosphate stimulated the release of calcium from intact vacuoles isolated from Acer cells. **FEBS Letters 230: 137-141**

Rincon M, Boss WF, (1987) Myo-Inositol trisphosphate mobilizes calcium from fusogenic carrot (Daucus carota L.) protoplasts. **Plant Physiol. 83: 395-398**

Salimath BP, Marmé D (1983) Protein phosphorylation and its regulation by calcium and calmodulin in membrane fractions from zucchini hypocotyl hooks. **Planta 158:560-568**

Sandelius AS, Sommarin M (1986) Phosphorylation of phosphatidylinositols in isolated plant membranes. **FEBS Letters 201: 282-286**

Schäfer A, Bygrave F, Matzenauer S, Marmé D (1985) Identification of a calcium and phospholipid dependent protein kinase in plant tissue. **FEBS Letters 187: 25-28**

Schumaker KS, Sze H (1987) Inositol 1,4,5-trisphosphate releases Ca^{2+} from vacuolar membrane vesicles of oat roots. **J. Biol. Chem. 262: 3944-3946**

Sommarin M, Sandelius AS (1988) Phosphatidylinositol and phosphatidylinositolphasphate kinase in plant plasma membranes. **Biochem. Biophys. Acta 958: 268-278**

Wheeler JJ, Boss WF (1987) Polyphosphoinositides are present in plasma membranes isolated from fusogenic carrot cells. **Plant Physiol. 85: 369-392**

Transgenic systems as a tool to probe essential components in physiological processes

Ramon Serrano[1], A. Spena[2], E. Weiler[3] and J. Morré[4]

1 EMBL, Postfach 10.22009, Meyerhofstrasse 1
D-6900 Heidelberg

2 Max Planck Institute
Abt. Genetische Grundlag. Pflanz D-5000 Köln 30

3 Rhur Universität Bochum
Lehrstuhl fur Pflanzen Physiol., Postfach 1021
D-4630 Bochum

4 Purdue University
Department of Medicinal Chemistry West Lafayette
In 47907 U.S.A.

Spena : Physiologists have used mutants obtained by classical genetics to study gene action on both metabolic and developmental processes. The dwarfing genes in Zea mays and their relation to the gibberellins is one of the most often cited examples of interaction between genetics and physiology. More recently auxin resistant Arabidopsis mutant lines have been used to correlate auxin resistance with morphological alterations. Tryptophan requiring mutants of Arabidopsis have confirmed and extended previous knowledge about the relationship between tryptophan and auxin. In

NATO ASI Series, Vol. H 47
Signal Perception and Transduction in Higher Plants
Edited by R. Ranjeva and A. M. Boudet

this meeting Dr. Hall has outlined the properties of ethylene-insensitive Arabidopsis plants and their use in studying ethylene mechanism of action. Another significant contribution to our understanding of biological processes in plants comes from microbial genetics. In particular the study of plant pathogens and the understanding of the pathogenesis of plant diseases caused by phytopathogens has greatly contributed to our under standing of the action of plant growth regulators. Dr. Weiler, what are, in your opinion, the major outcomes from studying the crown gall disease ?

Weiler : The understanding of the pathogenesis of the crown gall disease is an example of interaction between plant physiologists and plant molecular biologists. Tissue culture studies had already established that, by varying the ratio between auxin and cytokinin, we can trigger either root or shoot formation when it was observed that T-DNA mutants of Agrobacterium tumefaciens were eliciting either root or shoot formation. The cloning and expression of T-DNA genes in E. coli, combined with immunoassays for phytohormones were able to clarify that T-DNA genes are responsible for auxin and cytokinin overproduction. I wish to emphasize that hormonal measurements of crown gall tissue were not conclusive. This was probably due to the existence of inactive hormonal pools and compartmentalization in plant cells. The characterization of the T-DNA hormone biosynthetic genes gave us the key to understand the crown gall phenomenon.

Spena : Now that we have briefly celebrated the role that classical plant and microbial genetics play in plant science, I would like to concentrate our discussion on the new genetics. What are the novelties offered by the reverse genetics approach, and how plant genetic engineering can contribute to the study of physiological problems ?

Serrano : The so-called new genetics has, in my opinion, two aspects ; a descriptive one and a plastic one. The first one is concerned with the cloning of genes, with their structural description, with the dissection of their controlling regions. The second one, relies on already isolated genes and, allows us to test the biological effects of their gene products by a gain of function or by a loss of function. The gain of function is performed not merely by introducing the isolated genes under the control of its own promoter, but, by replacing the wild type 5 flanking sequences with heterologous promoters, we can pilot gene expression at our will. Using different promoters we can obtain different biological effects. In this respect, Angelo, can you tell us about your work with the rolC gene of *Agrobacterium rhizogenes* ?

Spena : The rolC gene *Agrobacterium rhizogenes* was identified for its capacity to affect root formation and growth in the hairy root disease. However this gene is able to affect not only root formation and growth but also plant morphogenesis and development as easily monitored in plants transgenic for the gene. Moreover when the rolC 5 flanking sequences are replaced by a promoter with different specificities like the 35s promoter from Cauliflower mosaic virus, new morphological and developmental alterations are established. Indeed a pletora of effects, including reduction in apical dominance are displayed by 35S-rolc transgenic plants.

Although most effects are reminiscent of an alteration in the activity of plant growth regulators, the rolC gene product acts in a cell-autonomous fashion. Most, but definitely not all, of the phenotypic alterations could be explained by a achieved in many different ways like auxin inactivation, inhibition of auxin transport, alteration of the auxin receptor system, etc.

Serrano : This is an example of the "gain of function" approach. The loss-of-function approach is performed by expressing the gene of interest in antisense orientation. Although the molecular details are not completely clear, antisense-RNA may produce a new type of mutants by inhibiting gene expression at the post-transcriptional level by forming double-stranded RNA.

Morré : It seems to me that genetic engineering will provide the tools to test the hypothesis that the plasmamembrane oxidoreductase system plays a crucial role in controlling growth both in animal and plant cells. We have purified the enzyme, and hope to clone the gene hopefully, we eventually construct transgenic systems overexpressing the oxidoreductase aiming to promote cell growth.

Spena : Ramon, you have worked for long time with yeast and now, while still engaged with this low eukaryote, you are approaching *Arabidopsis*. Is *Arabidopsis* like yeast, only a bit more green ?

Serrano : First, yeast is a good model for plant cells because they have the same basic transport mechanisms in their plasmamembrane (i.e. H^+-ATPase, H^+-symports, H^+-antiports, K^+-channels). In yeast, gene replacement is easy because of its high frequency of homologous recombination. More specifically, mutational analysis of yeast plasma membrane H^+-ATPase has demonstrated that the control coefficient of growth rate by ATPase activity is close to 1. This means that the enzyme is ratelimiting (i.e. pace-maker) for growth by auxin, which activates the plant H^+-ATPase. Increased ATPase activity correlates with increased intracellular pH and increased active nutrient transport. Lastly, the yeast proton pump expressed in animal cells behaves as an oncogene causing uncontrolled proliferation in tissue culture, and neoplastic growth in nude mice.

Spena : I would like to close this round table on transgenic plants as a tool to probe essential components in physiological responses by discussing the possible use of transposable elements. Obviously transposable elements are going to be used as tools to tag genes. However what I would like to consider now is their use as elements able to control gene expression. We can use transposable elements to selectively activate a gene in a cellular clone. This can be easily achieved by splitting any gene of interest with a transposable element. In such a way that somatic excision will reconstruct the gene and will allow clonal expression of the gene. This is the type of genetic switch that we have used with the rolC gene in order to show the cell autonomous behaviour of its product. The same type of switch could be peculiarly useful for genes affecting plant morphogenesis and growth, where to overcome the difficulties of pleiotropic effects of hormones one should try to localize the effect in one organ or tissue. For example, Dewulf et al. have presented a poster showing that expression of the yeast Ras2 gene kills plant cells. Of course this type of biological effect is bound to generate some troubles when trying to regenerate transgenic plants. One possibility to circumvent this type of problem is to use promoters which trigger expression only under certain environmental or developmental conditions. Unfortunately, many inducible promoters are leaky. An alternative is to deliver not the Ras2 gene as such, but a construction where the promoter is separated from the coding region of the gene by a transposable element. Gene expression can take place only after transposon excision and therefore this will probably allow to regenerate plants and then to observe the killing effects in clonal population of cells within an organ.

Weiler : Transgenic systems could be used also to express a gene specifically in one type of cells and then to monitor the effects on neighbouring cells in order to study cell-cell communication.

Serrano : In my opinion, transgenic systems will be essential to test the physiological role(s) of the several signal transducing pathways proposed in this meeting.

CONTROL OF GLYCOGEN METABOLISM AND PHOSPHORYLASE KINASE. A MODEL SYSTEM FOR STUDYING SIGNAL TRANSDUCTION MECHANISMS MEDIATED BY PROTEIN PHOSPHORYLATION AND Ca^{2+}

T.G. SOTIROUDIS, V.G. ZEVGOLIS, L.G. BALTAS, S.M. KYRIAKIDIS
Institute of Biological Research
The National Hellenic Research Foundation
48 Vassileos Constantinou Avenue
Athens 116 35
Greece

One of the best studied metabolic control systems regulates the formation and breakdown of glycogen in mammalian tissues (Huijing, 1975). Glycogen constitutes the main storage form of carbohydrate in mammalian cells and it is the major energy source which sustains ATP levels during muscle contraction. At least 95% of the polymer is located in skeletal muscle and liver, where it plays quite different roles (Cohen, 1976). The rate limiting enzymes in cytosolic glycogenolysis and glycogen synthesis are glycogen phosphorylase and glycogen synthase, while branching and debranching enzymes are considered to be present in excess in normal animals and are therefore not rate-limiting. In this respect, although the current convention regards phosphorylase as a purely degradative enzyme with no synthetic role, it seems possible that intracellular local concentrations of its substrates glucose-1-P and Pi may on occasions favour synthesis (Geddes, 1986). In addition, a potential cellular degradative pathway, concerning cellular glycogen associated with the lysosomal compartments, was also revealed; in this case, the enzyme 1,4-a-glycosidase is responsible for the degradation of the polysaccharide (Geddes, 1986).

It was through work on glycogen metabolism that the first examples of hormonal regulation of enzyme activity by reversible phosphorylation were discovered. The extracellular signal-generated second messenger (cAMP or Ca^{2+}) is decoded by a protein kinase (cAMP-dependent protein kinase or phosphorylase kinase) and further translated into the phosphorylation of the specific enzyme target, resulting in increase (phosphorylase kinase, glycogen phosphorylase) or decrease (glycogen synthase) of its catalytic action (Cohen, 1976). Today it is well established (in vitro or in vivo) that a number of enzymes and regulatory proteins involved in the control of

NATO ASI Series, Vol. H 47
Signal Perception and Transduction in Higher Plants
Edited by R. Ranjeva and A. M. Boudet

glycogen metabolism can be post-translationally modified by several phosphorylation processes, involving a number of protein kinases and phosphatases (Sotiroudis and Evangelopoulos, in press). Thus, this metabolic regulation cascade can be used as a model for the comparison with other cellular systems controlled by the same type of covalent modification, to reveal the molecular basis of signal transduction mechanisms (Chan and Krebs, 1986; Larner, 1988).

Among the enzymes involved in the control of glycogen metabolism, phosphorylase kinase has a central role, linking glycogen breakdown to both nervous and endocrine stimulation. It is regulated in a complex way by phosphorylation of various serine residues and by calcium (reviews: Carlson et al, 1979; Chan and Graves, 1984; Pickett-Gies and Walsh, 1986), while recently two other mechanisms of regulation have been suggested to be physiologically significant: One involves allosteric activation by ADP (Cheng and Carlson, 1988) and the other activation or inhibition by interaction with lipid molecules (Kyriakidis et al, 1986b; Baltas et al, 1989). The complexity of the system offers more than one avenue of signal transduction and thus, phosphorylase kinase can be used as a key target for studying the processes by which hormones and other extracellular agents exert their biological action.

In this article we briefly summarize currently available data on the structural and functional features of two isoenzymes of phosphorylase kinase: (i) The rabbit skeletal muscle type, which is the most extensively characterized heteropolymeric form and (ii) The smooth muscle homopolymeric isoenzyme. Furthermore, we will discuss research which has been directed toward an understanding of the interaction of phosphorylase kinase with membranes and membrane constituents.

Characterization of rabbit skeletal muscle phosphorylase kinase

Rabbit skeletal muscle phosphorylase kinase, a key regulatory enzyme in glycogen metabolism which catalyzes the phosphorylation of phosphorylase b to make phosphorylase a, is a multisubunit Ca^{2+}, calmodulin-dependent enzyme composed of four different subunits with a stoechiometry $(\alpha\beta\gamma\delta)_4$ and a native molecular weight of approximately 1.3×10^6. The isoenzymes of white skeletal muscle and liver are characterized by the 132-kDa α subunit (apparent size) which is replaced by the smaller α′ subunit (127 kDa) in red skeletal muscle and heart. The complete amino acid sequences of all four subunits are known. γ (Mr 44,673) belongs to the family of protein kinases and is catalytically active in its isolated form. δ (Mr 16,680) is identical to calmodulin and confers Ca^{2+} sensitivity to the enzyme. The two large subunits α and β are homologous proteins, (Mr 138,422 and 125,117, respectively), they cary all phosphorylation sites and at least one of their function is regulatory (Zaman et al, 1989). Recently we have demonstrated that fluorescein isothiocyanate (FITC) reacts specifically with the α and α′ subunits of phosphorylase kinase resulting in enzyme inactivation, while both FITC labelling and kinase inactivation are inhibited by ATP and ATP/Mg^{2+} (Sotiroudis and Nikolaropoulos, 1984; Zaman et al, 1989). These results point to a catalytic function of the α-subunit.

As isolated from rabbit skeletal muscle, phosphorylase kinase has little activity when assayed at neutral pH; however the enzyme can be stimulated by a variety of mechanisms including covalent and allosteric modifications. Phosphorylation represents probably the most important covalent regulation of phosphorylase kinase in vivo, while the activity of the kinase towards phosphorylase b can be also affected (directly or indirectly) by two other types of covalent modification. Partial proteolysis (the α and β subunits are only degraded) drastically increases the pH 6.8 activity of the enzyme, while ADP-ribosylation of α and β subunits diminishes both cAMP-dependent and autocatalytic phosphorylation and, as a result suppresses phosphorylation-dependent activation (Pickett-Gies and Walsh, 1986). The ability of adrenergic agents to stimulate in vivo the formation of cAMP and the activation of phosphorylase kinase is correlated with the phosphorylation in vitro of both α and β subunits by cAMP-dependent protein kinase.

In addition, phosphorylation of phosphorylase kinase catalyzed by a number of different kinases and by Ca^{2+}-dependent autocatalytic process induces enzyme activation (Pickett-Gies and Walsh, 1986). Both γ and δ subunits can not be phosphorylated. Phosphorylase kinase activity can be further activated by a variety of effectors that interact in a noncovalent, specific manner and presumably modulate activity by affecting the conformation of the kinase. This group of activators include Ca^{2+}, Mg^{2+}, exogenous calmodulin, troponin C, ADP, actin, glycogen, organic solvents acidic phospholipids, unsaturated fatty acids and gangliosides (Pickett-Gies and Walsh, 1986; Kyriakidis et al, 1986b; Baltas et al, 1989).

Although phosphorylase kinase is usually characterized as a dedicated Ca^{2+}-dependent protein kinase, that is a protein kinase with a limited substrate specificity (Schulman, 1988), several substrates of possible physiological significance in addition to phosphorylase b have been described. These include glycogen synthase, myelin basic protein, a number of myofibril proteins and several membrane protein components (Pickett-Gies and Walsh, 1986; Kobayashi et al, 1984; Schlender et al, 1988). In addition it has been reported that phosphorylase kinase exhibits phosphatidylinositol kinase activity (Georgoussi and Heilmeyer, 1986).

Smooth muscle phosphorylase kinase

Glycogen is the most conspicuous source of endogenous carbohydrate metabolism observed in the smooth muscle cells of different organs. Nevertheless, little is known on the catalytic and structural characteristics of the enzymes involved in the cascade of reactions associated with glycogen synthesis or degradation in smooth muscle (Sotiroudis et al, 1986). We have undertaken the purification and characterization of phosphorylase kinase from two types of smooth muscle: a primitive vertebrate (chicken gizzard) (Nikolaropoulos and Sotiroudis, 1985) and a mammalian smooth muscle type (bovine stomach) (Zevgolis et al, submitted). It was interesting as well as puzzling that both smooth muscle isoenzymes of phosphorylase kinase showed a completely different subunit pattern compared to that of rabbit skeletal muscle enzyme. Upon SDS-PAGE the gizzard kinase showed one main protein band of 61 kDa, while the bovine stomach enzyme migrated as a single protein band of 43 kDa, although the native Mr of both kinases were similar to that of skeletal muscle enzyme. These results suggest that smooth muscle phosphorylase kinase is an oligomer composed of identical subunits which reminds of the large holoenzyme of multifunctional Ca^{2+}/calmodulin-dependent protein kinase (Schulman, 1988). More important, the bovine stomach 43 kDa subunit exhibits a strong polydispersity, as revealed by glycerol density ultracentrifugation and non-denaturing PAGE, and weak immunoreactivity with anti-actin antibody.These properties may largely affect its localization within the contractile machinery of the cell. The above hypothesis is strengthened by the fact that there is experimental evidence for functional compartmentation of glycogen metabolism in smooth muscle cell and for the

Table 1. Comparative properties of rabbit skeletal muscle and bovine stomach smooth muscle phosphorylase kinase

	Rabbit muscle[c]	Bovine stomach
Mr	1.3×10^6	1×10^6
Subunit pattern	α, 138 kDa	43 kDa
	β, 125 kDa	
	γ, 45 kDa	
	δ, 17 kDa	
	$(\alpha_4\beta_4\gamma_4\delta_4)$	
Activity ratio (pH 6.8/8.2)	0.04-0.08	0.23
Km (pH 8.2), mM		
for ATP	0.06-0.31	1.6
for phosphorylase b	0.1-0.27	0.015
$c^{a}_{0.5}$ for Ca^{2+} (pH 8.2), µM	16	0.05
Activation by phosphorylation	yes	yes
Activation by calmodulin[b]	yes	yes
Activation by proteolysis	yes	yes

[a]$c_{0.5}$, concentration required for half-maximal activation
[b]Exogenous calmodulin, [c]Pickett-Gies and Walsh, 1986

association of glycogenolysis with contractile filaments (Paul, 1989). Furthermore, both smooth muscle kinases are Ca^{2+}-dependent protein kinases, are activated by calmodulin and limited proteolysis and are characterized by a high affinity for phosphorylase b. They did not appreciably incorporate ^{32}Pi from [γ-^{32}P]ATP donor into histone II-AS, casein and phosvitin, but the bovine stomach enzyme presented significant activity with protamine sulfate (10% activity compared to that with phosphorylase b), while at the same time the phosphorylase b to a conversion is drastically inhibited by protamine (IC_{50}=15µg/ml) (unpublished results).

Our data (Nikolaropoulos and Sotiroudis, 1985) indicated also, that gizzard phosphorylase kinase can neither be autophosphorylated nor activated by phosphorylation with cAMP-dependent protein kinase. In contrast, bovine stomach phosphorylase kinase is phosphorylated in presence of the catalytic subunit of protein kinase and this reaction is accompanied by enzyme activation at pH 6.8 and ^{32}P-incorporation into the 43 kDa protein band. These results imply that the hormonal control of glycogenolysis in mammalian smooth muscles can be exercised by a direct interaction between the cAMP-dependent protein kinase and phosphorylase kinase as in the case of mammalian skeletal muscle. This hypothesis is consistent with the results of Silver and Stull (1982) who have shown that β-adrenergic stimulation significantly increases phosphorylase a levels in bovine tracheal smooth muscle. Table 1 presents a comparison of significant properties of rabbit skeletal muscle and bovine stomach smooth muscle phosphorylase kinase.

Interaction of phosphorylase kinase with membranes and membrane constituents

A number of previous studies have reported the association of several activities of glycogen metabolism with sarcoplasmic reticulum(SR) membranes (reviewed:Dombradi et al, 1984). In this respect, phosphorylase kinase activity is well known to be associated with SR or plasma membranes, while recently molecular structures related to phosphorylase kinase have been localized at the SR of rabbit skeletal muscle employing polyclonal antibodies directed against the holoenzyme as well as monoclonal antibodies specific for its α-, β- or γ- subunits (Dombradi et al, 1984;Thieleczek et al, 1987).

In addition, several observations from our and other laboratories suggested that the hydrophobic properties of "soluble" phosphorylase kinase are important for its mode of action. Thus, it was found that: (i) The flavone quercetin, a bioflavonoid known to affect membrane-linked activities and inhibit several protein kinases including the Ca^{2+} and phospholipid-dependent protein kinase, was found to be an efficient inhibitor of ATP for the phosphorylation of phosphorylase b at pH 8.2 (Kyriakidis et al, 1986a). (ii) Polymyxin B, an amphiphilic cyclic decapeptide antibiotic, which strongly inhibits protein kinase C and myosin light chain kinase, has been shown to effectively inhibit "soluble" and SR-bound phosphorylase kinase and to stimulate its autophosphorylation (Ktenas et al, 1985; Ktenas et al, 1989). (iii) Organic solvents, unsaturated fatty acids (Singh and Wang, 1979) and the flavanone hesperetin (Kyriakidis et al, 1986a) stimulated the activity of nonactivated phosphorylase kinase. Moreover, we have demonstrated that acidic phospholipids are also activators and that their activating effect largely depends on the size of lipid vesicles which is connected with the procedure of their preparation (Kyriakidis et al, 1986b). (iv) Phosphorylase kinase exhibits phosphatidylinositol kinase activity (Georgoussi and Heilmeyer, 1986), while in parallel, the SR Ca^{2+}-ATPase, the sarcolemma Na^{+}, K^{+}-ATPase (reviewed by Pickett-Gies and Walsh, 1986) and a 40 kDa protein in synaptic plasma membranes (Browning et al, 1979) may be substrates of possible physiological significance.

In our effort to investigate the relation between cytoplasmic and membrane bound phosphorylase kinase we have chosen as a membrane system the inside-out human erythrocyte vesicles which have been proved to be a very useful tool for studying physiologically relevant protein-protein and protein-lipid interactions (Kyriakidis et al, 1988). We found that at pH 7.0, the kinase binds to the inner face of the erythrocyte membrane and that this binding is Ca^{2+} and Mg^{2+} dependent. In the presence of 3mM Mg^{2+}, the sharpest increase in the binding reaction occured at concentrations between 70 and 550 nM free Ca2+, agreeing favourably with the reported levels of intracellular Ca^{2+} (Kyriakidis et al, 1988). Furthermore, it was revealed that phosphorylase kinase binding leads to a preferred ^{32}P labelling of a 93 kDa membrane protein and that this phosphorylation increases in presence of Ca^{2+}(Fig.1). The 93 kDa protein possibly represents the band 3 polypeptide of red blood cell membrane; it shows a similar Mr with the anion transport protein, while after staining of the gel for carbohydrate it is visualized as a small shoulder on the trailing edge of the periodic acid-Schiff reagent (Steck, 1974). At this point we must emphasize that even 10-fold lower amount of phosphorylase kinase, than that of Fig.1, is also able to increase the level of phosphorylation of the 93 kDa band.

During the past few years several lines of evidence have accumulated that the breakdown products and metabolites of membrane lipids function predominantly in signal trasnduction as agonists or as second messengers (reviewed by Hannun and Bell, 1989). In this respect, sphingolipid breakdown products represent a recent addition to this growing family of biologically and physiologically active lipids. As it is known this class of lipid molecules has very profound effects on protein phosphorylation,

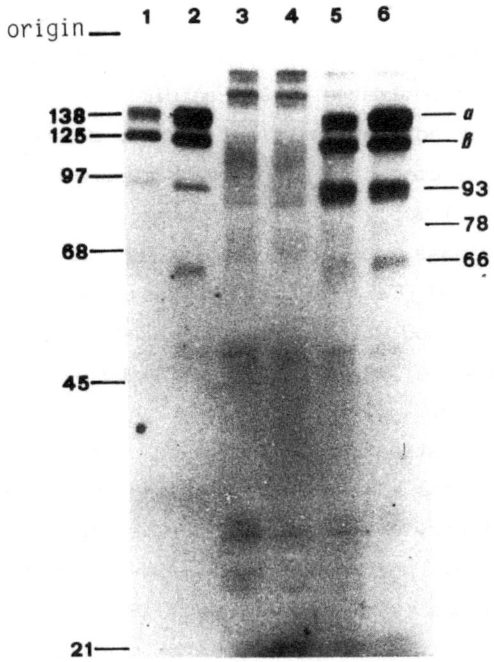

Fig. 1. ^{32}P incorporation into erythrocyte inside-out vesicles in presence and absence of exogenous phosphorylase kinase. The phosphorylation experiment was performed with 1 mM [^{32}P] ATP in a reaction mixture similar to the binding reaction mixture described by Kyriakidis et al (1988). SDS-PAGE (10%) acrylamide was performed as described by Baltas et al (1989). The concentrations of phosphorylase kinase, bovine serum albumin and Ca^{2+} (free), when present, were 50 µg/ml, 1mg/ml and 45µM respectively. Lanes 1, 2, 4, 6 with phosphorylase kinase; lanes 3-6 with human erythrocyte membrane; lanes 2, 4, 6 with Ca^{2+}; lanes 1, 2 with bovine serum albumin. Marker proteins: phosphorylase kinase α subunit (138 kDa), β subunit (125 kDa), γ subunit (45 kDa); phosphorylase b (97 kDa); bovine serum albumin (68 kDa); soybean trypsin inhibitor (21 kDa)

while recent results suggest that a number of sphingolipids potently affect the activity of protein kinase C and of several calmodulin dependent enzymes (reviewed by Baltas et al, 1989).

We found that psychosine (galactosyl sphingosine) (Fig.2) and sphingosine drastically inhibit the activity or rabbit skeletal muscle phosphorylase kinase, while a number of gangliosides stimulated phosphorylase b to a conversion by the nonactivated kinase. Among the individual ganglioside tested G_{D1a} was the most potent stimulator (Fig.2) activation potency:$G_{D1a}>G_{T1b}>G_{M1}$) and G_{M3} was without effect. Most important, G_{D1a} dramatically increases the activity of the kinase at low Ca^{2+} concentrations. On the other hand both psychosine and G_{D1a} increased

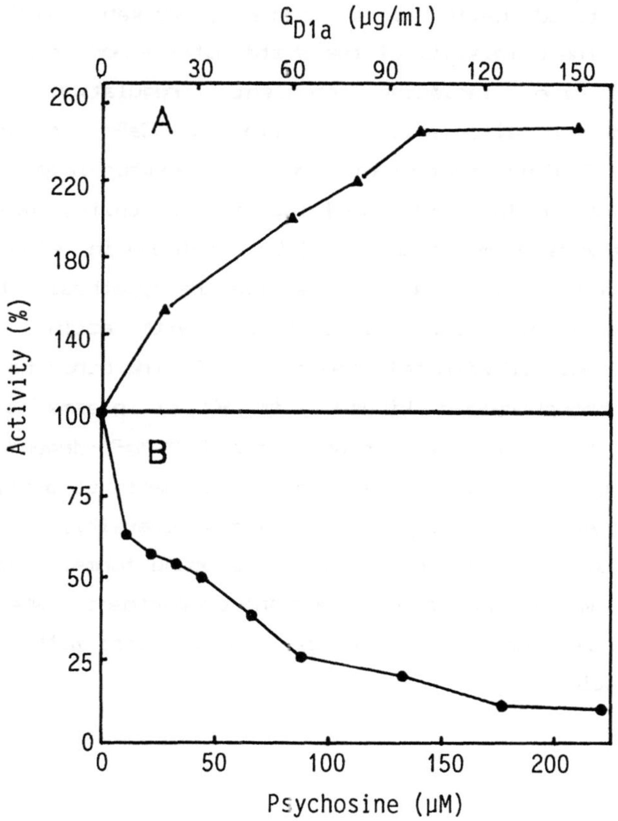

Fig. 2. Effect of G_{D1a} (A) and psychosine (B) on the pH 6.8 activity of phosphorylase kinase. Experimental conditions as described by Baltas et al, 1989

the rate of kinase autophosphorylation on α-subunit only. In this respect, ganglioside-induced stimulation of autophosphorylation was accompanied with an enhancement of the rate of autoactivation, while psychosine completely blocked autoactivation (Baltas et al, 1989).

Concluding remarks

The investigation of effects on glycogen metabolism could hold the key to explaining the molecular basis for the action of hormones and other extracellular signals. We suggest the phosphorylase kinase regulation cascade in cultured cells as a promising workable model system for investigation. The complexity of the system offers more than one avenue of signal transduction including covalent regulation by protein phosphorylation and allosteric regulation by Ca^{2+} or other messenger molecules. Our findings that smooth muscle phosphorylase kinase is an oligomeric enzyme complex composed of identical subunits, which reminds of the large holoenzyme of multifunctional Ca^{2+}/calmodulin kinase and shows similarities with actin, make attractive the hypothesis that there is a close interaction of phosphorylase kinase with actin microfilaments ubiquitously associated with cell membranes. On the other hand, the close association of phosphorylase kinase with SR or plasma membranes, its interaction with erythrocyte membranes in a Ca^{2+}/Mg^{2+}-dependent manner and its activation or inhibition by membrane constituents strongly favour the following hypothesis: Phosphorylase kinase may reversibly partition between kinetically distinct soluble and membrane-bound forms. Under mobilizing Ca^{2+} conditions may translocate to membrane compartments where it serves for a specialized function not necessarily connected with phosphorylase b to a transformation.

References

Baltas LG, Zevgolis VG, Kyriakidis SM, Sotiroudis TG, Evangelopoulos AE (1989) Modulation of phosphorylase kinase activity by sphingolipids. Biochem Inter, in press

Browning M, Bennett W, Lynch G (1979) Phosphorylase kinase phosphorylates a brain protein which is influenced by repetitive synaptic activation. Nature 278:273-275

Carlson GM, Bechtel PJ, Graves DJ (1979) Chemical and regulatory properties of phosphorylase kinase and cyclic AMP-dependent protein kinase. Adv Enzymol 50:41-115

Chan KFJ, Graves DJ (1984) Molecular properties of phosphorylase kinase. In:Cheung WY (ed) Calcium and cell function vol 5. Academic Press, New York, p 1

Chan CP, Krebs EG (1986) Effects of growth factors on carbohydrate metabolism in cultured mammalian cells. In:Belfrage P, Donner J, Stalfors P (eds) Mechanisms of insulin action. Elsevier, Amsterdam, p 13

Cheng A, Carlson GM (1988) Competition between nucleoside diphosphates and triphosphates at the catalytic and allosteric sites of phosphorylase kinase. J Biol Chem 263:5543-5549

Cohen P (1976) Control of enzyme activity. Chapman and Hall Ltd., London

Dombradi VK, Silberban SR, Lee EYC, Caswell AH, Brandt NR (1984) The association of phosphorylase kinase with rabbit muscle T-tubules. Arch Biochem Biophys 230:615-630

Geddes R (1986) Glycogen: a metabolic viewpoint. Biosci Rep 6:415-428

Georgoussi Z, Heilmeyer LMG Jr (1986) Evidence that phosphorylase kinase exhibits phosphatidylinositol kinase activity. Biochemistry 25:3867-3874

Hannun YA, Bell RM (1989) Functions of sphingolipids and sphingolipid breakdown products in cellular regulation. Science 243:500-507

Huijing F (1975) Glycogen metabolism and glycogen-storage diseases. Physiol Rev 55:609-658

Kobayashi T, Nakaza T, Negami A, Nakamura S, Yamamura H (1984) Phosphorylation of myelin basic protein by glycogen phosphorylase kinase. FEBS Lett 169:224-228

Ktenas TB, Sotiroudis TG, Nikolaropoulos S, Evangelopoulos AE (1985) Interaction of phosphorylase kinase with polymyxins. Biochem Biophys Res Commun 133:891-896

Ktenas TB, Sotiroudis TG, Evangelopoulos AE (1989) Effects of polymyxin B on the sarcoplasmic reticulum membrane. Biosci Rep, in press

Kyriakidis SM, Sotiroudis TG, Evangelopoulos AE (1986a) Interaction of flavonoids with rabbit muscle phosphorylase kinase. Biochim Biophys Acta 871:121-129

Kyriakidis SM, Sotiroudis TG, Evangelopoulos AE (1986b) Stimulation of glycogen phosphorylase kinase by phospholipids. Biochem Inter 13: 853-861

Kyriakidis SM, Sotiroudis TG, Evangelopoulos AE (1988) Ca^{2+}-and Mg^{2+}-dependent association of phosphorylase kinase with human erythrocyte membranes. Biochim Biophys Acta 972:347-352

Larner J (1988) Insulin-signaling mechanisms. Lessons from the old testament of glycogen metabolism and the new testament of molecular biology. Diabetes 37:262-275

Nikolaropoulos S, Sotiroudis TG (1985) Phosphorylase kinase from chicken gizzard. Partial purification and characterization. Eur J Biochem 151:467-473

Paul RJ (1989) Smooth muscle energetics. Annu Rev Physiol 51:331-349

Pickett-Gies CA, Walsh DA (1986) Phosphorylase kinase. In:Boyer P, Krebs

EG (eds) The Enzymes vol 17. Academic Press, New York, p 395

Schlender KK, Thysseril TJ, Hegazy MG (1988) Calcium-dependent phosphorylation of bovine cardiac C-protein by phosphorylase kinase. Biochem Biophys Res Commun 155:45-51

Schulman H (1988) The multifuntional Ca^{2+}/calmodulin dependent protein kinase. In:Greengard P, Robinson GA (eds) Advances in second messenger and phosphoprotein research vol 22. Raven Press, New York, p 39

Silver PJ, Stull JT (1982) Regulation of myosin light chain and phosphorylase phosphorylation in tracheal smooth muscle. J Biol Chem 257:6145-6150

Singh TJ, Wang JH (1979) Stimulation of glycogen phosphorylase kinase from rabbit skeletal muscle by organic solvents. J Biol Chem 254: 8466-8472

Sotiroudis TG, Nikolaropoulos S (1984) Selective labelling of phosphorylase kinase with fluorescein isothiocyanate. FEBS Lett 176: 421-425

Sotiroudis TG, Nikolaropoulos S, Evangelopoulos AE (1986) Glycogen metabolism in smooth muscle. In:Heilmeyer LMG (ed) Signal transduction and protein phosphorylation, Plenum Press, New York, p 243

Sotiroudis TG, Evangelopoulos AE. Protein kinases, protein phosphatases and the control of glycogen metabolism. In:Evangelopoulos AE, Changeux JP, Packer L, Sotiroudis TG and Wirtz KWA (eds) Receptors, membrane transport and signal transduction, Springer-Verlag, Berlin, in press

Steck TL (1974) The organization of proteins in the human red blood cell membrane. J Cell Biol 62:1-19

Thieleczek R,, Behle G, Messer A, Varsanyi M, Heilmeyer LMG Jr, Drenckhahn D (1987) Localization of phosphorylase kinase subunits at the sarcoplasmic reticulum of rabbit skeletal muscle by monoclonal and polyclonal antibodies. Eur J Cell Biol 44:333-340

Zaman N, Varsanyi M, Heilmeyer LMG Jr, Sotiroudis TG, Johnson CM, Crabb JW (1989) Reaction of fluorescein isothiocyanate with an ATP binding site on the phosphorylase kinase α-subunit. Eur J Biochem, in press

Zevgolis VG, Sotiroudis TG, Evangelopoulos AE. Phosphorylase kinase from bovine stomach smooth muscle. Purification and characterization. Submitted for publication

MOLECULAR CLONING OF PLANT PROTEIN KINASES

Mark S. Dixon, Shaohui Zhang, Robert T. Yamamoto, Christopher J. Lamb and Michael A. Lawton

Plant Biology Laboratory,
Salk Institute for Biological Studies,
PO Box 85800,
San Diego,
California 92138,
USA

INTRODUCTION

Many cellular proteins are subject to post-translational modifications which alter their activity. One of the most widespread and extensively studied modifications is that of reversible phosphorylation. First demonstrated in the control of glycogen metabolism, reversible phosphorylation is now recognized as a major mechanism for regulating the activities of enzymes and proteins.

The phosphorylation state of a target protein is determined by opposing phosphorylation and dephosphorylation reactions which are catalysed by protein kinases and phosphoprotein phosphatases, respectively (Krebs, 1985). The activities of these regulatory enzymes may in turn be regulated by reversible phosphorylation as well as by the binding of small effector molecules. These properties allow the elaboration of complex regulatory circuits through cascades of positively and negatively interacting protein kinases and phosphoprotein phosphatases. In this way, reversible phosphorylation allows not just the simple turning on or off of an enzyme or protein activity in response to a particular cellular perturbation, but the integration of these activities into a multidimensional regulatory network sensitive to changes in a diverse number of

NATO ASI Series, Vol. H 47
Signal Perception and Transduction in Higher Plants
Edited by R. Ranjeva and A. M. Boudet

metabolic, developmental and environmental signals (Hunter, 1987).

The number of reported protein kinases has grown exponentially since phosphorylase kinase was first purified by Krebs and co-workers in 1959. Since then, over 100 different protein kinases have been described falling broadly into two classes based on their ability to phosphorylate either serine/threonine or tyrosine. These groups can be further classified into families according to substrate specificity, regulatory ligand specificity, structural features and cellular function.

The biochemical properties of protein kinases make these enzymes particularly well-suited to function as transducers of specific cellular signals (Hunter, 1987); signal input is determined by the regulatory properties of the enzyme with respect to ligands such as cyclic nucleotides, lipids and calcium, while signal output is determined by the catalytic properties of the kinase with respect to substrate specificity. This notion is supported by extensive biochemical, molecular and genetic studies which have demonstrated the involvement of protein kinases in the regulation of transcription, translation, cell division, metabolic flux or the cellular transduction of external signals such as hormones and growth factors (Krebs, 1985; Hunter, 1987). Of particular significance to the present work is the finding that input (effector sensing) and output (phosphotransfer reaction) signals are associated with physically distinct domains or subunits of the protein. As we discuss later, this molecular division of labour forms the basis for the cloning of novel protein kinase homologs.

Although protein kinases have been shown to play a central role in cellular signal transduction in animals and simple eukaryotes such as yeast or *Dictyostelium*, little is known of the specific functions of individual plant protein kinases as signal transducers. Thus, although protein phosphorylation has been observed *in vivo* and protein kinase activity demonstrated *in vitro*, in only a few cases has reversible phosphorylation been shown to be associated with the regulation of plant proteins of known function (Ranjeva and Boudet, 1987). Several enzymes of intermediary metabolism have been shown to be regulated by phosphorylation in a similar manner to their mammalian counterparts, for example, pyruvate dehydrogenase and 3-hydroxymethyl glutaryl CoA reductase. Protein kinases have also been implicated in cellular processes specific to plants. For example, a redox-sensitive protein kinase has been shown to regulate the distribution of excitation energy between the two photosystems by phosphorylating the light-harvesting chlorophyll a/b-binding protein (Bennett, 1979). A second example concerns the enzyme quinate:NAD^+ oxidoreductase whose activity is regulated by a Ca^{2+}/calmodulin-dependent protein kinase (Ranjeva *et al.*, 1984). In many cases, however, the natural phosphorylation targets of plant protein kinases are unknown while the significance of stimulus-induced changes in the overall pattern of protein phosphorylation has not been established. We can expect that the significance of some of these activities will become apparent as the role of phosphorylation is examined in more plant systems (for example, Hiddinga *et al.*, 1988; Suzuki *et al.*, 1989).

We have adopted a molecular genetic approach to the isolation of novel protein kinases. The rationale for this approach is several fold: molecular cloning is independent of abundance and therefore is particularly appropriate for protein kinases which are expressed in a limited temporal or spatial manner; isolation of cloned sequences requires neither the stability of the protein nor the establishment of an *in vitro* enzymological assay; comparison of the derived amino acid sequence of cloned protein kinases with existant sequences may provide an indication of substrate specificity and regulatory properties, for example, the presence of canonical calcium or cyclic nucleotide-binding sites; availability of cloned sequences encoding protein kinases forms the basis for examining both activity and function through a combination of molecular and biochemical methods. In particular, the regulatory and catalytic properties of cloned protein kinases may be addressed through the expression of wild-type and mutant genes in heterologous systems. In concert, transgenic plants can be used to address function through the inappropriate expression of wild-type protein kinases or by expression of mutant proteins acting as dominant negative mutants.

PROTEIN KINASES: A GENE FAMILY OF MODULAR PROTEINS

The nucleotide and amino acid sequences of extant protein kinases forms a data base for delineating conserved regions likely to be important in the function of these enzymes. Sequence comparisons between existing members of the protein kinase family have established that all protein kinases contain a common region responsible for catalytic activity (Figure 1).

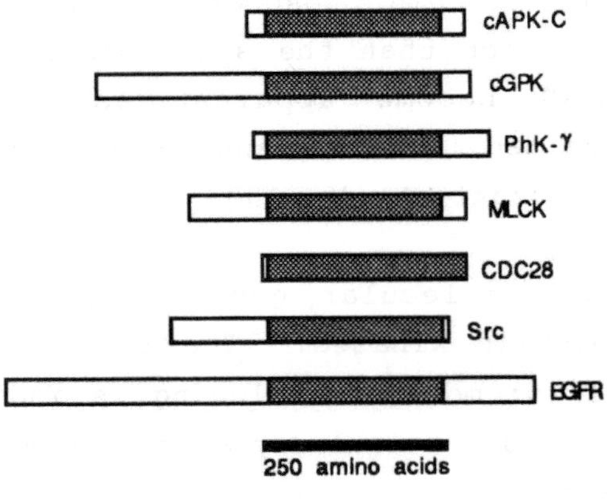

Figure 1. Conservation of the catalytic domain in protein kinases. The bars correspond to the following polypeptides: cAMP-C: cyclic AMP-dependent protein kinase, catalytic subunit; cGPK: cyclic GMP-dependent protein kinase; PhK-γ: phosphorylase kinase γ-subunit; MLCK: myosin light chain kinase; CDC28: P^{34} protein encoded by *S. cerevisiae*; Src: the *src* proto-oncogene; EGFR: epidermal growth factor receptor. The shaded bar indicates the conserved catalytic region

Sequences outside this catalytic domain have either been shown, or are presumed to be involved in regulating enzyme activity and in interactions with other proteins and effector molecules. Alignment of the amino acid sequences of the catalytic domains identifies short regions and single amino acids which are highly conserved (Hanks *et al.*, 1988). Some of these amino acids are characteristic of different sub-families, for example, distinguishing between serine/threonine and tyrosine kinases, or between cyclic nucleotide-dependent protein kinases and the protein kinase C family. DNA sequences encoding these short regions of amino acid homology can be identified by cross-hybridization with synthetic oligonucleotides designed to target such regions. In this manner, previously uncharacterized protein kinases can be identified from the nucleic acid sequences in which they are encoded. Oligonucleotide probes corresponding to conserved regions have been used to identify novel protein kinases in both mammals (Hanks, 1987) and in yeast (Levin *et al.*, 1987).

If genomic DNA sequences are screened using this approach, then potentially all protein kinases are sampled, as each sequence is equally represented. On the other hand, protein kinases which are expressed in specific tissues may be identified by screening sequences contained within cDNA libraries. In addition, the specificity of the screen can be tuned by tailoring the oligonucleotide probes to encompass particular protein kinase sub-families and by precisely controlling the conditions of hybridization.

ISOLATION OF PLANT PROTEIN KINASE HOMOLOGS

We have used this strategy to isolate plant cDNAs encoding potential protein kinases (Lawton *et al.*, 1989). Since this method requires only the conservation of the catalytic domain, it seems particularly well suited to the isolation of plant protein kinases, whose regulatory properties are generally not well characterized. Using the method described by Hanks

(1987), we have isolated cDNA clones encoding plant protein kinase homologs in the dicot bean (*Phaseolus vulgaris*) and the monocot rice (*Oryzae sativa*) (Lawton *et al.*, 1989). Two discrete amino acid sequences representing highly conserved regions within the catalytic domains of eukaryotic protein kinases were chosen as targets for screening (Figure 2). Two partially degenerate oligonucleotide probes were synthesized to represent codons most likely to encode the respective amino acid target regions. The degenerate oligonucleotides were used to screen a plasmid based cDNA library corresponding to polyadenylated RNA isolated from suspension cultured bean cells 38 hours after elicitor treatment. To select against false positives, clones were sequentially probed with the partially degenerate oligonucleotides corresponding to the amino acid sequences DLKPEN and GTPEYLAPE. Since cDNA synthesis proceeds in a 5' to 3' direction, true homologs isolated using the primary DLKPEN probe should also contain sequences for GTPEYLAPE, assuming conservation in these regions and complete second strand cDNA synthesis. From a sample of 19 colonies that hybridized with the DLKPEN probe, 4 hybridized to the secondary GTPEYLAPE probe. These colonies were identical as judged by the size of insert and restriction enzyme digestion pattern, and one clone PVPK-1 (*Phaseolus vulgaris* protein kinase-1) was chosen for further study.

Sequence analysis of the entire 2.2 kb PVPK-1 clone revealed a 609 amino acid open reading frame, from nucleotides 217 to 2036. The region upstream of this translational unit contained stop codons in all reading frames, suggesting that the PVPK-1 clone contained a complete polypeptide coding region. A 1.1 kb *Acc* I fragment from PVPK-1 containing the target regions was used to screen a monocot cDNA library made from leaves of dark-grown rice seedlings. One clone, G11A, hybridized strongly to the PVPK-1 probe and contained a 1.6 kb insert encoding a highly related polypeptide. G11A contains an open reading frame encoding a 536 amino acid polypeptide which

PANEL A

```
                PRIMARY TARGET                      SECONDARY TARGET

cAPK   (161) DLIYRDLKPENLLIDQ.......(195) TWTLCGTPEYLAPEIILSK
cGPK   (478) GIIYRDLKPENLILDH.......(514) TWTFCGTPEYVAPEIILNK
PhK-γ  (144) NIVHRDLKPENILLDD.......(180) LREVCGTPSYLAPEIIECS
MLCK   (177) RVLHLDLKPENILCVN.......(215) LKVNFGTPEFLSPEVVNYD
consensus         DLKPEN                          GTPEYLAPE
```

PANEL B

```
Consensus motif        -Asp-Leu-Lys-Pro-Glu-Asn-

Possible codons     5' -GAC-CUN-AAA-CCN-GAA-AAC- 3'
                          U UUA   G       G   U
                              G

Selected codons     5' -GAC-CUC-AAA-CCN-GAA-AA-  3'
                          U   G   G       G

cDNA probes         3' -CTG-GAG-TTT-GGN-CTT-TT-  5'
                          A   C   C       C
```

PANEL C

```
Consensus motif        -Gly-Thr-Pro-Glu-Tyr-Leu-Ala-Pro-Glu-

Possible codons     5' -GGN-ACN-CCN-GAA-UAC-CUN-GCN-CCN-GAA- 3'
                                      G   U UUA           G
                                              G

Selected codons     5' -GGC-ACC-CCC-GAG-UAC-CUC-GCA-CCN-GA-  3'
                                  U       U   G   C
                                                  U

cDNA probes         3' -CCG-TGG-GGG-CTC-ATG-GAG-CGT-GGN-CT-  5'
                                  A       A   C   G
                                                  A
```

Figure 2. (A) Alignment of amino acid sequences from the catalytic domains of four mammalian protein-serine kinases reveals two separate highly conserved regions. The sequences shown are; cAPK, cyclic AMP-dependent protein kinase from bovine cardiac muscle; cGPK, cyclic GMP-dependent kinase from bovine lung; PhK-γ, phosphorylase kinase γ-subunit from rabbit skeletal muscle; MLCK, myosin light chain kinase from rabbit skeletal muscle. The single-letter amino acid code is used and only the conserved stretches and the surrounding residues of each protein are shown. The two consensus sequences chosen for probe targeting are shown aligned below four kinase sequences. (B and C) Oligonucleotide probe design. Figure adapted from Hanks, (1987) extends throughout the cDNA, indicating that it is incomplete at both the 5' and 3' ends.

Although PVPK-1 and G11A contain the target sequence DLKPEN, in both cDNAs the regions homologous to the secondary probe corresponds to the sequence GTHEYLAPE rather than the target GTPEYLAPE. These sequences fall within a region of 410 amino acids that is highly conserved between PVPK-1 and G11A, however, both the C-terminal and N-flanking regions are divergent.

Much of the region conserved between PVPK-1 and G11A is homologous to the catalytic domains of eukaryotic protein kinases and the deduced plant polypeptides contain all but one of the characteristic features of these enzymes (Taylor, 1987). These include (relative to the residue numbers for the catalytic subunit for bovine cAMP-dependent protein kinase) amino acids corresponding to the motif Gly^{50}-Xaa-Gly^{52}-Xaa-Xaa-Gly^{55}, which forms part of the ATP binding region; Lys^{72}, which is implicated in ATP binding and the phosphotransfer reaction; Asp^{166} and Asn^{171}, which form part of the DLKPEN motif and are also implicated in ATP binding; Asp^{184}, whose chemical modification by carbodiimide destroys catalytic activity; Ala^{206} and Glu^{208}, which are diagnostic for the catalytic domain, and reside within the GTHEYLAPE motif.

Both cDNA sequences also contain the invariant or nearly invariant amino acids corresponding to Val^{57}, Ala^{70}, Glu^{91}, Asp^{220}, Gly^{225} and Arg^{280}, whose functions have not been defined. The one invariant amino acid not conserved in the plant homologs is Gly^{168}. In both PVPK-1 and G11A, Gly^{168}, which in other protein kinases forms part of the nearly invariant sequence $Asp^{184}Phe^{185}Gly^{186}$. In PVPK-1 and G11A, Gly^{186} is replaced by an aspartate residue. Additionally the residue corresponding to Thr^{197}, which in cAMP-dependent kinase is autophosphorylated, and whose mutation to a glycine leads to loss of cAMP regulation, is replaced by a serine, which could fulfill a similar function.

The most striking differences between the catalytic domains of the plant homologs and the catalytic domains of known and suspected protein kinases is the presence of an approximately 80 amino acid insert between the regions encoding DLKPEN and GTHEYLAPE. This region, which is conserved between the two plant homologs, is rich in lysine and contains five tandem copies of a motif whose consensus is C(V/I)(Q/E)P(S/T). The protein kinase encoded by the *Saccharomyces cerevisiae* *CDC7* gene contains an unrelated insert of similar size, indicating that additional sequences can be accommodated in this region without destroying catalytic activity. Presumably, these sequences do not alter the configuration of the bordering conserved regions implicated in catalysis. It is impelling to speculate that such an insert region may have some role in the regulating the catalytic activity of the PVPK-1 protein kinase. In particular, the presence of regularly spaced cysteine residues is reminiscent of metal-coordinating proteins, such as ferredoxin.

Comparison of the putative catalytic domains of PVPK-1 and G11A with the catalytic domains of other protein kinases using an unrooted tree building concept (Hanks *et al.*, 1988) revealed that the plant homologs are more related to each other than to any other known protein kinase, and that both sequences fall into a branch cluster of the tree that includes both the cyclic nucleotide dependent protein kinases and the protein kinase C family. In contrast, the regions flanking the putative catalytic domains of PVPK-1 and G11A are unrelated to each other and to known protein kinases. By analogy with other eukaryotic protein kinases the regions flanking the catalytic regions may be involved in regulating kinase activity.

A clue to the possible functions of PVPK-1 and G11A comes from the close similarity of the catalytic domains to those of the cyclic nucleotide-dependent protein kinases, which are involved in the transduction of extracellular signals. The PVPK-1 and G11A kinases may have equivalent functions in plant cells, with similar outputs in terms of target substrate specificities. However, physiological studies to date indicate that cAMP and cGMP do not have major regulatory roles in higher plants (Ranjeva *et al.*, 1987). This information is consistent with the high conservation of the catalytic domains but not the regulatory domains to the animal enzymes. These results suggest that plant protein kinases, like their counterparts in mammals, insects and lower eukaryotes, are modular proteins consisting of a conserved catalytic domain attached to divergent regulatory domains specialized for binding specific effector ligands or interacting with other macromolecules.

Southern blots of restriction enzyme digested genomic DNA from bean and rice were probed, respectively, with the cDNA inserts isolated from PVPK-1 and G11A. Hybridization at high stringencies revealed only one strongly hybridizing band and one or two weakly hybridizing bands depending on the restriction enzyme used. At low stringencies, additional bands hybridized to the PVPK-1 probe, suggesting that these cDNAs correspond to single copy genes that form part of a larger gene family.

ISOLATION OF PROTEIN KINASE HOMOLOGS USING THE POLYMERASE CHAIN REACTION

The modular nature of protein kinases allowed the isolation of plant genes encoding serine/threonine kinase homologs independent of their regulatory properties by use of sequential probing with partially degenerate oligonucleotides that span large evolutionary distances. Using a variation of this approach, based on amplification by the polymerase chain reaction, several additional sequences related to protein kinases have been isolated. This approach again requires the use of two degenerate oligonucleotide probes corresponding to the conserved amino acid regions within the protein kinase catalytic domain. To function in a polymerase chain reaction, the oligonucleotide probes must be homologous to opposite strands of a kinase clone and hybridize in such a manner that their 3' ends are proximal to each other. This gives a second level of specificity in the screening for protein kinase sequences as selection is based not only on sequence specificity but also on oligonucleotide orientation. In consecutive rounds of the polymerase chain reaction, the protein kinase sequences flanked by the degenerate oligonucleotide probes are amplified greatly, substantially reducing the background of sequences that recognize only a single oligonucleotide. A third level of specificity is provided by selecting for products that are of a size expected for the amplified product.

This approach was applied to total genomic DNA derived from *Arabidopsis thaliana,* chosen for its small genome size. Figure 3 shows the results of an amplification using two sets of oligonucleotides corresponding to different classes of protein kinase. Blotting of the gel and probing the filter with PVPK-1 revealed a highly homologous sequence predominant within one band amplified with oligonucleotides A and B. Since these oligonucleotides incorporate the sequences present in PVPK-1 this suggests firstly that *Arabidopsis,* like rice, contains sequences homologous to PVPK-1, and secondly, that the conditions of the reaction are appropriate for the amplification of protein kinase homologs. Oligonucleotides A and C, which should amplify sequences corresponding to the mammalian protein kinase C family, gave rise to four major bands. A small amount of PVPK-1 cross-hybridizing material was present in the amplification products and could represent either the low efficiency generation of the PVPK-1-like product by these oligonucleotides or the amplification of a weakly related sequence.

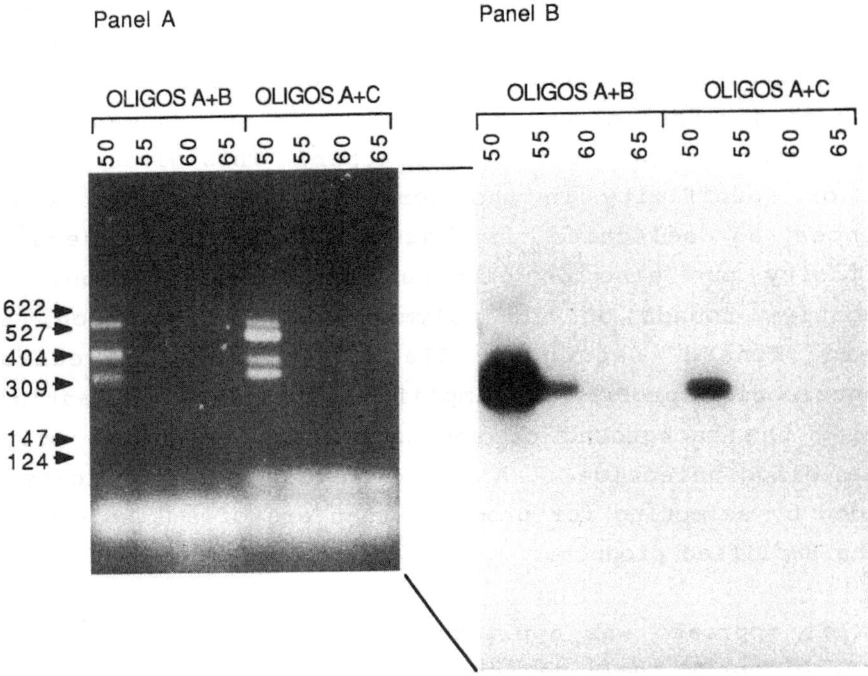

Figure 3. Direct amplification of protein kinase homologs using the polymerase chain reaction. Panel A shows the ethidium bromide stained gel of products amplified from *Arabidopsis thaliana* DNA using the indicated combinations of oligonucleotides and annealing temperatures. Panel B shows an autoradiogram of the corresponding filter probed with radiolabeled PVPK-1 sequences

We have now cloned and sequenced some of these products and have established that the size heterogeneity of the amplification products results from the presence or absence of either introns or coding sequences (as per PVPK-1). Using this approach, we have isolated protein kinase homologs that are:(i) related to additional members of the PVPK-1/G11A class; (ii) closely related to the protein kinase C and cyclic nucleotide dependent protein kinases (definitive identification of these products awaits the sequence determination of flanking regions); (c) related to the budding yeast *CDC28* and human cdc2hs P^{34} gene products that form part of the mitotic oscillator responsible for regulating the progression of the cell cycle (Dunphy and Newport, 1988). A similar approach has been adopted for the isolation of other regulatory proteins, including plant sequences homologous to the *Schizosaccharomyces pombe CDC25* and *Drosophila melanogaster String* genes, which encode positive regulators of the mitotic promoting factor (Russell *et al.*, 1989)

CONCLUSIONS AND PROSPECTS

The isolation of cloned plant sequences homologous to known eukaryotic protein kinases now extends the family of related protein kinases to include the plant kingdom. Clearly, those features that typify the larger family of plant protein kinases will become apparent as additional sequences are isolated and their sequences determined. Although homology probing provides a powerful tool for the isolation of novel protein kinase sequences, there are certain drawbacks associated with this approach. The independence of the method with respect to expression necessarily means that nothing is initially known about tissue specificity of the cloned kinases sequences. Further, no information is available concerning the regulatory and catalytic properties of the enzyme or its phosphorylation targets within the cell. Nonetheless, the availability of cloned sequences encoding these enzymes will facilitate their genetic manipulation and expression in both heterologous systems and in transgenic plants. Analogous approaches have proven extremely powerful in the molecular dissection of regulatory proteins in mammals and there is every reason to suppose that similar methods will provide insight into the functions and activities of plant protein kinases.

Accordingly, protein kinases can be expressed in bacterial, yeast or insect systems to produce both active enzyme suitable for biochemical studies, and antigen for generating monospecific antiserum for isolation of the native protein. Similarly, the expression of particular protein kinases *in planta* can be altered by placing these genes under the control of strong constitutive or regulated promoters. Alternatively, it should be possible to down-regulate endogenous activities through the expression of either anti-sense RNA or dominant negative mutants (Herskowitz, 1987). The use of the latter is particularly appropriate for the dissection of protein kinase function since the modular construction of the protein kinases potentially allows the expression and manipulation of individual catalytic and regulatory domains, while the wealth of biochemical information concerning the catalytic domains of mammalian protein kinases (especially cAMP-dependent protein kinase) can be used to express mutant proteins deficient in catalysis but still able to interact with other substrates and effectors. Phenotypes associated with the altered expression of endogenous protein kinases may provide important clues as to the natural function of the enzyme. By a combination of these and other approaches we can begin to determine the regulatory circuits in which these enzymes serve, and define the molecular components with which they interact.

ACKNOWLEDGEMENTS

This work was supported by NSF grant DCB-8945970, The Samuel Roberts Noble Foundation and The Rockefeller Foundation. S.H.Z. is in the Graduate Program in Chemistry, University of California, San Diego, La Jolla, CA 92037 USA.

REFERENCES

Bennett J (1979) Chloroplast phosphoproteins. The protein kinase of thylakoid membranes is light-dependent. FEBS Lett. 103:342-344.

Dunphy WG and Newport JW (1988) Unraveling mitotic control mechanisms. Cell 55:925-928.

Hanks SK (1987) Homology probing: Identification of cDNA clones encoding members of the protein-serine kinases family. Proc. Natl. Acad. Sci. USA. 84:388-392.

Hanks SK, Quinn AM and Hunter T (1988) The protein kinase family: Conserved features and deduced phylogeny of the catalytic domains. Science 241:42-52.

Herskowitz I (1987) Functional inactivation of genes by dominant negative mutations. Nature 329:219-222.

Hiddinga HJ, Jessen Crum C, Hu J, Roth DA (1988) Viroid - induced phosphorylation of a host protein related to a ds-RNA dependent protein kinase. Science 241:451-453.

Hunter T (1987) A thousand and one protein kinases. Cell 50:823-829.

Krebs EG (1985) The phosphorylation of proteins: a mechanism for biological regulation. Biochem. Soc. Trans. 12:813.

Lawton MA, Yamamoto RT, Hanks SK and Lamb CJ (1989) Molecular cloning of plant transcripts encoding protein kinase homologs. Proc. Natl. Acad. Sci. USA. 86:3140-3144.

Levin DFE, Hammond CI, Ralston RO and Bishop JM. (1987) Two genes that encode unusual protein kinases. Proc. Natl. Acad. Sci. USA. 84:6035-6039.

Ranjeva R, Graziana A, Ranty B, Cavalie G and Boudet AM (1984) Phosphorylation of proteins in plants: A step in integration of extra and intracellular stimuli? Physiol. Veg. 22:365-376.

Ranjeva R and Boudet AM (1987) Phosphorylation of proteins in plants: Regulatory effects and potential involvement in stimulus/response coupling. Ann. Rev. Plant Physiol. 38:73-93.

Russell P, Moreno S, Reed SI (1989) Conservation of mitotic controls in fission and budding yeasts. Cell 57:295-303.

Suzuki H and Verma DPS (1989) Nodule-specific kinase phosphorylating nuclear factors present in isolated nuclei. The Plant Cell 1:373-379.

Taylor SS (1987) Protein kinases: a diverse family of related proteins. Bioessays 7:24-29.

SIGNAL TRANSDUCTION MECHANISMS IN PHYTOCHROME ACTION

C. B. Johnson
Plant Science Laboratories
University of Reading
Whiteknights
Reading RG6 2AS
United Kingdom

Phytochrome is a chromoprotein of molecular weight approximately 120-125 000 (depending on the source) with an open chain tetrapyrrole chromophore. There are two stable forms of the molecule _in vivo_. One, known as Pr, has an A_{max} at around 660nm: phytochrome is synthesised in this form. Absorption of light by Pr results in phototransformation via a series of intermediates to an equilibrium mixture in which the other stable form, Pfr, which has an A_{max} at ca. 730nm predominates. This form in turn is photoconvertible back to the Pr form so that light of any spectral quality, monochromatic or broad band, results in the formation of an equilibrium mixture of the two forms which depends on the spectral quality. The range of variation in natural daylight results in maximal variation in this equilibrium, so that phytochrome is an excellent sensor of daylight quality (Figure 1). The function of phytochrome is the detection of such changes and redirection of plant development accordingly: the collective name given to these responses is photomorphogenesis.

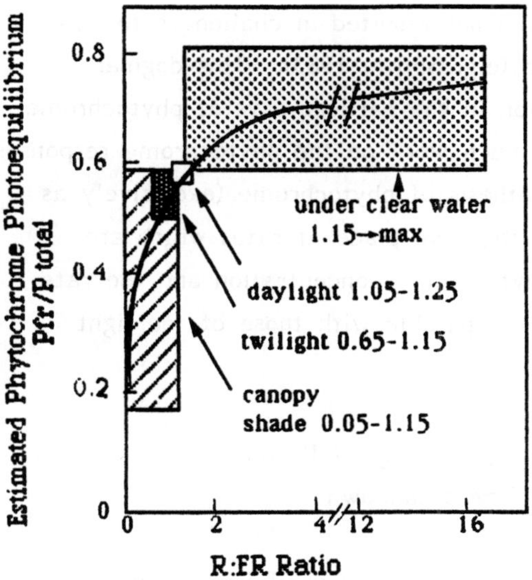

Figure 1. The relationship between light quality, defined here in terms of the red:far-red ratio, and the estimated phytochrome photoequilibrium, indicating some of the ranges found in nature under the conditions indicated. (From Smith, 1982)

So much is clear but from this point on, from the nature of the active form of phytochrome to the mechanisms underlying the phytochrome control of gene expression in photomorphogenesis, almost every step involves some degree of controversy; a sign of course of lack of definitive evidence concerning each stage of the chain of signal perception and transduction.

NATO ASI Series, Vol. H 47
Signal Perception and Transduction in Higher Plants
Edited by R. Ranjeva and A. M. Boudet

THE ACTIVE FORM OF PHYTOCHROME

Biological responses are usually but not always functions of an increasing concentration or proportion of Pfr. For example, seedlings previously grown in the dark and given short pulses of light typically produce greater responses as the Pfr concentration is increased. The response function is often quite complex; a substantial response to low Pfr levels typically followed by one or more phases of lower responsivity (Figure 2a). After prolonged irradiation the action maximum for the responses shifts towards the far-red and becomes irradiance dependent; the so-called 'High Irradiance Response' (Figure 2b). In green plants which have been grown for prolonged periods in the light the response characteristics are often different and usually an approximately linear relationship obtains between the extent of the response and the photoequilibrium of phytochrome, defined as [Pfr]/[Pr]+[Pfr] and modified by changing the ratio of red to far red light in the incident radiation (Figure 2c).

From the earliest days, phytochrome research has been based on the notion of Pfr as the active form of phytochrome, Pr as inactive, thus;

$$\text{Pr} \rightleftharpoons \text{Pfr} \longrightarrow \text{Biological Responses}$$

The strong correlations frequently found in light-grown plants between response and phytochrome photoequilibrium has resulted in challenges to this 'central dogma' of phytochrome research (e.g. Smith, 1983). This dogma, originating as no more than an assumption in the earliest days of phytochrome research had been largely borne out by much work on the phytochrome responses of etiolated seedlings. For example, synthesis of phytochrome (exclusively as Pr) and turnover (almost exclusively of Pfr) take place at rates which are significant as a function of the total phytochrome concentration and the rate constants for these processes are often comparable with those of the light reactions themselves, thus;

$$\text{synthesis} \longrightarrow \text{Pr} \rightleftharpoons \text{Pfr} \begin{cases} \nearrow \text{Biological Responses} \\ \searrow \text{breakdown} \end{cases}$$

As a result, after prolonged irradiation, wavelengths establishing the highest phytochrome photoequilibria do not yield the highest values of Pfr. However, the wavelengths that do establish the highest Pfr values do become

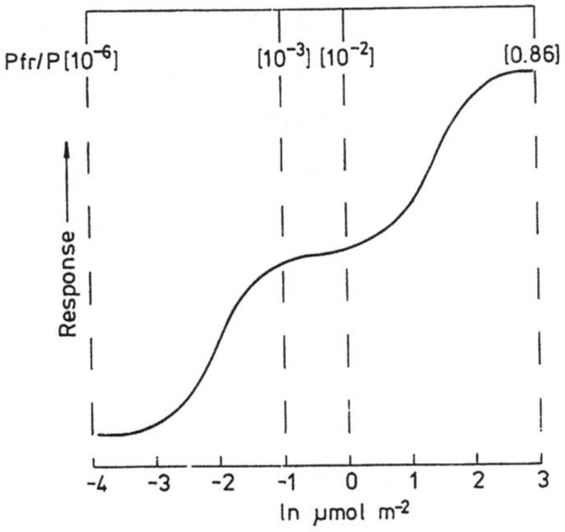

Figure 2a. Illustration of the typical phytochrome fluence responses to short irradiations with red light. On the lower x-axis is given the fluence, plotted on a logarithmic scale: an indication of the Pfr formed (expresssed as the ratio Pfr/Pr+Pfr) is given above (from Kendrick & Kronenberg, 1986)

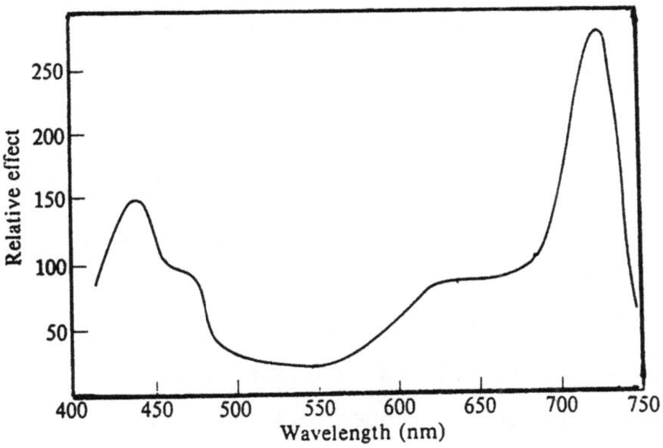

Figure 2b. Generalized 'High Irradiance Response' of phytochrome. The relative effectiveness of each wavelength is presented as the reciprocal of the fluence rate required to promote 50% of the maximum response (from Mohr, 1964)

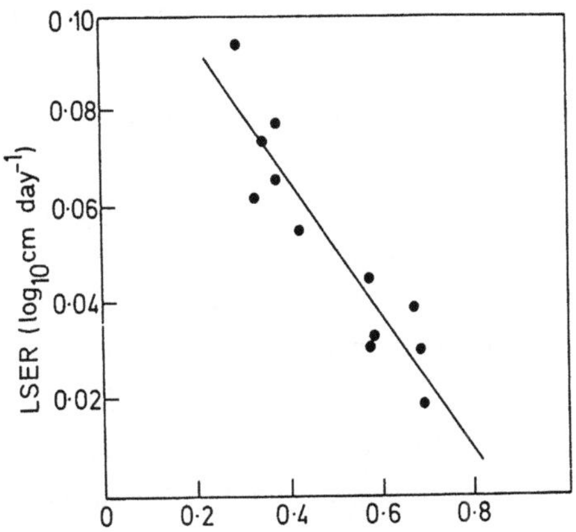

Figure 2c. The relationship between internode elongation and the estimated phytochrome photo-equilibrium (Pfr/Pr+Pr) in *Chenopodium album* L. The photo-equilibrium is altered by progressive addition of far-red light to a constant background of white light provided by fluorescent tubes (from Morgan & Smith, 1976)

optimal for phytochrome action in the High Irradiance Response (even though it is not possible to explain the High Irradiance Response entirely in terms of [Pfr]; see later for a discussion of this). Perhaps, it might be argued, Pfr is the active component in etiolated seedlings whereas light-grown plants monitor the ratio Pfr/Pr. This idea had seemingly little in its favour until it received a revival with the discovery that phytochrome in light-grown plants was not only immunologically different from 'dark' phytochrome but was a different gene product. The two forms of phytochrome are now commonly referred to as 'Type I' (dark) and 'Type II' (green plant) phytochrome. As well as relatively small differences in the spectral properties between the two types, the kinetics of Type II phytochrome differ significantly in that rates of synthesis and degradation are negligible, both Pr and Pfr apparently being very stable. Since these differences offer a ready explanation for the different response characteristics of seedlings and light-grown, green plants it is usually assumed that Type I phytochrome is responsible for the characteristic responses of etiolated seedlings, Type II for those of green plants. I shall return to this at the end of the paper.

Could the active component of the two types of phytochrome be different? Are Type I responses restricted to seedlings, Type II to green plants? How justified are we in seeking a single route of signal transduction from phytochrome to photoresponse? The enormous diversity of reported responses has encouraged generalizations to be made about these questions, which are worth examining in more detail. First, let us consider the exclusivity or otherwise of the relationship between Type I phytochrome and etiolated seedlings, and Type II with green plants. Such a distinction would itself help to clarify the advisability of looking for a single primary mechanism of action of phytochrome, a much vaunted but so far elusive goal of phytochrome research. Two examples which argue against such exclusivity are discussed below.

EVIDENCE FOR ACTION OF TYPE I PHYTOCHROME IN LIGHT GROWN PLANTS

Flowering in wheat is dependent on long days. The photoreceptor is phytochrome. Plants grown with 8h days do not flower; those given 8h white light together with an 8h day extension of monochromatic light do so to varying degrees according to the wavelength and fluence rate of the day-extension light treatment (Carr-Smith, Johnson & Thomas, 1989). The wavelength and fluence rate dependence of the response are identical in characteristics with the High Irradiance Response of etiolated seedlings (Figure 3).

Furthermore, immunological assay of phytochrome using monoclonal antibodies specific for Type I phytochrome reveals a pool of this type of phytochrome with the expected kinetic properties, e.g. fast synthesis of Pr, rapid turnover of Pfr (unpublished results of H. Carr-Smith). It is difficult to reach any conclusion other than that this response in a mature green plant is controlled by Type I (seedling) phytochrome.

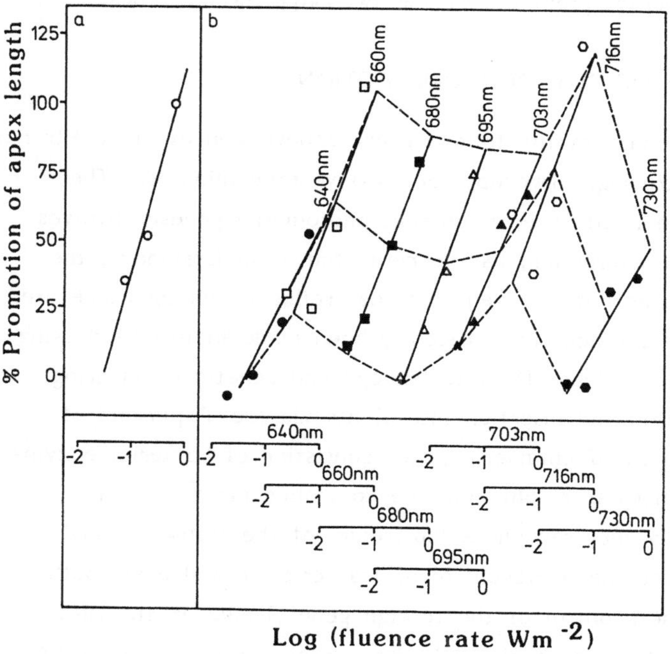

Figure 3.

Fluence-rate response curves for promotion of floral development in wheat by day-extensions of (a) white light (b) monochromatic light. The broken lines join points of equal fluence rates. The plants had been grown in short-days and were given 4 day-extensions of the wavelengths and fluence rates indicated before being returned to short days until harvest

EVIDENCE FOR ACTION OF TYPE II PHYTOCHROME IN SEEDLINGS

It is much harder to be sure that Type II phytochrome is active in seedlings even though it can clearly be shown to be present in the seed. However, one interesting case suggests that it may be. Bartley and Frankland (1982) have shown that germination in Sinapis arvensis is inhibited in a fluence- and wavelength-dependent manner by prolonged irradiation, the inhibition showing the characteristic action maximum and irradiance dependence of the High Irradiance Response. This photo-inhibition of germination is compounded with photo-promotion of germination which is a function of the photo-equilibrium established by the incident light; i.e. identical to the characteristics of light-grown plants discussed above. The simplest interpretation of these data (in the view of the author) is that it is the Type II phytochrome which is responsible for the promotion, Type I for inhibition.

Taken together these two sets of experiments support the view that Type I and Type II phytochrome are not exclusive in their action to etiolated and green plants respectively. It may be noted that this interpretation allows the retention of the notion that Pfr is the active component for Type I promoted responses, and that somehow the ratio Pfr/Pr+Pr acts as the sensor for Type II. This in turn leaves open the possibility that a different mechanism of action applies to responses promoted by the two phytochrome types.

PRIMARY MECHANISMS OF PHYTOCHROME ACTION

The search for the primary events following photoconversion of Pr to Pfr has been intriguing but frustrating. Two approaches have predominated. The first, based on the self-evident observation that photomorphogenesis involves selective control of gene expression, has concentrated on understanding the mechanisms whereby the activities of enzymes controlled by phytochrome are regulated. Considerable advances have recently been made in this field using the tools provided by recombinant DNA technology and genetic engineering. In particular, the construction and insertion into plants via the Ti-plasmid of *Agrobacterium tumefaciens*, of chimaeric genes consisting of bacterial enzymes joined to flanking sequences from phytochrome controlled plant genes has demonstrated conclusively that sequences 'upstream' of the transcriptional start site are required for the photocontrol of transcription and exert both positive and negative photocontrol of the foreign gene (Timko et al., 1985; Morelli et al., 1985; Castresana et al, 1988: for a review see Jenkins, 1988). *Trans*-acting protein factors which bind conserved portions of the upstream regulatory DNA sequences have been isolated (Green, Kay and Chua, 1987; Giuliano et al., 1988) but no evidence has emerged to support the view that phytochrome itself could be one of these. The possibility remains that there is a direct interaction between phytochrome and the DNA-binding proteins but, all in all, it now seems unlikely that *qualitative* photocontrol of gene expression, exerted through the action of these 'upstream' regulatory sequences, is a direct phytochrome-genomic interaction.

On the other hand, the alternative approach, a study of the events most immediately following Pfr formation has not yet led to the discovery of any factors which themselves might interact with the *cis*-acting regulatory DNA sequences. In this regard the situation has not changed fundamentally since Quail produced an extensive review of the most rapid effects attributable to phytochrome (Quail, 1983). He concluded that although the earliest effects specifically controlled by phytochrome appeared to affect membrane properties, "there was no rigorous evidence that the alterations result from direct, physical interaction of phytochrome with these membranes." Still less is there direct evidence that such alterations lead to changes in gene expression via the mechanisms described in the previous section, though, of course, the difficulties involved in obtaining decisive evidence on this sequence of events should not be overestimated. Here, I have concentrated on some early effects which either appear to have the potential to lead to morphogenetic effects or, in one unusual case, where there is evidence of a more or less complete signal transduction chain. Omission of some of the fastest known phytochrome-mediated effects on membranes is justified only on the grounds that in several years that have elapsed since the initial observations nothing has yet emerged linking them to possible subsequent events in a signal transduction chain. Included in this category are the various reports of phytochrome effect on membrane potential, the fastest of which (described by Racusen, 1976) takes place at a rate which is almost inseparable from Pfr formation.

PELLETABILITY OF PHYTOCHROME

More difficult to eliminate as a candidate for a primary role in phytochrome action, even though it too has not been decisively associated with the final display of any phytochrome response, is the phenomenon of Pfr-mediated pelletability of phytochrome (Quail, Marme & Schäfer, 1973). The essential observation of this phenomenon, that phytochrome, predominantly soluble as Pr, becomes associated with a pelletable, possibly membranous, fraction after conversion to Pfr, enables a rational explanation to be made of the problems arising from the contradiction between the numerous observations of early phytochrome-mediated events involving changes in membrane properties, and the readily demonstrable solubility of the majority of the phytochrome in the cell. Furthermore, the kinetics of the event, rapid Pfr promoted pelletability, slow Pr promoted release, have been shown to be consistent with the wavelength and fluence rate optima of the High Irradiance Response (Schäfer, 1975). Unlike many of the other 'early' responses, pelletability has so far been found to occur in all species in which it has been sought. Finally, electron microscopy, combined with various immunological techniques has similarly confirmed a redistribution of phytochrome in the cell from more diffuse to more discrete following a short saturating red-light pulse (for example, see Epel, Butler, Pratt & Tokuyashu, 1980), showing that the response observed in the test tube has a counterpart in vivo. Despite all these observations, the lack of decisive evidence linking pelletability with any physiological response must leave open the question of its biological significance. The elusive process, if it exists, must rapidly precede the post-translational modification of Pfr which occurs immediately following aggregation, leading to formation of ubiquitin-phytochrome conjugates prior to the well-documented breakdown of Pfr (Shanklin, Jabben & Vierstra, 1987). It is difficult to believe however that the sole purpose of Pfr aggregation is to instigate its own breakdown.

PHYTOCHROME CONTROL OF GIBBERELLIN EFFLUX

Another phenomenon linking phytochrome with regulation of membrane properties is that of the photocontrol in vitro of gibberellin efflux from etioplasts (Evans & Smith, 1975; Hilton & Smith, 1980). Leaves of dark-grown cereals unroll when placed either in the light, in the dark following a pulse of red light (far-red reversible) or following addition of GA_3. Hilton & Smith showed that a purified etioplast fraction from the dark-grown cereal leaves released gibberellin-like substances within minutes of a pulse of red light, a result which had previously been found with 'crude' leaf extracts. The interest of this example lies with the similarity of the photoresponse to the effect of applied gibberellin. The notion that phytochrome might act through the inter-mediary of plant growth substances has often been advanced and these findings are consistent with that view, one which has not, however, been supported by two-factor analysis experiments (e.g. Bühler, Drumm & Mohr, 1978a, 1978b; Warner, Ross & Coombs, 1981). Smith, Evans & Hilton (1978) also found that a small portion of the total cellular phytochrome was associated with the etioplast fraction and, in an interesting analogy to the 'pelletability' phenomenon discussed above, found that the etioplast membrane-associated phytochrome doubled after irradiation with red light. It is not known whether this phytochrome is Type I or Type II, nor that it is involved in photocontrol of this response. Recently however, Viner, Whitelam & Smith (1988) have shown that calcium is involved in the response; the promotion by red light being blocked by EGTA but restored on addition of calcium. Calcium links this example with the next, in which definitive evidence for the involvement of both calcium and membrane-associated phytochrome is available.

PHYTOCHROME LOCALIZATION AND PHYTOCHROME ACTION IN MOUGEOTIA

The photocontrol of chloroplast orientation in the filamentous alga Mougeotia, first reported in 1958 by Haupt (Haupt, 1958) is an almost unique example of a phytochrome-mediated physiological response known to involve control by membrane-associated phytochrome. In Mougeotia, the orientation of the single flat chloroplast is controlled by phytochrome in such a way that the flat surface of the chloroplast faces the light following a pulse of red light, whereas a far-red light pulse leads to an edge-on orientation. By using very short irradiations with microbeams of light, Haupt has shown that the response depends on the polarity of the light in the irradiating beam. This must be parallel to the long axis of the cell for promotion and at right angles for reversal, demonstrating that the phytochrome controlling this response is located in or bound rigidly to the plasma membranes and undergoes a change of orientation upon conversion from Pr to Pfr and vice versa (Haupt & Weisenseel, 1976). Again, calcium is implicated in the response: it has been shown (Dreyer & Weisenseel, 1979) that Pfr formation is quickly (<3min.) followed by substantial accumulation of Ca^{2+}, either by selective transport through the plasma-membrane or possibly from calcium-binding vesicles (Grolig & Wagner, 1989). Wagner and Rossbacher (1980) have demonstrated that the chloroplast is connected to the plasma membrane by actomyosin filaments and it seems likely that orientation is accomplished by selective action of calcium on these filaments following a the establishment of a Pfr gradient. The calcium may act via calmodulin, which is known to activate myosin light chain kinase and is itself activated by Ca^{2+}. Wagner, Grolig and Altmüller (1987) have shown that the response is inhibited by inhibitors of calmodulin and actin. On the other hand, the involvement of calmodulin has not been unequivocally established and recent work (Kohno & Shimmen, 1987, 1988; Kachar & Reese, 1988) has shown that organellar movement in both pollen tubes and characean algal cells may be the result of the direct [Ca^{2+}] dependent activation or inactivation of actin microfilaments attached to the organelles.

Either way, it seems that the complete transduction chain for this phytochrome reponse is not far from elucidation. Thus, we reach one of the first certainties in our quest for phytochrome signal transduction chains: it is clear from these results that there is a phytochrome response involving calcium, membrane bound (and presumably Type II) phytochrome, and that this response does not involve changes in gene expression. How much of this chain, if any, could be common to other responses?

PHYTOCHROME AND RUN-ON TRANSCRIPTION

No less equivocal than the absence of gene expression in the previous example is the certainty that differential regulation of gene expression is at the heart of many photomorphogenetic responses. As discussed earlier, not all of the effects of phytochrome on gene transcription (and of course none of those involving post-transcriptional control) need involve direct interaction between phytochrome and the genome and the primary mechanism of action of these could in principle be the same as in Mougeotia. However, there have been several recent albeit controversial reports which suggest that a more direct action between phytochrome and the genome may occur. In the first of these, Ernst and Oesterhelt (1983) showed that purified phytochrome added to purified nuclei isolated from dark-grown rye seedlings in vitro, stimulated the rate of run-on transcription, measured by incorporation of ^{3}H-UTP into mRNA. Specificity for Pfr (compared with Pr) and for high-molecular weight, intact phytochrome was demonstrated. In 1985, Mösinger et al. showed that overall transcription in vitro was markedly stimulated by irradiation in vivo but that when transcription rates of two genes differential regulated by light in vivo were compared one of them, the light-harvesting chlorophyll a/b protein (LHCP) was enhanced in vitro as well as in vivo whereas NADPH protochlorophyllide oxidoreductase (reductase), which was decreased in vivo by light treatment, showed also a decreased rate of transcription in vitro. Mösinger et al. (1987) subsequently confirmed the findings of Ernst & Oesterhelt by showing that there is an overall stimulation of transcription in vitro in nuclei isolated from barley following addition of oat or rye Pfr to isolated nuclei. They were further able to show that, when purified oat Pfr was added, the transcription rates of LHCP were enhanced above the mean level whereas transcription of the reductase was depressed below the mean, confirming that the specificity of the direction of the photoresponse found in vivo, was fully preserved in vitro (Figure 4a).

Unlike Ernst & Oesterhelt they were only able to obtain successful in vitro control with previously red-irradiated nuclei. A surprising omission from these experiments was confirmation of the specificity of Pfr compared with Pr. For this reason we repeated the experiments in our own laboratory. Using purified high molecular weight wheat phytochrome and nuclei isolated from wheat we were readily able to confirm the basic observations (Figure 4b).

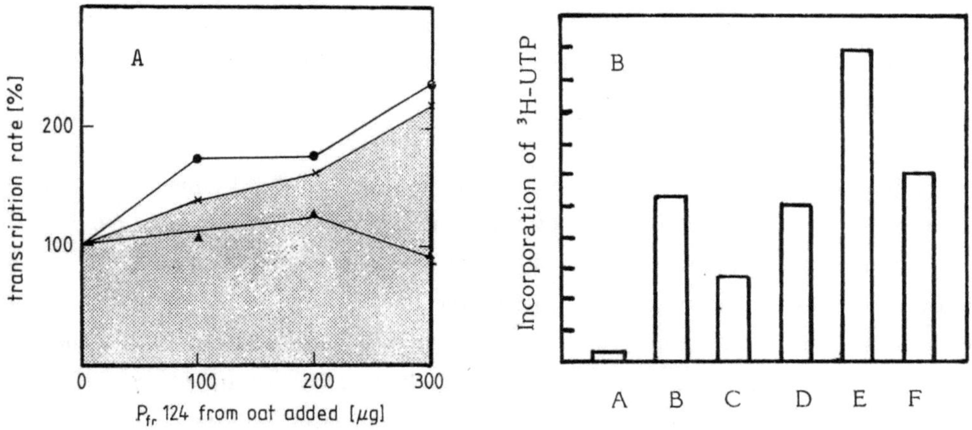

Figure 4. A: _In vitro_ transcription rate of isolated barley nuclei as affected by addition of Pfr. Crosses, overall transcription (shaded); circles, LHCP mRNA; triangles, reductase mRNA (from Mösinger et al.,1987). B: _In vitro_ transcription rate of isolated wheat nuclei after A, darkness; B & D, 10'R prior to extraction; C, 10'R:10FR; E, 10'R prior to extraction + 100μg wheat Pfr added _in vitro_; F, 10'R + 100μg Pr (unpublished results of C. Messham)

The specificity of the response, both with regard to Pfr compared with Pr, and with regard to the preservation _in vitro_ of both positive and negative control suggests that a specific interaction (not necessarily direct) between Pfr and the nucleus is occurring. The nature of the interaction is unclear at present. However, since transcription _in vitro_ is thought not to involve initiation of new transcripts but only completion of existing RNA chains, this probably eliminates any interaction between phytochrome and the upstream regulatatory sequences discussed earlier and represents a subsequent _quantitative_ phase in the regulation. This conclusion is consistent with results of Mösinger et al (1987) who found two phases to the photo-responses; a rapid and transient change promoted only by irradiation _in vivo_ (presumed to be related to 'upstream' regulation) and change in the duration and maximum of the response promoted both by irradiation _in vivo_ and by Pfr addition _in vitro_[2]. Recently Dean et al. (1989) have shown that in addition to the upstream regulatory sequences required for photocontrol of the small subunit of ribulose bisphosphate carboxylase, sequences 'downstream' of the translation 'stop' signal regulate quantitative expression of the gene. This regulation is retained in nuclear run-off experiments. These sequences are obvious candidates for investigation in respect of the _in vitro_ induced promotion of

2 Recent work of Lissemore & Quail (1989) using run-on transcription assays has shown that phytochrome regulation of transcription _in vitro_ differs in mechanism for two genes studied: induction of LHCP occurs much more slowly than repression of the phytochrome gene itself (phyt). Furthermore LHCP induction _in vitro_ requires much higher Pfr levels than phyt repression. Finally phyt repression is not inhibited _in vitro_ by inhibitors of protein synthesis, whereas LHCP induction is.

transcription, though it is not yet clear whether the sequences, which are similar to downstream 'enhancer' sequences found in animal cells, are subject to photocontrol.

Clearly, much more work is required in order to establish whether these sequences or any other components of the transcription machinery are involved in direct or indirect interaction with phytochrome within the nucleus. However, whatever the outcome of these experiments, one of the most important conclusions based on a comparison the two examples discussed above, and the justification for their otherwise disproportionately lengthy treatment, is the fact that their signal transducing pathways share virtually nothing in common. This makes it all but impossible to retain the notion of a single primary mode of action of phytochrome.

DIFFERENT MODES OF ACTION FOR TYPE I AND TYPE II PHYTOCHROME?

One possibility not ruled out by these experiments is that the transduction chains for Type I and Type II phytochrome are different. It is obvious that many of the characteristic changes which occur in dark grown plants upon exposure to light involve changes in gene expression, and many of these changes have been shown to involve transcriptional regulation. Indeed, Mohr (1983) attempted to restrict the term 'photomorphogenesis' to such responses, while those phenomena which were readily reversible and presumed not to involve differential gene expression he defined as 'photomodulations'. Chloroplast rotation in Mougeotia he included in this latter class, together with the most widely researched of the responses of light-grown plants known to be promoted by Type II phytochrome, the photo-inhibition of stem extension. This response has been shown to be extremely rapid and it has often been presumed that, like the Mougeotia response, it does not involve differential gene expression. Could 'photomorphogenesis' be controlled by Type I phytochrome and involve one primary transduction mechanism, with 'photomodulations' involving Type II phytochrome and a completely different transduction chain?

This distinction does not survive scrutiny. Firstly, with the advances in our understanding of gene regulation in recent years has come the appreciation that transcriptional regulation can be much more rapid than had hitherto been assumed. Hence, rapidity of response can no longer be assumed to rule out phytochrome action at the genome. Secondly, the photocontrol of elongation growth has proved to be a much more complicated phenomenon than originally

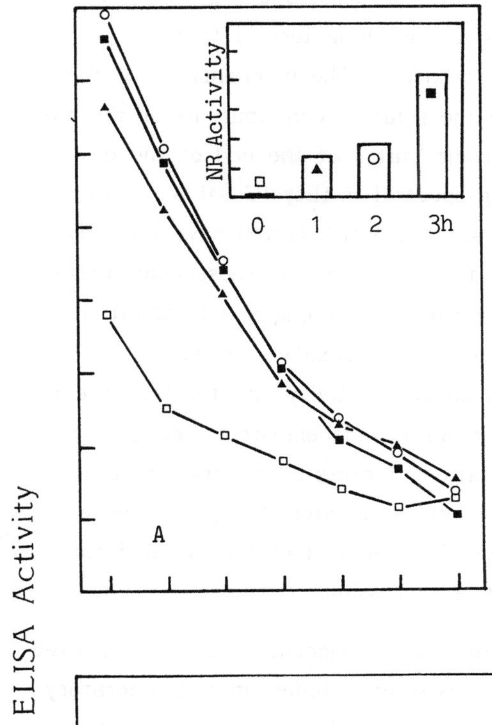

Figure 5. A comparison of nitrate reductase (NR) activity (inset) with NR activity measured by enzyme-linked immunosorbent assay (ELISA). Seedlings of Sinapis alba were grown for 66h on 1% agar. They were given three 10' pulses of red light 0, 8 & 16h prior to this. At 66h they were treated with 100mM KNO_3 except where otherwise stated. Both light and nitrate are required for NR induction at this stage

Fig 5a; Time course of induction following addition of KNO_3

Fig 5b; NR and ELISA activity 3h after addition of KNO_3. Where indicated, far-red light was given immediately after each 10' red pulse. Note that this reverses the effect of the red pulses with respect to NR activity but not NR protein

Fig 5c; As Fig 5b except that the red pulses were given at a high fluence rate for only 2min each. Induction of both NR activity and NR protein is reversed

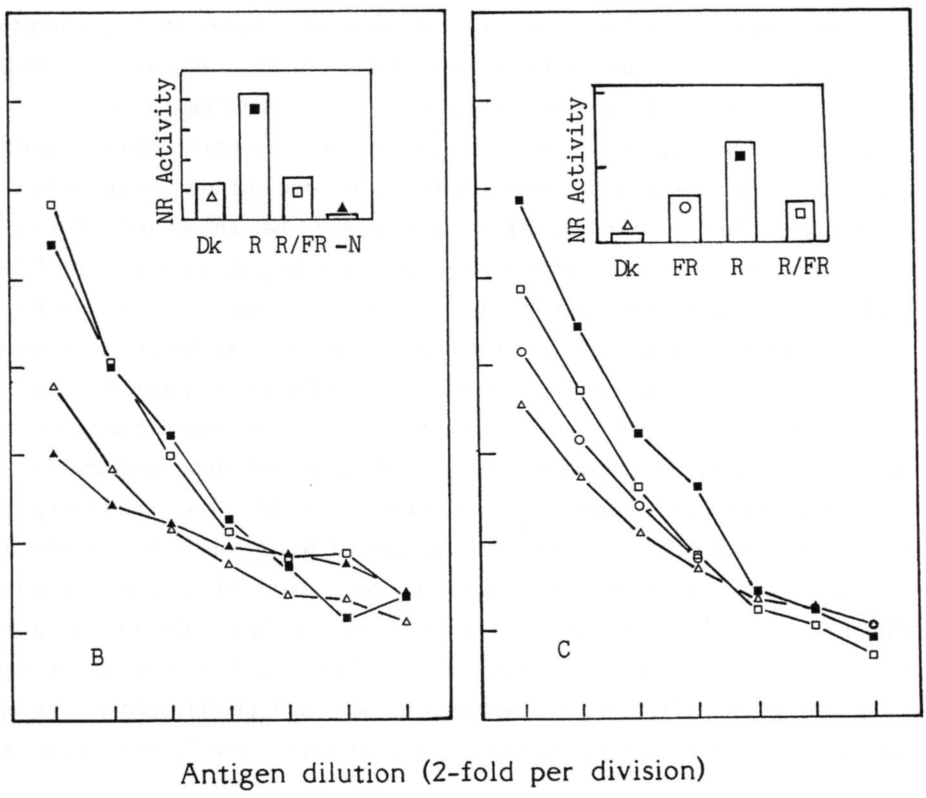

thought. Smith's group have shown that as well as a rapid effect of light quality, perceived in the internode there is a more long term effect, perceived in the leaves as well as in the internode. The mechanisms for the two components may not be the same and the relative contributions of the two differ for responses to changes in phytochrome status at the end of the day compared with continuous light of differing spectral quality (Casal & Smith, 1988). For the former case, Casal, Whitelam and Smith (1989) have recently demonstrated that the end-of-day effects of phytochrome on subsequent internode growth rates in the dark are correlated both in timing and localization with rapid changes in the activity of extracellular peroxidase activity. A causal relationship between the changes in enzyme activity and the response is not yet proven but, as the authors argue, there is independent evidence that extracellular peroxidases catalyse the formation of cross-links between cell wall matrix components, thereby regulating cell wall extensibility and hence extension growth. These data suggest that Type II phytochrome is able to promote responses involving differential gene expression.

Even more convincing evidence that phytochrome responses are not restricted to one species of phytochrome comes from extensive studies in this laboratory on the photocontrol of nitrate reductase. Photocontrol occurs at every stage in development from the barely imbibed seed, to the seedling, with a characteristic High Irradiance Response (Whitelam & Johnson, 1981), to the mature plant, with a typical linear Pfr/Pr+Pfr-related, irradiance-independent response (Whitelam & Johnson, 1982). Recent work has shown that these changes all involve photocontrol of enzyme synthesis, indicating again that both Type I and Type II phytochrome are involved in the regulation of gene expression (Clarke, Fido & Johnson, 1989). Furthermore the same approach has revealed that, like several of the responses discussed already, the response has two components. In this case, the second component of the response involves enzyme activation, this separate phytochrome-controlled step following the phytochrome-mediated synthesis of new enzyme protein. The second step can be resolved experimentally from the first because it has a much shorter escape time; the interval between a promoting red pulse after which a far-red pulse will no longer reverse the response is more than 10min for the first step but much less than 10min for the activation step (Johnson, Allsebrook & Thomas, 1989) (Figure 5). These results support the view that photoresponses sharing the same mechanism can be controlled by both Type I and Type II phytochrome.

Recently, Keller, Shanklin, Vierstra and Hershey (1989) have succeeded in transferring the gene coding for Type I phytochrome from oat via the Ti-plasmid to tobacco. When functionally linked to a promoter from cauliflower mosaic virus the oat gene was expressed in tobacco, producing spectrophotometrically detectable phytochrome, recognizable immunologically as being from oat. This Type I oat phytochrome was physiologically active in green plants, where it was shown to increase substantially the photoinhibition of internode extension, producing semi-dwarf plants. Thus, not only can Type I and Type II phytochrome control the same response, as with the previous examples, but a mixture of Type I and Type II phytochrome can apparently control the same response at the same time in the same plant; i.e. added Type I phytochrome can participate in a response originally being promoted by Type II. It will be interesting to see a kinetic analysis of the responses of these transgenic green plants. These experiments indicate more decisively than had previously been possible that the active component of phytochrome involved in this response, typically controlled by Type II phytochrome, must be Pfr, the concentration of which is increased in the transgenic plants, rather than the ratio Pfr/Pr+Pfr which would be unchanged by the added phytochrome. It therefore seems likely that Pfr is the active component of both Type I and Type II phytochrome.

TWO ACTIVE COMPONENTS IN PHYTOCHROME ACTION

There is a further factor which needs to be considered in relation to mechanisms of phytochrome action. As mentioned briefly before, there is now a substantial body of evidence supporting the view that responses promoted by Type I phytochrome at least, involve a multiplicative interaction between Pfr and a factor generated at a rate which is a function of the phytochrome cycling rate (Johnson & Tasker, 1979; Johnson, 1980). The two factors can be resolved temporally by exploiting the period prior to the time at which phytochrome responses in Sinapis alba become 'competent' to respond to Pfr (Mohr, 1976). Light treatments given during this period (which differs for each response) have no effect provided that any Pfr is removed by the competence time. However, the effectiveness of Pfr subsequent to 'competence' is strongly dependent on the quality and fluence rate of light treatments given previously, in a manner which is a function of the phytochrome cycling rate (Whitelam & Johnson, 1981). Moreover, the effects of such pretreatments can be stored and decay away with a half-life of many hours (much more than that of Pfr), again depending on the individual response (Johnson & Whitelam,

1983). Thus, even in dealing with what I have so far referred to as single responses, we know that two components are involved in each; Pfr and an unknown factor or factors related to the rate of phytochrome cycling. Responses to Type II phytochrome which are irradiance dependent have not so far been found and it may be that the two-factor responses are unique to Type I phytochrome. Recent work of Briggs and his colleagues (Eisinger, Short and Briggs, 1989; and W.R. Briggs, personal communication) has shown, using monoclonal antibodies, that plasma membrane-associated phytochrome in etiolated pea stems is of Type I and shows characteristic Type I kinetics. They also found that isolated membrane vesicles pump calcium: right-side-out vesicles released calcium in red light and took it up in far-red light; inside-out vesicles showed the opposite effects. In each case the transport of calcium occurred only when the light was on, not during an ensuing dark period (in this, very different from phytochrome-mediated calcium transport in Mougeotia). Thus, it is possible that such calcium pumping could be the cycling dependent factor (or one of them) in Type I phytochrome action. However, as with a number of other examples of calcium pumping, it is too soon to be sure that this one is related to any phytochrome-mediated physiological responses.

CONCLUSIONS

I have attempted to show that our understanding of the processes of signal transduction mediated by phytochrome has been hampered by an initially sensible but increasingly untenable attempt to identify a single primary mechanism of phytochrome action. Several potential candidates have emerged but decisive evidence has been restricted to somewhat unusual examples. However, from these examples it is possible to conclude that there are phytochrome responses with mutually incompatible signal transduction chains, such as the photocontrol of chloroplast movement in Mougeotia compared with control of transcription in isolated nuclei by exogenously supplied Pfr. In addition, there are responses themselves compounded of two or more different phytochrome-mediated responses, possibly involving completely different response chains. Photocontrol of transcription again provides one such example, the photocontrol of nitrate reductase another, but many of the phytochrome responses having complex dose-response characteristics may also fall into this category. All of this should encourage us to be less than surprised that no single candidate for the role of primary effector of phytochrome action has emerged: in all probability one never will.

REFERENCES

Bartley MR, Frankland B (1982) Analysis of the dual role of phytochrome in the photoinhibition of seed germination. Nature **300**:750-752

Bühler B, Drumm H, Mohr H (1978) Investigations on the role of ethylene in phytochrome-mediated photomorphogenesis. I. Anthocyanin synthesis. Planta **142**:109-117

Bühler B, Drumm H, Mohr H (1978) Investigations on the role of ethylene in phytochrome-mediated photomorphogenesis. II. Enzyme levels and chlorophyll synthesis. Planta **142**:119-122

Carr-Smith H, Johnson CB, Thomas B (1989) Action spectrum for floral induction and apex elongation in green, light-grown wheat. Planta (in press)

Casal JJ, Smith (1988) Persistent effects of changes in phytochrome status on internode growth in light-grown mustard: occurrence, kinetics and locus of perception. Planta **175**:214-220

Casal JJ, Whitelam GC, Smith H (1989) Phytochrome control of extra-cellular peroxidase activity in mustard internodes: correlation with growth and comparison with the effect of wounding. Plant Physiol (in press)

Castresana C, Garcia-Luque I, Alonso E, Malik VS, Cashmore AR (1988) Both positive and negative regulatory elements mediate expression of a photoregulated CAB gene from Nicotiana plumbaginifolia. EMBO J **7**:1929-1936

Clarke IE, Fido R, Johnson CB (1989) Nitrate and phytochrome control of nitrate reductase synthesis in seedlings and mature plants of Sinapis alba. Phytochemistry (in press)

Dean C, Favreau M, Bond-Nutter D, Bedbrook J, Dunsmuir P (1989) Sequences downstream of translation start regulate quantitative expression of two petunia rbcS genes. Plant Cell **1**:201-208

Eisinger W, Short TW, Briggs WR (1989) Light regulation of calcium fluxes in isolated membrane vesicles from pea (Pisum sativum) seedlings. In "Signal Perception and Signal Transduction in Higher Plants" Book of Abstracts of the NATO symposium, Toulouse 1989)

Epel BL, Butler WL, Pratt LH, Tokuyasu KT (1980) Immunofluorescence localization studies of the Pr and Pfr forms of phytochrome in the coleoptile tips of oats, corn and wheat. In: De Greef JA (ed) Photoreceptors and plant development. Antwerpen University Press, Antwerpen, pp121-133

Ernst D, Oesterhelt D (1984) Purified phytochrome influences in vitro transcription in rye nuclei. EMBO J **3**:3075-3078

Evans A, Smith H (1976) Localization of phytochrome in etioplasts and its regulation in vitro of gibberellin levels. Proc Natl Acad Sci USA **73**: 138-142

Giuliano G, Pichersky E, Malik, VS, Timko MP, Scolnik PA, Cashmore RA (1988) An evolutionarily conserved protein binding sequence upstream of a plant light-regulated gene. Proc Natl Acad Sci USA **85**:7089-7093

Green PJ, Kay SA, Chua N-H (1987) Sequence-specific interactions of a pea nuclear factor with light-responsive elements upstream of the rbcS-3A gene. EMBO J **6**:2543-2549

Grolig F, Wagner G (1989) Characterization of the isolated calcium-binding vesicles from the green alga Mougeotia scalaris, and their relevance to chloroplast movement. Planta **177**:169-177

Haupt W (1958) Hellrot-dunkelrot-Antagonismus bei der Auflösung der Chloroplastenbewegung. Naturwissenschaften **45**:273-274

Haupt W, Weisenseel MH (1976) Physiological evidence and some thoughts on lo localized responses, intracellular localization and action of phytochrome. In: H. Smith (ed) Light and Plant Development. Butterworths London Boston, pp63-74

Hilton JR, Smith H (1980) The presence of phytochrome in purified barley etioplasts and its in vitro regulation of biologically active gibberellin levels in etioplasts. Planta **148:**312-318

Jenkins GI (1988) Photoregulation of gene expression in plants. Photochem Photobiol **48:**821-832

Johnson CB (1980) The effect of red light in the high irradiance reaction of phytochrome: evidence for an interaction between Pfr and a phytochrome cycling-driven process. Plant Cell Environ **3:**45-51

Johnson CB, Tasker R (1979) A scheme to account quantitatively for the action of phytochrome in etiolated and light-grown plants. Plant Cell Environ **2:**259-265

Johnson CB, Whitelam GC (1983) Enhancement of Pfr-mediated responses by light pretreatments: persistence of the pretreatment effects. Plant Cell Environ **6:**89-93

Johnson CB, Allsebrook SM, Thomas B (1989) Sequential phytochrome-mediated synthesis and activation of nitrate reductase in seedlings of Sinapis alba. Phytochemistry (in press)

Kachar B, Reese TS (1988) The mechanism of cytoplasmic streaming in characean algal cells: sliding of endoplasmic reticulum along actin filaments. J Cell Biol **106:**1545-1552

Keller JM, Shanklin J, Vierstra RD, Hershey HP (1989) Expression of a functional monocotyledonous phytochrome in transgenic tobacco. EMBO J **8:**1005-1012

Kohno T, Shimmen T (1987) Ca^{2+}-induced fragmentation of actin filaments in pollen tubes. Protoplasma **141:**177-179

Kohno T, Shimmen T (1988) Accelerated sliding of pollen tube organelles along characeae bundles regulated by Ca^{2+}. J Cell Biol **106:**1539-1543

Lissemore JL, Quail PH (1988) Rapid transcriptional regulation by phytochrome of the genes for phytochrome and chlorophyll a/b-binding protein in Avena sativa. Molecular and Cellular Biology **8:**4840-4850

Mohr H (1964) The control of plant growth and development by light. Biol Rev **39:**87-112

Mohr H (1978) Pattern specification and realization in photomorphogenesis. Bot Mag Tokyo (special issue) **1:** 199-217

Mohr H (1983) An introduction to photomorphogenesis for the general reader. In:Shropshire W Jr, Mohr H (eds) Encyclopedia of Plant Physiology New Series Vol 16. Springer, Berlin Heidelberg New York, pp24-38

Morelli G, Nagy F, Fraley RT, Rogers SG, Chua NH (1985) A short conserved sequence is involved in the light inducibility of a gene encoding ribulose 1,5-bisphosphate carboxylase in higher plants. Nature 315:200-204

Morgan DC, Smith H (1976) Linear relationship between phytochrome photo-equilibrium and growth in plants under simulated natural radiation. Nature **262:**210-212

Mösinger E, Batschauer A, Schäfer E, Apel K (1985) Phytochrome control of in vitro transcription of specific genes in isolated nuclei from barley (Hordeum vulgare). Eir J Biochem **147:**137-142

Mösinger E, Batschauer A, Vierstra R, Apel K, Schäfer E (1987) Comparison of the effects of exogenous native phytochrome and in vivo irradiation on in vitro transcription in isolated nuclei from barley (Hordeum vulgare). Planta **170:**505-514

Quail PH (1983) Rapid action of phytochrome in photomorphogenesis. In Encyclopedia of Plant Physiology, New Series, Vol. 16; Shropshire W & Mohr H, eds; Springer-Verlag, Berlin-Heidelberg, pp178-212.

Quail PH, Marme D, Schäfer E (1973) Particle-bound phytochrome from maize and pumpkin. Nature New Biol **245:**189-191

Racusen RH (1976) Phytochrome control of electrical potentials and intercellular coupling in oat coleoptile tissue. Planta **132**:25-29

Schäfer E (1975) A new approach to explain the "high irradiance responses" of photomorphogenesis on the basis of phytochrome. J. Math Biol **2**:41-56

Shanklin J, Jabben M, Vierstra RD (1987) Red light induced formation of ubiquitin-phytochrome conjugates: identification of possible intermediates of phytochrome degradation. Proc Natl Acad Sci USA **84**:359-363

Smith H (1982) Light quality, photoperception, and plant strategy. Ann Rev Plant Physiol **33**:481-518

Smith H (1983) Is Pfr the active form of phytochrome? Phil Trans R Soc Lond B **303**:443-452

Smith H, Evans A, Hilton JR (1978) An in vitro association of soluble phytochrome with a partially purified organelle fraction from barley leaves. Planta **141**:71-76

Timko MP, Kausch AP, Castresana C, Fassler J, Herrera-Estrella L, Van den Broeck G, Van Montagu M, Schell J, Cashmore AR (1985) Light regulation of plant gene expression by an upstream enhancer-like element. Nature **318**: 579-582

Viner N, Whitelam GC, Smith H (1988) Calcium and phytochrome control of leaf unrolling in dark-grown barley seedlings. Planta **175**:209-213

Wagner G, Rossbacher R (1980) X-ray microanalysis and chlorotetracycline staining of calcium vesicles in the green alga Mougeotia. Planta **149**: 298-305

Wagner G, Grolig F, Altmüller D (1987) Transduction chain of low irradiance response of chloroplast reorientation in Mougeotia in blue or red light. Photochem Photobiol Suppl 183-189

Warner TJ, Ross JD, Coombs J (1981) Phytochrome control of maize coleoptile section elongation: a rapid loss of photoreversibility. Plant Physiol **67**:355-357

Whitelam GC, Johnson CB (1981) Temporal separation of two components of phytochrome action. Plant Cell Environ **4**:53-57

Whitelam GC, Johnson CB (1982) Photomorphogenesis in Impatiens parviflora and other plant species under simulated natural canopy conditions. New Phytol **90**: 611-618

LIGHT MODULATION OF CHLOROPLASTIC ENZYMES BY THE FERREDOXIN-THIOREDOXIN SYSTEM : THE CASE OF NADP-MALATE DEHYDROGENASE

P. Gadal, J.P. Jacquot, M. Miginiac-Maslow, M. Droux, P. Decottignies, C. Crétin

Physiologie Végétale Moléculaire, Université Paris Sud, URA CNRS D1128, Bâtiment 430, 91405 Orsay-Cedex, France

INTRODUCTION

The different reactions that occur in photosynthesis are usually divided in two broad categories, the light dependent reactions and the dark reactions.

In the light dependent reactions, energy derived from sunlight energizes an electron in chlorophyll enabling it to move along an oxidation chain in the thylakoid membrane. This electron transport process is used to pump protons across the thylakoid membranes and the resulting protonmotrice force drives the phosphorylation of ADP to produce ATP. At the same time, the processes generate high energy electrons that reduce NADP.

In the dark reactions, otherwise designated as the synthetic phase of photosynthesis the ATP and the NADPH generated in the light dependent reactions serve as the sources of energy for conversion of CO_2 to carbohydrates, ammonia assimilation and sulfate reduction; in other words, light had no other effect but the production of ATP and NADPH as suggested by the early experiments of Trebst et al (1958). Subsequent experiments however led to the conclusion that light also functions in the regulation of the synthetic phase. (Buchanan, 1980). In fulfilling its regulatory role light modulates the activity of selected enzymes. Such regulation is of paramount importance because enzymes for degradating carbohydrates coexist in the chloroplast with the enzymes implicated in carbohydrate synthesis and because many shuttle systems are involved in connecting the chloroplast to other cell compartments and especially the cytosol; very precise mechanisms have

NATO ASI Series, Vol. H 47
Signal Perception and Transduction in Higher Plants
Edited by R. Ranjeva and A. M. Boudet

to be worked out in order to keep the right concentrations of the substrates in the right compartment.

Evidences from different laboratories indicated that enzymes of the Calvin Cycle such as RuBP carboxylase/oxygenase (RuBPC EC 4.1.39), fructose-1,6- bisphosphatase (FBPase EC 3.1.3.11), sedoheptulose-1,7-bisphosphatase (SBPase EC 3.1.3.37), phosphoribulokinase (PRK EC 2.7.1.19) and NADP-glyceraldehyde-3 phosphate dehydrogenase (NADP-G3PDH EC 1.2.1.) are light activated. This proved to be also true for enzymes outside the Calvin cycle, including the NADP-malate dehydrogenase (NADP-MDH EC 1.1.1.82), pyruvate Pi dikinase (EC 2.7.9.1), phenylalanine ammonia lyase (PAL 4.3.1.5) and the ATPase activity of the coupling factor (Buchanan, 1980).

This paper will be focused on the mechanisms which lead to the rapid light dependent activation of one enzyme of the C_4 dicarboxylic acid cycle the NADP-malate dehydrogenase taken as a model system.

GENERAL CHARACTERISTICS OF THE NADP-MALATE DEHYDROGENASE.

NAD-malate dehydrogenases are very active enzymes present in all living organisms, bacteria, animals, plants (Banaszak and Bradshaw, 1975) A number isoforms have been identified and shown to be present in the cytosol, mitochondria, microbodies,(Mc Millin and Scandalios,1980).

Polyclonal monospecific antibodies directed against the NADP-MDH from Sorghum leaves purified to homogeneity have been raised in rabbits (Jacquot et al, 1981). On a crude extract, as expected, these antibodies immunoprecipitate quantitatively the NADP-dependent enzyme but not the NAD dependent malate dehydrogenase, demonstrating that in higher plants, the native NAD dependent and NADP-dependent malate dehydrogenases are immunologically distinct proteins bearing only a few common epitopes (Figure 1).

In contrast to the NAD-MDHs, NADP-MDH is strictly restricted to chloroplasts (Hatch,1971, Perrot-Rechenmann 1983) and consequently is absent from non chlorophyllous tissues such as roots, it can then be considered as an organelle-specific enzyme.

The structure of NADP-MDH varies according to the authors. In C_3 plants it was considered either as a dimer or as a tetramer composed of apparently identical subunits from 28 kDa to 40 kDa depending on the preparations. In C_4 plants several structures have also been proposed such as two non identical subunits of 33 kDa and 45 kDa (Geetha et al, 1981), a tetramer of four identical subunits of 38 kDa each (Ashton and Hatch ,1983) or a dimer of two identical subunits of 38kDa each (Jacquot et al., 1984). As these data appear very heterogeneous a complete comparison of NADP-malate dehydrogenases from two

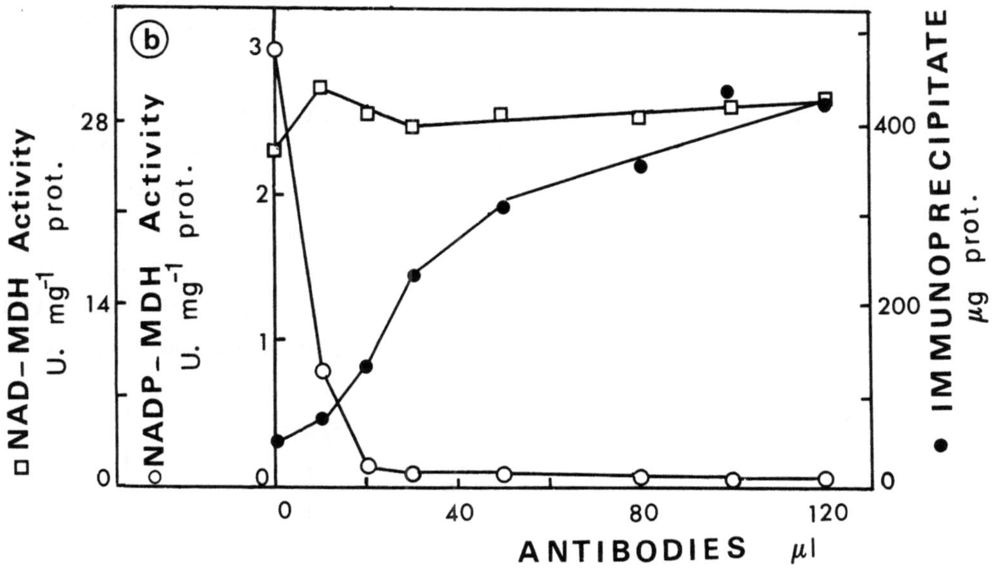

Figure 1 . Immunoprecipitation of NADP-MDH and NAD-MDH by specific polyclonal antibodies raised against NADP-MDH. Extracts from corn leaves were precipitated by ammonium sulfate (35-55% saturation), collected by centrifugation and redisolved in Tris-HCl. After incubation overnight at 4°C, the immunoprecipitates were centrifuged and the activities determined on the supernatants

different plants, Corn and Spinach, was made (Ferté et al. 1986). From this study it was deduced that both enzymes have actually the same dimeric structure made of two subunits exhibiting similar molecular weights whatever their oxidoreduction status. Moreover, Corn and Spinach NADP-MDH show very similar amino acid composition indicating that there might exist a strong homology between them. The estimation of the molecular mass of these proteins gave variable results, higher values being obtained by polyacrylamide gel electrophoresis. Depending on which method is used the molecular masses of NADP-MDH subunits vary by 20 %. More recent results (unpublished) have shown that the corn enzyme is very susceptible to proteolytic degradation and from a full length cDNA,theMW of the subunit has actually been determined to be 42.2 KDa.

The functions of the NADP-MDH are specially well documented in photosynthetic C_4 type plants such as Corn of Sorghum. The mesophyll cells of these plants fix CO2 into oxaloacetate through the cytosolic enzyme phosphoenolpyruvate carboxylase (Hatch and Slack, 1970). Oxaloacetate is then reduced to malate by the NADP-malate dehydrogenase located in the chloroplasts of the mesophyll cells. In photosynthetic C_3 plants the enzyme is believed to be involved in the translocation of reducing power from the stroma to the cytosol in the form of malate through the envelope.(Anderson and House 1979).

It was discovered by Johnson and Hatch(1970) that in C_4 plants that NADP-MDH is inactive in leaves kept in the dark and activated subsequent to tranfer of the organs to light, another major contribution

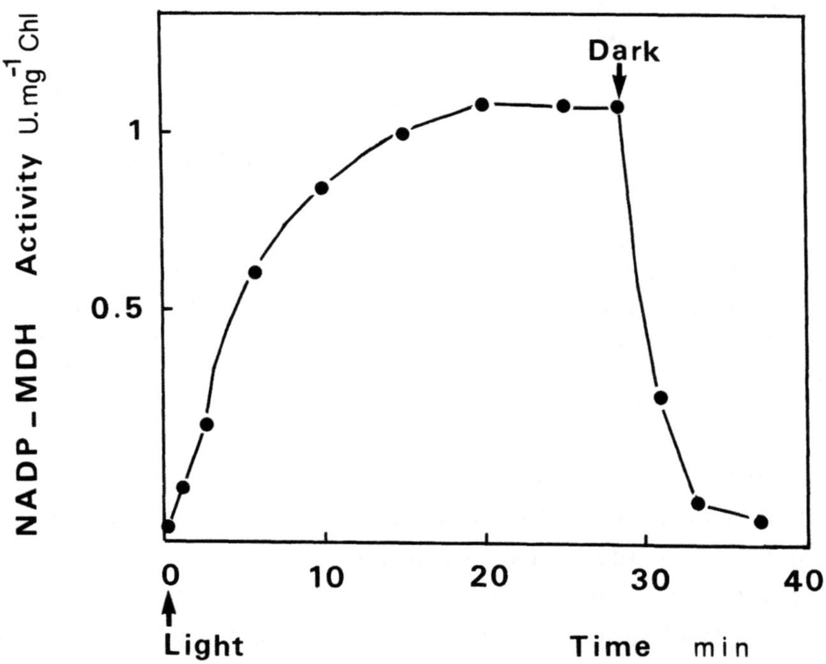

Figure 2 . Evolution of NADP-MDH activity in isolated pea chloroplast during light-dark transition

of this work was the finding that the activitation can be performed *in vitro* by a strong reductant such as dithiothreitol (DTT) suggesting that the activation proceeds throught the reduction of disulfide bridges. Figure 2 illustrates the light activation of the enzyme in isolated pea chloroplasts, similar results have been published for other C_3 plants such as spinach (Vidal et al 1980). As a consequence it can be concluded that light modulation is a general characteristic of all NADP-MDH whatever the photosynthetic metabolic type C_4 or C_3.

COMPONENTS OF THE PHOTOMODULATION SYSTEM

Three systems namely :

- ferredoxin-thioredoxin (Buchanan, 1980),
- light effect mediators or LEM (Anderson, 1979)
- ferralterin (De La Torre et al., 1982)

capable of promoting the light activation of enzymes in higher plants chloroplasts have been described; they have been compared in detail by Jacquot in1984.

From the inhibition of activation of different enzymes, based on the use of electron transfer inhibitors like DCMU or DSPD, or uncouplers like FCCP and also of arsenite or sulfite which react with disulfides and vicinal thiols respectively Anderson and Avron (1976) and Heuer et al., (1982),have concluded that in pea chloroplast, membrane bound proteins termed LEM were essential for light modulation. This hypothesis is now severely questioned as the activation results probably from a partial binding of ferredoxin, thioredoxin, and ferredoxin-thioredoxin reductase (FTR) to the thylacoids as shown in figure 3.a et 3b.

With preparation of once purified thylakoids, the light activation of NADP-MDH is stimulated by the addition of FTR and ferredoxin, the addition of thioredoxin is effectless.(Figure 3a). The experiment was repeated with purified thylakoids washed with a buffer containing 200mM NaCl and 0.02% Triton X 100 and resuspended. Using these washed thylakoids, the light activation of NADP-MDH is then strictly dependent on the addition of Fd, FTR and TRX (Figure 3b) These data strongly suggest that those components when still associated to the membranes may constitute the LEM system as described by Anderson and coworkers.

Ferralterin has now been shown to be synonymous to FTR (Droux et al., 1987).

The ferredoxin-thioredoxin (FTR) system what first described by

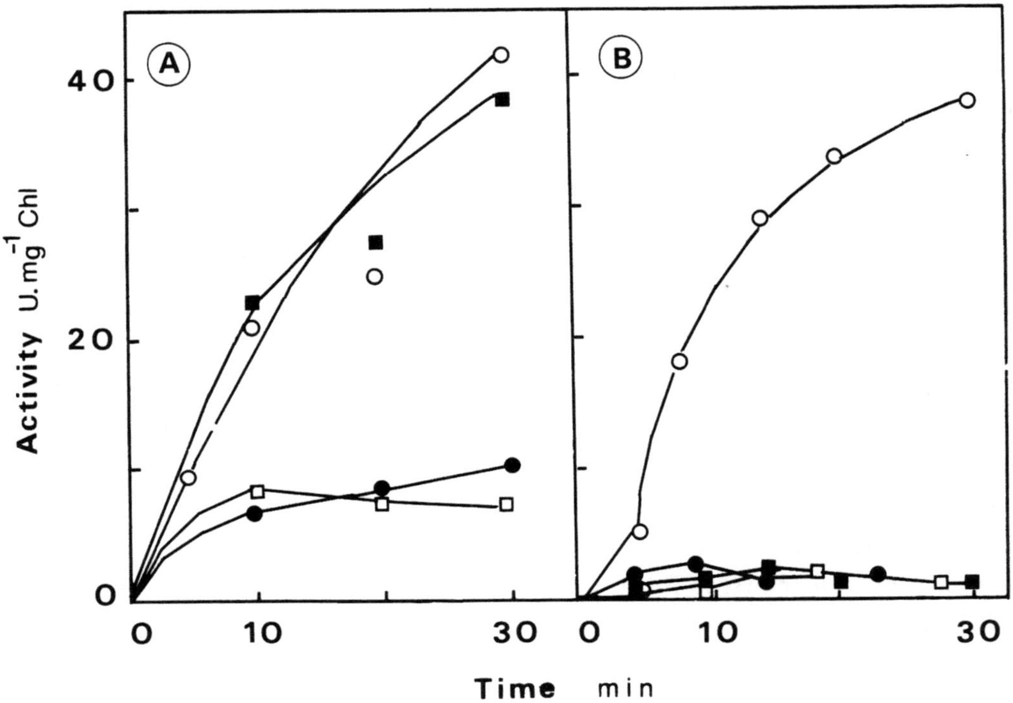

Figure 3. Light activation of NADP-MDH in a Pea chloroplast reconstituted system.

The reconstituted light activation system was obtained by combining : thylakoids prepared as described in Jacquot et al (1984) : (30-50 μg chlorophyll), ferredoxin(15-25 μg), thioredoxin (20-30μg), ferredoxin thioredoxin reductase(2-5 μg) and NADP-MDH (2-6μg) the components were prepared as described below.

3a Thylakoids

3b Thylakoids washed with NaCl and Triton.

(O) complete, (●) minus FTR, (■) minus TRX, (□) minus Fd

Buchanan and colleagues (Wolosiuk and Buchanan, 1977, Holmgren et al 1977); it is composed of ferredoxin, ferredoxin-thioredoxin reductase, thioredoxin and the target enzyme. Taken into account the considerations presented above it is now widely accepted that this system represents the only system involved in light activation, through thiol-disulfide interchange, of chloroplast enzymes.

ACTIVATION OF NADP-MDH IN A RECONSTITUTED SYSTEM

Dithiothreitol (DTT) can replace light for photoactivation of the NADP-MDH but as DTT is a strong reductant, the possibility does exist that DTT reduces not only the regulatory -S-S- bonds but also -S-S- groups unrelated to the regulation and for maintaining the enzyme structure.

In order to overcome this problem we set up a chloroplast reconstituted system prepared from pure components as follow In a total volume of 100 ml, the activation mixture contained in Tris-HCl pH 7.9 100 mM, 25 Units of catalase, 2 mM sodium ascorbate, 0.2mM 2,6 dichlorophenol indophenol, twice washed thylakoids (30mg chlorophyll), 5-10 mM ferredoxin, 10-15 mM spinach thioredoxin *m*, 1mM NADP-MDH. Assays were performed in Eppendorf tubes that were equilibrated with nitrogen and then illuminated from below a water bath at 20°C with 300 W/m^2 white light. NADP-MDH activated this manner was assayed as in Ferté et al. 1986.

Preparation of the components of the reconstituted system

The most labile component of the system are the thylakoids, the preparation of which had to be repeated daily. Freezing the thylakoids in small aliquots in liquid nitrogen and storing them at -90°C in a buffer containing 50% glycerol resulted in preparations which are stable over a long period of time (Jacquot et al. 1987). Before freezing the thylakoids were treated at 55°C for 5 min. in order to destroy the oxygen evolving photosystem 2 activity, PS2 was replaced by ascorbate and dichlorophenolindophenol.

The ferredoxin, and the thioredoxin *m* component of the system, specific for the photoactivation of the NADP-MDH, and NADP-MDH were purified according to published procedures (Mayhew, 1971, Schürmann et al., 1981, Ferté et al.1986).

In constrast to all those proteins, FTR remained a poorly understood; because of this gap we carried out intensive investigations in order to characterize this enzyme in oxygenic photosynthetic organisms. FTR was assayed routinely by measuring its capacity to promote the activation of NADP-MDH in the reconstituted thylakoid system described above.

FTR extracted from Spinach, Corn and *Nostoc muscorum in* a Tris-HCl pH7,9 30 mM buffer supplemented with EDTA, 2-mercaptoethanol $MgCl_2$ polyvinyl pyrrolidone was purifed to homogeneity through the following steps : ammonium sulfate precipitation, acid precipitation,

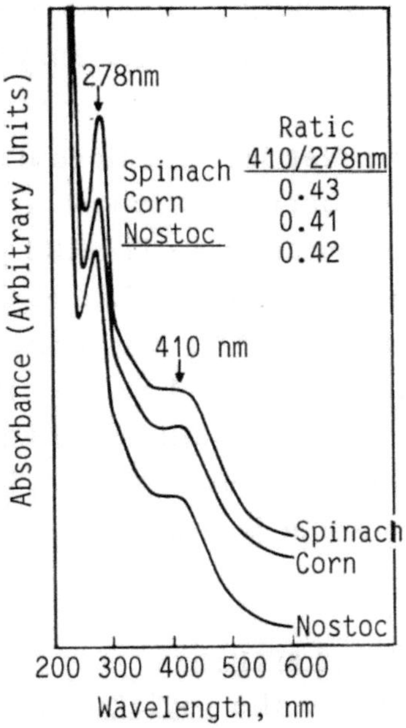

Figure 4. Absorption spectra of FTR from corn and spinach chloroplasts and from *Nostoc muscorum*. Spectra were measured in 30 mM Tris-HCl buffer pH 7.9 with *ca* 0.5 mg protein/ml

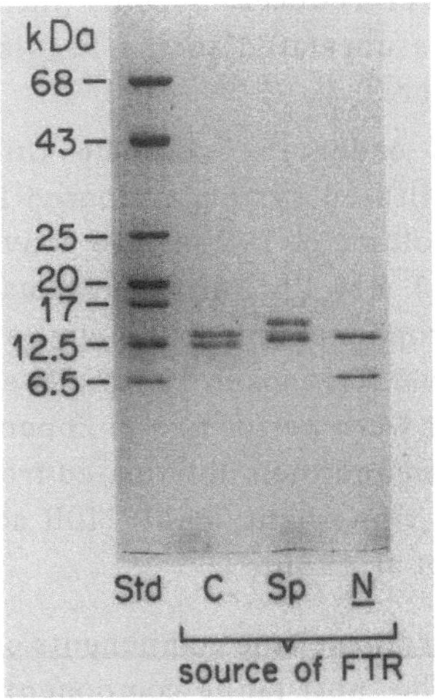

Figure 5 Molecular weight of FTR from corn and spinach chloroplasts *Nostoc muscorum* as determined by SDS-PAGE. In each case 4µg of FTR was used

DEAE-Sephacel chromatography, Ultrogel ACA 44 filtration, affinity chromatography on ferredoxin-sepharose for the Corn and the spinach enzyme. For the *Nostoc* cells, the final steps are hydroxylapatite chromatography and fast protein liquid chromatography; for further detail see Droux et al.1987.

The FTR preparations were yellowshish brown when concentrated. The color of the FTR is also reflected by its absorption spectrum as illustrated in figure 4. In each case, FTR showed showed a chromophore peak in the visible region (410 nm) and a protein peak in the ultra-violet (278 nm), this spectrum is typical of a four iron-four sulfur protein if one assume an extinction coefficient of 4000 M^{-1} cm^{-1} per iron-sulfur pair.

The thiol content of FTR was examined in several ways. The aminoacid composition yielded 10 cysteine plus cystine residues per mole for the Spinach enzyme. The results from the different types of analysis such as reactions with p chloromercuribenzoate, iodoaceate, 5-5'dithiobisnitrobenzoate before and after denaturation with urea suggest that FTR has the thiol content presented in Table 1 : four free cysteines, four cysteines in the iron-sulfur cluster, and one cystine.

The molecular weight of the FTR was estimated by several methods: sedimentation equilibrium, amino-acid analysis, gel electrophoresis, it was found to be about 30000 irrespective of the analytical method. When analyzed by SDS-PAGE, each FTR showed two subunits different in molecular weight. The subunit profile was unique for a given FTR but each showed a common subunit with amolecular weight around 13000. (Figure 5).

As demonstrated in figure 6, the 13 kDa subunits also resemble one another immunologically. This common subunit to the various enzymes was immunologically cross-reactive in Western blot, the only exception in this case being the absence of reaction between *Nostoc* FTR and the spinach anti-FTR antibodies.

Table 1
Proposed thiol content of native spinach FTR

Thiol	number per mole
Free cysteine	4
Cysteine in the iron-sulfur cluster	4
Half-cystine	2

An additional common feature of the 13 kDa subunit was it capability to react with ^{14}C iodoacetate following reduction with DTT. As shown in figure 7 the 13kDa subunit of *Nostoc* and spinach reacted with ^{14}C iodoacetate following a 15 min. reduction with DTT. There was no reaction with the variable subunit suggesting that the disulfide bridge in FTR is located in the similar subunit.

In conclusion FTR is an iron sulfur protein containing four labile sulfide groups has a molecular weight of 30 000 and an absorption peak at 410 mm. It contains two non-identical subunits, one subunit of 13kDa common to each of the purified FTR and a second subunit of variable molecular weight. The similar subunit was immunologically related in the three FTR and contained a reducible -S-S bridge.

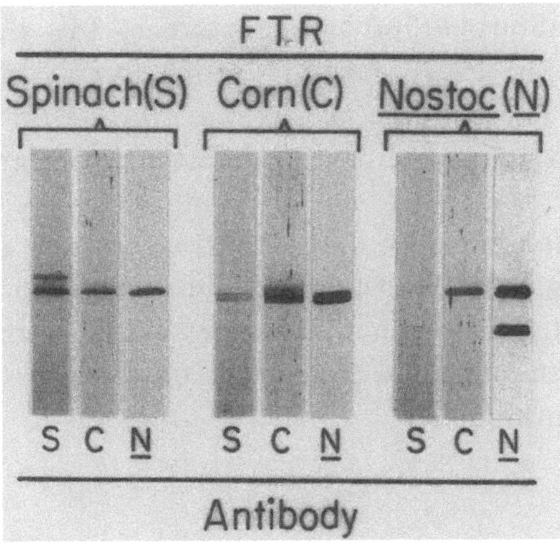

Figure 6.Identification of similar and variable subunits in Spinach, Corn and *Nostoc* muscorum FTR with corresponding antibodies (Western blot). The respective amounts of spinach, corn and *Nostoc* FTR applied µg : 2,4 and 9 for anti-spinach FTR antibody, 30,2, 20 for corn FTR antibody, 40,4 and 1 from anti *Nostoc* antibodies

MECHANISM OF LIGHT ACTIVATION IN THE RECONSTITUTED SYSTEM

It has been demonstrated that DTT reduces in tandem the disulfide groups on thioredoxin and NADP-MDH to the -SH level (Jacquot et al. 1984). To determine whether a change of this type takes place in a physiological light activation system we examined the reduction of thioredoxin *m*, and NADP-MDH as well as FTR in the chloroplast system reconstituted from pure components; The thiols content of the different soluble proteins of the system was determined by following the incorporation of ^{14}C iodoacetate in the light and in the dark. By measuring the differences in these treatments we were able to deduce the light generated -SH groups.(Table 2).

The results of table 2 agree with known structural properties of TRX m and with earlier findings made with DTT for both NADP-MDH and FTR. In agreement with its known structure ferredoxin showed no light reducible disulfide groups.

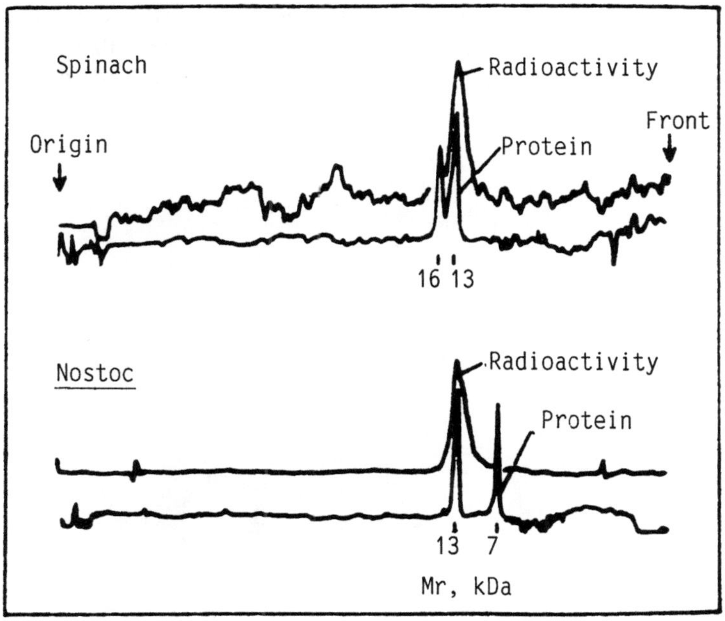

Figure 7. Identification of FTR subunit labeled with ^{14}C iodoacetate after reduction with DTT. Superimposed densitometric tracings of radioactivity and protein are presented

Table 2
Light generated sulhydryls in components of the ferredoxin-thioredoxin system and in NADP-MDH

	SH per mol Protein	
	No addition	+8M Urea
Ferredoxin	0.0	0.0
FTR	2.0	2.0
TRX m	2.0	2.0
NADP-MDH (subunit)	2.0	2.0

The ^{14}C iodoacetate labeling of each component of the ferredoxin-thioredoxin system was confirmed in accompanying SDS-PAGE experiments by coupling protein staining with fluorography-autoradiography (Figure 8). Radioactivity was recoverd almost entirely in the added soluble FTR, TRXm, and NADP-MDH. Interestingly,the major labeling on FTR seemed to be localized on the 13kDa subunit. This agrees with above mentioned results showing that the DTT reducible -S-S- also resides on this subunit.

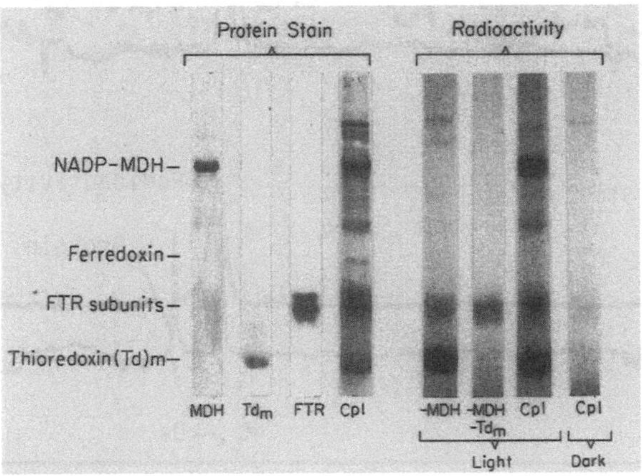

Figure 8. Localization of light generated sulfhydryls in components of the ferredoxin-thioredoxin system and the NADP-MDH. Carboxymethylated proteins were separated on a 12 to 18 linear gradient polyacrylamide SDS gel and vizualized by staining of the proteins and by autoradiography

IDENTIFICATION OF THE REGULATORY PEPTIDE OF THE CORN NADP-MDH

As shown previously, light activation of NADP-MDH was accompanied by the appearance of -2SH per subunit representing a net reduction. The next goal was to determine which part of the protein was involved in this modification. For this purpose, light activated NADP-MDH in thereconstituted system was ^{14}C carboxymethylated purified to homogeneity and then cleaved with trypsin and the peptide mixture purified on a gel filtration column and by reverse-phase chromatography. The distribution of the radioactivity was determined on a C_{18} column from which 80 % of the applied radioactivity was recovered thus allowing precise quantification The radioactivity was recovered in three fractions. A set of evidences presented in Decottignies et al 1988 demonstrated that it contains the regulatory -S-S- site. Sequence analysis of the peptide gave the following results : Lys-glu-Cmcys-phe-gly-val-phe-Cmcys-thr-thr-tyr-asp- leu-lys.

This sequence was compared to a full length cDNA of NADP-MDH from Sorghum recently sequenced in our laboratory. The radioactive peptide fits a postulated tryptic sorghum peptide located in the N-terminus part of the enzyme which is as follows :

ser-val-asp-ala-ala-lys-gln-val-gln-asp-gly-val-ala-thr-ala-glu-ala-pro-ala-thr-arg-lys-asp-**cys-phe-gly-val-phe-cys**-thr-thr-tyr-asp-leu-lys-

the two cysteineresidues implicated in the light regulation of the enzyme are located at positions 24 and 29, from this sequence it can be seen that there are small differences between the sorghum and the corn enzyme, a glu being replaced by an asp in position 23.

The possibility for the existence of an interchain -S-S-bridge in the oxidized enzyme has been ruled out. Indeed, in SDS-gel electrophoresis performed in the absence of any reductant NADP-MDH migrates like its subunits which excludes the existence of an intersubunit disulfide bond.

The presented results demonstrate that light activation of corn NADP-MDH proceeds via the reduction of a disulfide bridge located at the N-terminus part of the enzyme between cys ^{24} and ^{29}. The most stricking feature of this short portion of NADP-MDH is the presence of hydrophobic residues inside the -S-S-bridge (2 phe, 1 val). It is well known that in thioredoxin the active site forms a hydrophobic protrusion out of the molecule comprising the following sequence : phe-trp-ala-pro-trp-cys-gly-pro-cys-. Thus hydrophobic reactions can occur between the regulatory site of the enzyme and the active site of the thioredoxin. Another interesting characteristic of the N-terminal sequence is the presence of many charged and hydrophilic aminoacids outside the bridge, the functional significance of this composition is unknown at the moment.

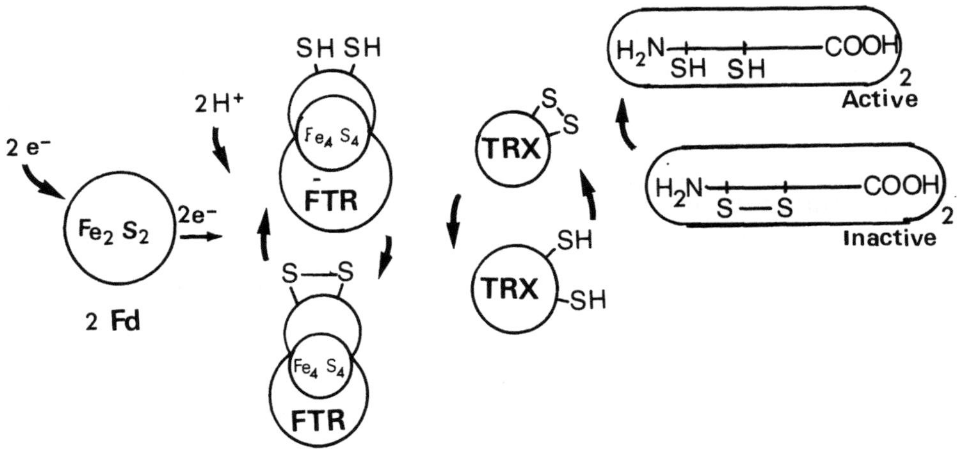

Figure 9. Schematic representation of the light dependent activation of NADP-MDH

CONCLUSION

The results summarized in this paper provide chemical evidence for the existence in oxygenic photosynthesis of a regulatory thiol chain presented in figure 9.

The transfer of thiol reducing equivalents leading from FTR to one target enzyme is now defined However little is known of the events leading from ferredoxin to the FTR disulfide bridge. This step is an especially interesting one as it involves the conversion of an electronic signal (reduced ferredoxin) to a thiol signal (SH groups on FTR). Similarly the function of the iron-sulfur cluster detected in FTR has to be discovered. Evidences on these points as well as on the way by which FTR promotes the photoreduction of other thioredoxin linked enzymes are now under investigation.

REFERENCES

Anderson LE (1979) Interaction between photochemistry and activation of enzymes. In : Gibbs M, Latzko E eds. Photosynthesis II. Springer Berlin Heidelberg New York, p 278-281

Anderson JW, House CM (1979) Polarographic studies of dicarboxylic acid dependent export of reducing equivalents from illuminated chloroplasts. Plant Physiol 64 : 1064-1069

Ashton RA, Hatch MD (1983) Regulation of C_4 photosynthesis : physical and kinetic properties of active (dithiol) and inactive disulfide NADP-malate dehydrogenase from Zea mays. Arch Biochem Biophys 227 : 406-415

Banaszak LJ, Bradshaw RA (1975) Malate dehydrogenases. In : Boyer P ed The Enzymes Academic Press New York San Francisco London p 369-396

Buchanan BB (1980) Role of light in the regulation of chloroplast enzymes. Annu Rev Plant Physiol 31 : 341-374

Decottignies P, Schmitter JM, Miginiac-Maslow M, Le Maréchal P, Jacquot JP, Gadal P (1988) Primary structure of the light dependent regulatory site of corn NADP-Malate dehydrogenase. J Biol Chem 263 : 11780-11785

De La Torre A, Lara C, Yee BC, Malkin R, Buchanan BB (1982) Physicochemical properties of ferralterin, a regulatory iron-sulfur protein functional in oxygenic photosynthesis. Arch Biochem Biophys 213 : 545-550

Droux M, Jacquot JP, Miginiac-Maslow M, Gadal P, Huet JC, Crawford N, Yee BC, Buchanan BB (1987) Ferredoxin-thioredoxin reductase, an iron-sulfur enzyme linking light to enzyme regulation in oxygenic photosynthesis : purification and properties of the enzyme from C_3, C_4, and cyanobacterial species. Arch Biochem Biophys 252 : 426-439

Droux M, Jacquot JP, Miginiac-Maslow M, Gadal P, Crawford N, Kosower NS, Buchanan BB (1987) Ferredoxin-thioredoxin reductase : a catalytically active dithiol groups links photoreduced ferredoxin to thioredoxin functional in photosynthetic enzyme regulation. Arch Biochem and Biophys 256 : 372-380

Ferté N, Meunier JC, Ricard J, Buc J, Sauve P.(1982) Molecular properties and thioredoxin mediated activation of spinach chloroplastic NADP-malate dehydrogenase. FEBS Lett. 146 : 133-139

Geetha V, Mohamed AH, Gnanam A (1981) Identification of chloroplast NADP-Malate dehydrogenase as a product of chloroplast protein synthesis. FEBS Lett. 126 : 223-226

Hatch MD (1971) Mechanism and function of the C_4 pathway of photosynthesis. In : Hatch MD, Osmond CB, Slatyer RO (eds) Photosynthesis and photorespiration. Wiley-Interscience New-York London Sydney Toronto, p 139-152.

Holmgren A, Buchanan BB, Wolosiuk RA (1977) Photosynthetic regulatory protein from rabbit liver is identical with thioredoxin. FEBS Lett 82 : 351-354

Jacquot JP (1984) Post-translation modifications of proteins in higher plant chloroplasts : enzyme regulation by thiol-disulfide interchange. Physiol. Vég. 22 : 487-507

Jacquot JP, Buchanan BB, Martin F, Vidal J (1981) Enzyme regulation in C_4 Photosynthesis : purification and properties of thioredoxin-linked NADP malate dehydrogenase from Corn leaves. Plant Physiol 68 : 300-304

Jacquot JP, Droux M, Miginiac-Maslow M, Joly C, Gadal P (1984) Frozen thylakoids : an improvement for reconstituted chloroplast enzyme light activation systems. Plant Science Lett 35 : 181-185

Jacquot JP, Gadal P, Nizhizawa A, Yee BC, Crawford N, Buchanan BB (1984) Enzyme regulation in C_4 photosynthesis : Mechanism of activation of NADP-Malate dehydrogenase by reduced thioredoxin. Arch Biochem Biophys 228 : 170-178

Johnson HS, Hatch MD (1970) Properties and regulation of leaf nicotinamide dinucleotide phosphate malate dehydrogenase and malic enzyme in plants with the C_4 dicarboxylic acid pathway of photosynthesis. Biochem. J. 119 : 273-280.

Mayhew SG (1971) Non denaturing procedure for rapid preparation of ferredoxin from *Clostridium pasteurianum*. Anal Biochem 42 : 191-194

Mc Millin DE, Scandalios JG (1980) Duplicated cytosolic malate dehydrogenases genes in *Zea mays*. Proc Natl Acad Sci USA 77 : 4866-4870

Perrot-Rechenmann C, Jacquot J-P, Gadal P, Weeden N-F, Cseke C, Buchanan B.B.(1983) Localization of NADP-Malate dehydrogenase of corn leaves by immunological methods. Plant Sci Lett 30 : 219-226

Schürmann P, Maeda K, Tsugita A (1981) Isomers in thioredoxins of Spinach chloroplasts. Eur J Biochem 116 : 37-45

Trebst AV, Tsujimoto HY, Arnon DI (1958) Separation of light and dark phases in the photosynthesis of isolated chloroplasts. Nature 182 : 351-355

Vidal J, Jacquot JP, Gadal P.(1980) Light activation of NADP-malate dehydrogenase in a reconstituted chloroplastic system. Phytochemistry 19 : 1919-1924

Wolosiuk RA, Buchanan BB (1977) Thioredoxin and glutathione regulate photosynthesis in chloroplasts. Nature 266 : 565-567

PROTEIN PHOSPHORYLATION AND THE CONTROL OF PHOTOSYNTHETIC CO_2 ASSIMILATION IN C4 AND CAM PLANTS

H.G. Nimmo*, P.J. Carter*+, C.A. Fewson*, G.A.L. McNaughton*+, G.A.Nimmo*+ and M.B. Wilkins+

Departments of *Biochemistry and +Botany
University of Glasgow
Glasgow G12 8QQ
Scotland

Introduction

Over the last thirty years it has become clear that protein phosphorylation/ dephosphorylation plays a central role in the regulation of many processes and pathways in both eukaryotic and prokaryotic cells (e.g. Edelman et al., 1987; Cohen, 1988). Relatively little attention has been paid to protein phosphorylation in higher plants, but several important targets of phosphorylation have now been identified (see e.g. Budde & Chollet, 1988, Ranjeva & Boudet, 1987 for reviews). Among the plant enzymes that have been shown to be regulated by phosphorylation are pyruvate dehydrogenase (e.g. Randall et al., 1981), pyruvate, phosphate dikinase (Burnell & Hatch, 1985), quinate:NAD^+ oxidoreductase (Refeno et al., 1982), sucrose phosphate synthase (Huber et al., 1989) and phosphoenolpyruvate carboxylase (PEPc) (see below). In this article we focus on the control of PEPc by phosphorylation in leaf tissue of C4 and Crassulacean acid metabolism (CAM) plants.

Regulation of PEPc

PEPc catalyses the fixation of CO_2 (as HCO_3^-) into oxalo-acetate. It plays an anaplerotic role in C3 plants and in non-photosynthetic tissue in C4 and CAM plants (e.g. Andreo et al., 1987). However the leaves of C4 and CAM plants contain distinct isoenzymes of PEPc with specialized roles in CO_2 fixation. In C4 plants, PEPc is located in leaf mesophyll cells and catalyses the first committed step in the C4 pathway of CO_2 fixation. CAM plants fix external CO_2 into malate at night via PEPc, and during the day the CO_2 is released from malate and re-fixed photosynthetically. Mechanisms must therefore exist to reduce or

NATO ASI Series, Vol. H 47
Signal Perception and Transduction in Higher Plants
Edited by R. Ranjeva and A.M. Boudet

eliminate the flux through PEPc at night in C4 plants and during the day in CAM plants.

The PEPcs of both C4 and CAM plants can be regulated allosterically. Effectors include glucose 6-phosphate, which activates the enzyme, and malate, which inhibits it. It is widely thought that malate may act as a feedback inhibitor that regulates flux through PEPc *in vivo*. However, some years ago, it became clear from the work of several groups that CAM PEPc must be controlled at an additional level, in that the enzyme is more sensitive to inhibition by malate, or less active, during the day than at night (e.g. Kluge et al., 1981; Winter, 1982; Buchanan-Bollig & Smith, 1984).

Using freshly prepared and desalted extracts of the CAM plants *Mesembryanthemum crystallinum* (Winter, 1982) and *Bryophyllum fedtschenkoi* (Nimmo et al., 1984) it was found that the PEPc was significantly more sensitive to inhibition by malate (i.e. the K_i for malate was lower) during the light period than during the middle of the dark period. For the *Mesembryanthemum* enzyme the K_m for phosphoenolpyruvate was higher in the light period than in the dark period. Results for the *Bryophyllum* enzyme are shown in Fig 1. This reveals that the conversions of the "night" to the "day" form of PEPc and *vice versa* coincided with the cessation and onset of malate accumulation respectively. Both conversions occurred during the dark period, suggesting that they are controlled by a circadian rhythm rather than by light/dark changes.

Phosphorylation of CAM PEPc in intact tissue

The mechanism responsible for this interconversion of two kinetically distinct forms of PEPc was investigated by Nimmo et al. (1984, 1986). Immunoprecipitation of PEPc from detached leaves that had been prelabelled with $^{32}P_i$ revealed that the "night" form of the enzyme contained ^{32}P whereas the "day" form did not. Purification of the two forms of the enzyme allowed us to show that the "night" form was phosphorylated on one or more serine residues. The phosphoserine was not a covalent intermediate in the reaction mechanism. Removal of the phosphate group by alkaline phosphatase increased the malate sensitivity of the purified "night" form to that of the "day" form. These data showed that the conversion of the

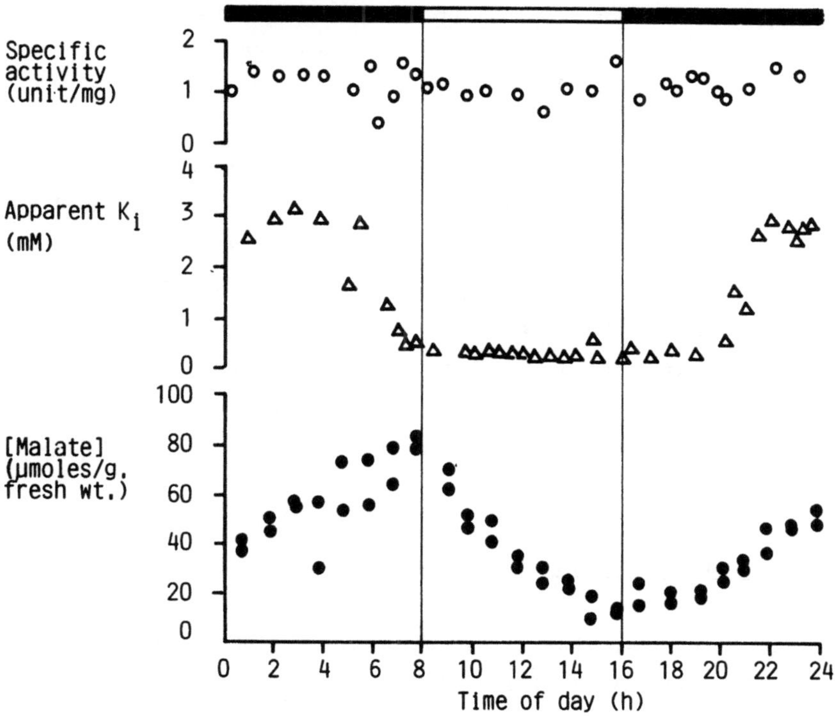

Fig 1. Diurnal rhythm in the malate sensitivity of PEPc in B. fedtschenkoi. Symbols show the specific activity (open circles) and apparent K_i for malate (open triangles) of PEPc and the malate content (closed circles) of leaves throughout the day. The dark and light periods are shown by the closed and open bars respectively. From Nimmo et al. (1984) with permission

"day" to the "night" form of Bryophyllum PEPc correlated with phosphorylation of the enzyme. A similar conclusion was reported for the PEPc of Kalanchoe species (Brulfert et al., 1986).

Detached leaves of B. fedtschenkoi can show persistent circadian rhythms of CO_2 metabolism in continuous darkness or continuous light (e.g. Wilkins 1959, 1984). In some conditions, such as in continuous darkness and CO_2-free air at 15°C, PEPc exhibited a persistent circadian rhythm of interconversions between a malate-sensitive, dephosphorylated form and a less sensitive, phosphorylated form. The changes in the K_i of PEPc for malate were in phase with the rhythm of CO_2 output (Nimmo et al., 1987b). These results strengthen the views that the phosphorylation of Bryophyllum PEPc is controlled by an endogenous rhythm rather than by changes in illumination, and that it plays a significant role in regulation of the flux through PEPc.

The "day" and "night" forms of Bryophyllum PEPc were indistinguishable by gel filtration or gel electrophoresis in non-denaturing conditions (Nimmo et al., 1986). Both forms appeared to be tetramers that could dissociate to dimers. In contrast, the "day" and "night" forms of PEPc from Crassula argentea could be resolved by non-denaturing gel electrophoresis (Wu & Wedding, 1985a). The former appeared to be a dimer whereas the latter behaved as a tetramer (Wu & Wedding, 1985b). It remains to be seen whether this represents a species difference and whether phosphorylation of PEPc occurs in C. argentea.

Phosphorylation of C4 PEPc in intact tissue

Several groups have shown that illumination of C4 plants results in a reduction of the sensitivity of leaf PEPc to inhibition by malate (e.g. Karabourniotis et al., 1983; Huber & Sugiyama, 1986; Nimmo et al., 1987a; Doncaster & Leegood, 1987; Jiao & Chollet, 1988). The K_i of the maize enzyme for malate changes only some two- to three-fold, a much less dramatic effect than the ten-fold change seen for the Bryophyllum enzyme (Nimmo et al., 1987a; McNaughton et al., 1989). The change requires light intensities of above $400uE/m^2/sec$, is complete in 30-60min and is reversed in darkness over 30-60min (Nimmo et al., 1987a). Blue, red and white light are equally effective in producing the change in K_i for malate, and the change is blocked by DCMU (data not shown). These results suggest that the photoreceptor involved in this effect is the photosynthetic apparatus itself.

The mechanism responsible for the reduction in the malate sensitivity of maize leaf PEPc upon illumination was investigated by immuno-precipitation of the enzyme from $^{32}P_i$-labelled tissue (Nimmo et al., 1987a). This showed clearly that phosphorylation and dephosphorylation of serine residues of PEPc correlated with the decrease and increase in malate sensitivity observed upon illumination and darkening respectively. In agreement with this, Jiao & Chollet (1988) purified PEPc from darkened and illuminated maize leaves and found that the enzyme from the latter tissue was significantly more phosphorylated than that from the former. Treatment of the "day" form of the enzyme with alkaline phosphatase both dephosphorylated it and increased its malate sensitivity. Thus, like the CAM enzyme, PEPc in C4 plants seems to be controlled by phosphorylation/ dephosphorylation. In C4 species the interconversions are triggered by light/dark changes and there is no evidence that a circadian rhythm is involved. The significance of these interconversions in regulation of flux through the C4 pathway is not yet clear.

PEPc kinases

Protein kinases capable of phosphorylating PEPc at the expense of ATP have been identified in several C4 and CAM plants including maize (Budde & Chollet, 1986; Jiao & Chollet, 1989; G.A.L. McNaughton et al., unpublished data), Sorghum (Echevarria et al., 1988), B. fedtschenkoi (see below) and several Kalanchoe species (Kluge et al., 1988). Typical results for Bryophyllum are shown in Fig 2. Incubation of the purified "day"-form of PEPc with a desalted "night" extract, ATP and Mg^{2+} results in both a time-dependent increase in the K_i of the PEPc for malate and transfer of the gamma-phosphoryl group of ATP to PEPc.

PEPc kinase activity has been partially purified from maize leaves by ammonium sulphate fractionation and chromatography on blue dextran-agarose (Jiao & Chollet, 1989). The extent of the phosphorylation of PEPc by this preparation correlated directly with changes in the activity of PEPc. However the maximal phosphorylation of PEPc that was observed corresponded to only 0.25mol phosphate/mol of PEPc subunits (Jiao & Chollet, 1989). Several explanations for this low stoichiometry are possible. One is that limited proteolysis of purified PEPc may decrease its ability to be phosphorylated (McNaughton et al., 1989).

Little is yet known about the regulation of PEPc kinase activity. Jiao & Chollet (1989) could detect no effect of thioredoxin h on the kinase. Echevarria et al. (1988) reported that the phosphorylation of PEPc in Sorghum leaf extracts by endogenous protein kinase activity was dependent on Ca^{2+}/calmodulin. This work must clearly be repeated with purified components. However, the possibility that Ca^{2+} is involved in the light-dependent phosphorylation of PEPc is potentially very significant.

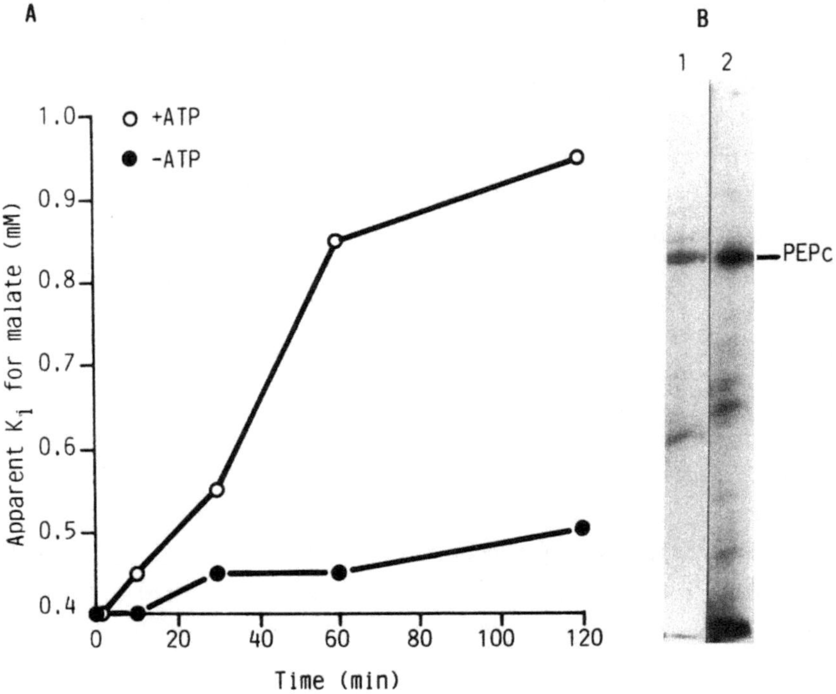

Fig 2. Phosphorylation and changes in the kinetic properties of B. fedtschenkoi PEPc. A) Purified "day"-form PEPc was incubated in 50mM Tris/HCl pH7.5, 1mM dithiothreitol, 10mM $MgCl_2$ at 25°C with a desalted "night" extract in the presence (open circles) or absence (closed circles) of 5mM ATP. The K_i of the PEP_c for malate was measured at the indicated times. B) From a duplicate incubation with [gamma-^{32}P]ATP, a sample was taken at 60min and subjected to SDS-polyacrylamide gel electrophoresis. The gel was stained with Coomassie blue (track 1) and autoradiographed (track 2)

PEPc phosphatases

Very little is known about protein phosphatases that dephosphorylate PEPc, but some progress has recently been achieved by comparing plant protein phosphatases with their mammalian counterparts. Mammalian tissues contain four main types of protein phosphatase (1, 2A, 2B and 2C) that dephosphorylate protein phosphoserine or phosphothreonine residues. These enzymes can be distinguished by their substrate specificities, sensitivities to certain inhibitors and requirements for activity (e.g. Cohen, 1988; Cohen et al., 1988; MacKintosh & Cohen, 1989). Phosphatase 1 is inhibited by two heat-stable protein inhibitors, termed inhibitors 1 and 2, and is fairly sensitive to inhibition by the tumour promoter okadaic acid (IC_{50} approx. 10nM). Phosphatase 2A is exquisitely sensitive to okadaic acid (IC_{50} approx. 0.1nM) but is unaffected by inhibitors 1 and 2. Phosphatase 2B is Ca^{2+}/calmodulin-dependent and phosphatase 2C requires Mg^{2+} for activity.

MacKintosh & Cohen (1989) showed that the phosphorylase phosphatase activity of oilseed rape seed extracts was entirely accounted for by two activities very similar to the mammalian phosphatases 1 and 2A. By using inhibitor 1 and okadaic acid, we found that 50-60% of the phosphorylase phosphatase activity of maize and Bryophyllum leaf extracts is contributed by a phosphatase 1-like activity and the remainder by a phosphatase 2A-like activity. Purified rabbit muscle phosphatase 2A (Cohen et al., 1988) was able to dephosphorylate the "night" form of Bryophyllum PEPc, as judged by changes in its K_i for malate. In contrast, phosphatase 1 was without effect (P.J. Carter et al., unpublished data). These preliminary results suggest promising experimental approaches to the identification of plant PEPc phosphatases.

Conclusions

There is now good evidence that the malate sensitivity of PEPc is regulated by phosphorylation/dephosphorylation in C4 and CAM plants. However much more work on the molecular details of these systems is required. It will be necessary to define the stoichiometry and sites of phosphorylation of PEPc both *in vivo* and *in vitro* and to investigate the regulatory properties of the protein kinase(s) and protein phosphatase(s) involved. This should allow us to understand the roles and regulation of the phosphorylation of PEPc and the signalling systems that affect phosphorylation of the enzyme in C4 and CAM plants.

Acknowledgements

Work from this laboratory was supported by the Agricultural and Food

Research Council. We thank Dr C. MacKintosh and Prof P. Cohen (Dundee University) for their help in the experiments on protein phosphatases.

References

Andreo CS, Gonzalez DH, Iglesias AA (1987) Higher plant phosphoenolpyruvate carboxylase. FEBS Lett 213:1-8

Brulfert J, Vidal J, Le Marechal P, Gadal P, Queiroz O (1986) Phosphorylation-dephosphorylation process as a probable mechanism for the diurnal regulatory changes of phosphoenolpyruvate carboxylase in CAM plants. Biochem Biophys Res Commun 136:151-159

Buchanan-Bollig IC, Smith JAC (1984) Circadian rhythms in Crassulacean acid metabolism: phase relationships between gas exchange, leaf water relations and malate metabolism in Kalanchoe daigremontiana. Planta 161:314-319

Budde RJA, Chollet R (1986) In vitro phosphorylation of maize leaf phosphoenolpyruvate carboxylase. Plant Physiol 82:1107-1114

Cohen P, Alemany S, Hemmings BA, Resink TJ, Stralfors P, Lim Tung HY (1988) Protein phosphatase 1 and protein phosphatase 2A from rabbit skeletal muscle. Methods Enzymol 159:390-408

Doncaster HD, Leegood RC (1987) Regulation of phosphoenolpyruvate carboxylase activity in maize leaves. Plant Physiol 84:82-87

Echevarria C, Vidal J, Le Marechal P, Brulfert J, Ranjeva R, Gadal P (1988) The phosphorylation of Sorghum leaf phosphoenolpyruvate carboxylase is a Ca^{++}-calmodulin dependent process. Biochem Biophys Res Commun 155: 835-840

Edelman AM, Blumethanl DK, Krebs EG (1987) Protein serine/threonine kinases. Annu Rev Biochem 56:567-617

Huber JLA, Huber SC, Nielsen TH (1989) Protein phosphorylation as a mechanism for regulation of spinach leaf sucrose-phosphate synthase activity. Arch Biochem Biophys 270:681-690

Huber SC, Sugiyama T (1986) Changes in sensitivity to effectors of maize leaf phosphoenolpyruvate carboxylase during light/dark transitions. Plant Physiol 81:674-677

Jiao JA, Chollet R (1988) Light/dark regulation of maize leaf phosphoenolpyruvate carboxylase by in vivo phosphorylation. Arch Biochem Biophys 261:409-417

Jiao JA, Chollet R (1989) Regulatory seryl-phosphorylation of C4 phosphoenolpyruvate carboxylase by a soluble protein kinase from maize leaves. Arch Biochem Biophys 269:526-535

Karabourniotis G, Manetas Y, Gavalas NA (1983) Photoregulation of phosphoenolpyruvate carboxylase in Salsola soda L. and other C4 plants. Plant Physiol 73:735-739

Kluge M, Brulfert J, Queiroz O (1981) Diurnal changes in the regulatory properties of PEP-carboxylase in Crassulacean Acid Metabolism (CAM) Plant Cell Environ 4:251-256

Kluge M, Maier P, Brulfert J, Faist K, Wollny E (1988) Regulation of phosphoenolpyruvate carboxylase in Carssulacean acid metabolism: in vitro phosphorylation of the enzyme. J Plant Physiol 133:252-256

MacKintosh C, Cohen P (to be published). Identification of high levels of type 1 and type 2A protein phosphatases in higher plants. Biochem J

McNaughton GAL, Fewson CA, Wilkins MB, Nimmo HG (1989) Purification, oligomerization state and malate sensitivity of maize leaf phosphoenolpyruvate carboxylase. Biochem J 261:349-355

Nimmo GA, Nimmo HG, Fewson CA, Wilkins MB (1984) Diurnal changes in the properties of phosphoenolpyruvate carboxylase in Bryophyllum leaves: a possible covalent modification. FEBS Lett 178:199-203

Nimmo GA, Nimmo HG, Hamilton ID, Fewson CA, Wilkins MB (1986) Purification of the phosphorylated night form and dephosphorylated day form of phosphoenolpyruvate carboxylase from Bryophyllum fedtschenkoi. Biochem J 239:213-220

Nimmo GA, McNaughton GAL, Fewson CA, Wilkins MB, Nimmo HG (1987a) Changes in the kinetic properties and phosphorylation state of phosphoenolpyruvate carboxylase in Zea mays leaves in response to light and dark. FEBS Lett 213:18-22

Nimmo GA, Wilkins MB, Fewson CA, Nimmo HG (1987b) Persistent circadian rhythms in the phosphorylation state of phosphoenolpyruvate carboxylase from Bryophyllum fedtschenkoi leaves and in its sensitivity to inhibition by malate. Planta 170:408-415

Randall DD, Williams M, Rapp BJ (1981) Phosphorylation-dephosphorylation of pyruvate dehydrogenase complex from pea leaf mitochondria. Arch Biochem Biophys 207:437-444

Ranjeva R, Boudet AM (1987) Phosphorylation of proteins in plants. Annu Rev Plant Physiol 38:73-93

Refeno G, Ranjeva R, Boudet AM (1982) Modulation of quinate:NAD^+ oxidoreductase activity through reversible phosphorylation in carrot cell suspensions. Planta 154:193-198

Wilkins MB (1959) An endogenous rhythm in the rate of carbon dioxide output of Bryophyllum I. Some preliminary experiments. J Exp Bot 10:377-390

Wilkins MB (1984) A rapid circadian rhythm of carbon-dioxide metabolism in Bryophyllum fedtschenkoi. Planta 161:381-384

Winter K (1982) Properties of phosphoenolpyruvate carboxylase in rapidly prepared, desalted leaf extracts of the Crassulacean acid metabolism plant Mesembryanthemum crystallinum L. Planta 154:298-308

Wu MX, Wedding RT (1985a) Diurnal regulation of phosphoenolpyruvate carboxylase from Crassula. Plant Physiol 77:667-675

Wu MX, Wedding RT (1985b) Diurnal regulation of phosphoenolpyruvate carboxylase from Crassula by interconversion of oligomeric forms. Arch Biochem Biophys 240:655-662

CELLULAR AND MOLECULAR APPROACHES OF DEFENSE IN PLANTS

M.T. Esquerré-Tugayé, J. Fournier, D. Mazau, A. Mouly, M.L. Pouénat, M.Rickauer, D. Rumeau and J.P. Sancan

Centre de Physiologie Végétale
Université Paul Sabatier
118, route de Narbonne
31062 Toulouse cédex
France

The signals exchanged during plant-microorganism interactions account for a three-step phenomenon leading from signal generation to cell responses, *via* transduction pathways. Recent reviews on signalling molecules and on their effects have been published (Darvill and Albersheim, 1984 ; Ryan, 1988). Attempts to integrate general features and more specific data from our work are presented.

SIGNAL GENERATION AND RECOGNITION

Signalling is not a one way phenomenon : the signals exchanged at the plant-microorganisms interface induce defense responses in plants and pathogenic traits in microorganisms (Table 1). In the past twenty years, it has been established that cell wall fragments of either pathogen or plant origin have the ability to induce the same defense reactions as pathogen attack. Such fragments, also called elicitors, are supposed to be among the earliest signals of the interactions. Several classes of elicitors (glycans, peptides, glycoproteins, lipids) have thus been identified. On the other hand, it has been shown that plant cell wall components and cell exudates trigger the synthesis of several hydrolytic enzymes in microorganisms and induce the expression of genes essential for pathogenicity. In this case, pectic polymers, cutin monomers and phenolics play the role of "elicitor" molecules active on pathogens. Thus, it was demonstrated that cutinase gene expression is induced in *Fusarium solani* f.sp. *pisi* within 1 hour or even less when cutin or its derivatives are added to the culture medium (Woloshuk and Kolattukudi, 1986). Similarly, it is well known that the production of pectinases is increased when microorganisms are grown in the presence of pectic polysaccharides. The recent findings that flavonoid molecules present in cell exudates induce the vir-genes of *Agrobacterium* and the Nod-genes of

NATO ASI Series, Vol. H 47
Signal Perception and Transduction in Higher Plants
Edited by R. Ranjeva and A.M. Boudet

Rhizobium provide excellent evidence that host signals are essential for microbial pathogenicity (Stachel et al., 1985 ; Redmond et al., 1986).
It is important to note that another class of signals, structurally related to elicitors, may exert opposite effects. Although most are not fully characterized (Table 1), this type of suppressor molecules lowers the induction of pathogenicity in pathogens and of defense responses in plants.

Table 1 SIGNALLING MOLECULES IN CELL SURFACES AND EXUDATES

SIGNALS	ORIGIN	RESPONSE
1. ELICITORS	MICROBE	HOST-DEFENSES
Oligosaccharides (glyco)peptides PUFAs		Phytoalexins PRs, HRGP, Lignin,...
2. INDUCERS	PLANT	PATHOGENICITY
Cutin monomers Pectic polymers Flavonoïds		Cutinases Pectinases VIR-, NOD-genes
3. SUPPRESSORS		
Polysaccharides	MICROBE	HOST DEFENSES
Phenolics	PLANT	PATHOGENICITY

It is likely that transduction of the signals is mediated through their interaction with membrane receptors. Recently, Schmidt and Ebel (1987) have characterized membrane proteins from soybean with a high affinity to a β-glucan elicitor preparation obtained from Phytophthora megasperma glycinea. Transduction events possibly related to the induction of defense responses in plants are now considered.

SIGNAL TRANSDUCTION

From the host point of view, signal transduction refers to a series of complex events leading from elicitor/pathogen recognition to defense gene expression. Understanding signal transduction always requires the availability of model systems on which cellular and molecular approaches can be achieved. One such system is provided by the tobacco-Phytophthora parasitica nicotianae interaction. This fungal pathogen interacts in a race-cultivar specific manner with its host plant Nicotiana tabacum (De Zoeten et al., 1982). Zoospore inoculation of the root system leads to the appearance of symptoms two to three days after inoculation. Browning of the tissues then rapidly extends in compatible interactions while it remains localized when the interaction is incompatible. Preliminary reports have shown that the interaction can be partly reproduced by using tobacco cell culture and elicitors of Phytophthora (Pélissier et al., 1986). The glycopeptide elicitors prepared from Phytophthora induce a slight browning of tobacco cells upon incubation. Using this system, we have looked for membrane related signals induced by elicitors, or infection, or both.

The very first event that was observed affected membrane potential : fungal elicitors depolorize plant membranes. This effect is instantaneous, reversible and non specific. The elicitors of Phytophthora, as well as of Colletotrichum lagenarium, another fungal pathogen, depolarize the membranes of host and non-host plants (Pélissier et al., 1986).

The second early event recorded in the tobacco-Phytophthora system is an increased production of ethylene. Elicitation of ethylene starts 15 to 30 minutes after addition of the elicitor, and results from the rapid stimulation of the methionine pathway of ethylene biosynthesis through ACC-synthase and other membrane-bound enzymes. The enhancement of ACC synthase activity occurs within the first five minutes ; it does not require transcription and is a transient phenomenon (Berthalon, 1986). A rapid and transient increase of ethylene has also been reported in infected tobacco and in other plant-elicitor and host-pathogen systems (Toppan et al.,1982 ; Chappell et al., 1984).

The third effect that we have studied is the induction of lipoxygenase. Lipoxygenases are active on polyunsaturated fatty acids (PUFAs). A number of reports indicate that PUFAs and derived products might play a significant role in plant-pathogen interactions. Thus, arachidonic acid induces phytoalexin biosynthesis in potato (Bostock et al., 1981), through the possible involvement of lipoxygenase-mediated degradation products. Lipoxygenase might also favour the last step of ethylene biosynthesis under certain conditions (Bousquet and Thimann, 1984). This hormone, in its turn, induces several defense responses in plants (Roby et al., 1986 ; Ecker and Davis, 1987 ; Rumeau et al., 1988).

The level of lipoxygenase is low in tobacco cell suspensions and is highly increased upon elicitor treatment (Fournier et al., 1986). Induction of the enzyme starts 1.5 hour after addition of the elicitor preparation, and shows a maximum at about 24 hours. In whole plants, lipoxygenase is increased upon infection, with a maximum two days after inoculation in the incompatible interaction, while it starts to increase at the time of symptom appearance (after 3 days) when the interaction is compatible. The effect of infection on lipoxygenase activity has also been reported in wheat infected by Puccinia graminis (Ocampo et al., 1986), and in tobacco infected by Erisyphe cichoracearum (Lupu et al., 1980).

The lipoxygenase that was purified from elicitor treated cells is mainly a 9-lipoxygenase, according to the hydroperoxides that it forms from linoleic and linolenic acids, which are the two main PUFAs in tobacco. The hydroperoxides are damaging compounds by themselves : they affect membrane integrity. They can also give rise to variety of products which exert deleterious effects on membranes, or which have regulatory activities (Vick and Zimmerman, 1987). For these reasons, PUFAs-derived compounds might be involved in the hypersensitive response which typifies incompatibility. Some of them might also act as secondary signals.

In animals, PUFA-derived products are part of a phospholipid breakdown phenomenon which involves several enzymes, most notably phospholipase C and A_2, neutral lipases, lipoxygenases and other oxygenases. Whether or not all these membrane related effects are part of elicitor-induced signal cascade(s) is an open question. Considering the timing of elicitor-, or infection-induced events in tobacco, it appears that one can distinguish early membrane steps and late defense responses. As a working hypothesis (Fig. 1), we propose that elicitor recognition induces membrane depolarization, which in its turn affects several membrane related events, notably ethylene synthesis. Ethylene and membrane depolarization might play the role of rapidly exported, cell to cell, signals. Intracellular signals might involve phospholipid metabolites, such as compounds resulting from lipoxygenase action and/or inositol phosphates. Ion fluxes, most notably calcium, and phosphorylation events mediated by protein kinases (Farmer et al., 1989) might play a role in this general scheme.

Preliminary data suggest that these putative cascades are controlled by complex cross relatory mechanisms.

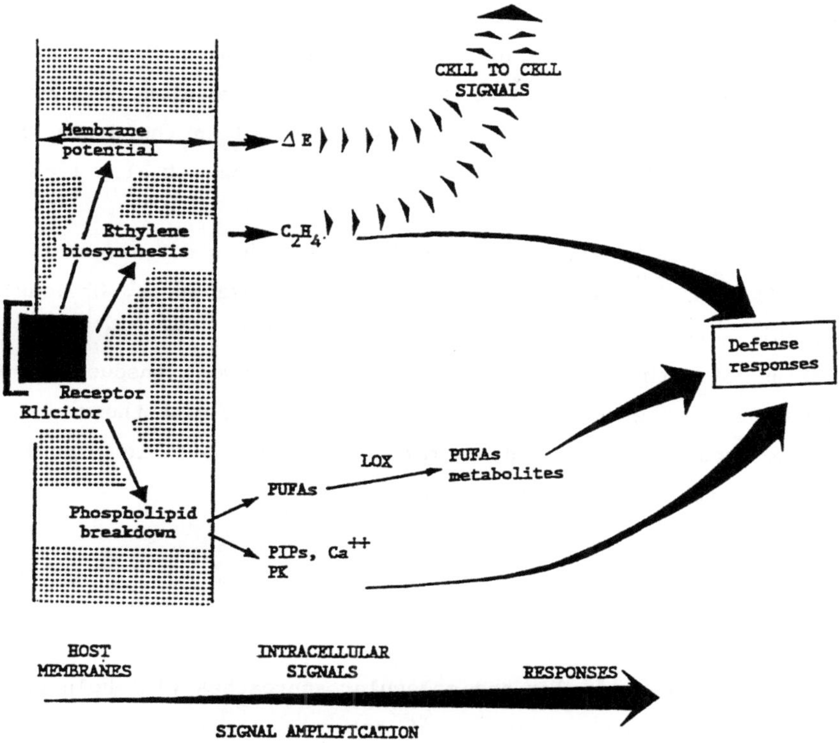

Figure 1. SIGNAL TRANSDUCTION IN RESPONSE TO ELICITORS : A PROPOSED SCHEME
(ΔE = variations of membrane potential : PUFAs = polyunsaturated fatty acids ; PIPs = phosphoinositides ; PK = protein kinases ; LOX = lipoxygenase)

RESPONSE TO MICROBIAL SIGNALS

The ultimate response to microbial signals is concerned with the induction of defense. Plants have developed a wide array of defense responses. In tobacco, it has been shown that elicitors and infection induce the synthesis of phytoalexins, lignin, hydroxyproline-rich glycoproteins (HRGPs), pathogenesis-related proteins (PRs), and proteinase inhibitors (Budde and Helgeson, 1981 ; Jamet and Fritig, 1986 ; Rickauer et al., 1989). In the tobacco-Phytophthora system, proteinase inhibitors, for instance, are induced at about 10 hours after addition of the elicitor to the cell suspension. In whole infected plants, they increase earlier in the incompatible than in the compatible interaction.

Molecular probes are now available for most of the defense genes. Using these probes for measuring gene expression should allow to answer a few questions such as : 1) are phospholipid metabolites involved in the regulation of defense responses ? 2) is ethylene a secondary signal ? Concerning the possible role of ethylene, it has been shown that treating plants with this hormone induces defense gene expression, as demonstrated for chitinase, HRGPs, and several enzymes involved in the synthesis of the flavonoid phytoalexins (Broglie et al., 1986 ; Rumeau et al., 1988 ; Eckert and Davis, 1987).

However, inhibition of elicitor-induced ethylene does not necessarily result in a significant decrease of the products of these genes. Obviously, more biochemical work is needed on ethylene itself, and on the possible interaction of ethylene with the other putative transduction pathways described above. The availability of defense genes and their use in gene constructs should allow to discriminate between possible signals and to clarify their role in gene regulation.

CONCLUSION

In the future, cellular and molecular approaches will help unravelling the internal transduction cascade of external signals such as elicitors. This is a prerequisite to understanding the higly localized and/or systemic responses observed in plants during plant-microorganism interactions.

REFERENCES

BERTHALON E (1986) Régulation de la voie de biosynthèse de l'éthylène dans les cellules de tabac en culture sous l'effet d'éliciteurs fongiques. Thèse Doct Spécialité Biologie et Physiologie Végétale, Toulouse.

BOSTOK RM, KUC JA, LAINE RA (1981) Eicosapentaenoic and arachidonic acids from Phytophthora infestans elicit fungitoxic sesquiterpenes in the potato. Science 212 : 67-69.

BOUQUET JF, THIMANN KV (1984) Lipid peroxidation forms ethylene from 1-amino cyclopropane-1-carboxylic acid and may operate in leaf senescence. Proc Natl Acad Sci USA 81 : 1724-1727.

BROGLIE KE, GAYNOR JJ, BROGLIE RM (1986) Ethylene-regulated gene expression: molecular cloning of the genes encoding an endochitinase from Phaseolus vulgaris. Proc Natl Acad Sci USA 83 : 6820-6824.

BUDDE AD, HELGESON JP (1981) Phytoalexins in tobacco callus tissues challenged with zoospores of Phytophthora parasitica var. nicotianae. Phytopathol 71 : 206.

CHAPELL J, HAHLBROCK K, BOLLER T (1984). Rapid induction of ethylene biosynthesis in cultured parsley cells by fungal elicitor and its relationship to the induction of PAL. Planta 161 : 475-480.

COLLINGE DB, SLUSARENKO AJ (1987) Plant gene expression in response to

pathogens. Plant Mol Biol 9 : 389-410.
DARVILL AG, ALBERSHEIM P (1984) Phytoalexins and their elicitors. A defense against microbial infection in plants. Annu Rev Plant Physiol 35 : 243-275.
De ZOETEN GA, GAARD G, HABERLACH GT, HELGESON JP (1982) Infection of tobacco callus by Phytophthora parasitica nicotianae. Phytopathol 72 : 743-746.
ECKER JR, DAVIS RW (1987) Plant defense genes are regulated by ethylene. Proc Natl Acad Sci USA 84 : 5202-5206.
FARMER EE, PEARCE G, RYAN CA (1989) In vitro phosphorylation of plant plasma membrane proteins in response to the proteinase inhibitor inducing factor. Proc Natl Acad Sci USA 86 : 1539-1542.
FOURNIER J, PELISSIER B, ESQUERRE-TUGAYE MT (1986) Induction of lipoxygenase activity in cultured tobacco (Nicotiana tabacum) cells by elicitor of ethylene from Phytophthora parasitica var. nicotianae. CR Acad Sc Paris t 303, série III, n° 15 : 651-654.
JAMET E, FRITIG B (1986) Purification and characterization of 8 of the pathogenesis-related proteins in tobacco leaves reacting hypersensitively to tobacco mosaic virus. Plant Mol Biol 6 : 69-80.
LUPU R, GROSSMAN S, COHEN Y (1980) The involvement of lipoxygenase and antioxydants in pathogenesis of powdery mildew on tobacco plants. Physiol Plant Pathol 16 : 241-248.
OCAMPO CA, MOERSCHBACHER B, GRAMBOW HJ (1986) Increased lipoxygenase activity is involved in the hypersensitive response of wheat leaf cells infected with avirulent rust fungi or treated with fungal elicitor. Z Naturforsch 41c : 559-563.
PELISSIER B, THIBAUD JB, GRIGNON C, ESQUERRE-TUGAYE MT (1986) Elicitor preparations from fungal pathogens depolarize plant membranes. Plant Sci 46 : 103-109.
REDMOND, JW, BAYLEY M, DJORDJEVIC MA, INNES RW, KUEMPEL PL, ROLFE BG (1986) Flavones induce expression of nodulation genes in Rhizobium. Nature 323: 632-635.
RICKAUER M, FOURNIER J, ESQUERRE-TUGAYE MT (1989) Induction of proteinase inhibitors in tobacco cell suspension culture by elicitors of Phytophthora parasitica var. nicotianae. Plant Physiol 90 : 1065-1070.
ROBY D, TOPPAN A, ESQUERRE-TUGAYE MT (1986) Elicitors of ethylene from Colletotrichum lagennarium trigger chitinase activity in melon plants. Plant Physiol 81 : 228-233.
RUMEAU D, MAZAU D, PANABIERES F, DELSENY M, ESQUERRE-TUGAYE MT (1988) Accumulation of hydroxyproline-rich glycoprotein mRNAs in infected or ethylene treated melon plants. Physiol Mol Plant Physiol 33 : 419-428.
RYAN CA (1988) Oligosaccharides as recognition signals for the expression of defensive genes in plants. Biochem 27 : 8879-8883.
STACHEL SE, MESSENS E, VAN MONTAGU M, ZAMBRYSKI P (1985) Identification of the signal molecules produced by wounded plant cells that activate T-DNA transfer in Agrobacterium tumefaciens. Nature 318 : 624-629.
SCHMIDT WE, EBEL J (1987) Specific binding of a fungal glucan phytoalexin elicitor to membrane fractions from soybean Glycine max. Proc Natl Acad Sci USA 84 : 4117-4121.
TOPPAN A, ROBY D, ESQUERRE-TUGAYE MT (1982) In vivo effect of ethylene on hydroxyproline-rich glycoprotein accumulation in the cell wall of diseased plants. Plant Physiol 70 : 82-86.
VICK BA, ZIMMERMAN DC (1987) Oxidative systems for modification of fatty acids : the lipoxygenase pathway. In STUMPF PK (ed) The Biochemistry of Plants : a comprehensive treatise. Academic Press, New York, pp 53-90.
WOLOSHUK CP, KOLATTUKUDY PE (1986) Mechanism by which contact with plant cuticle triggers cutinase gene expression in the spores of Fusarium solani f. sp. pisi. Proc Natl Acad Sci USA 83 : 1704-1708.

ELICITORS AND DEFENSE GENE ACTIVATION IN CULTURED CELLS

Richard A. Dixon[1], Arvind D. Choudhary[1], Robert Edwards[1], Maria J. Harrison[1], Christopher J. Lamb[2], Michael A. Lawton[2], Mehrdad Mavandad[1], Bruce A. Stermer[1] and Lloyd Yu[1,2].

[1]Plant Biology Division, Samuel Roberts Noble Foundation, P. O. Box 2180, Ardmore, Oklahoma 73402, USA

[2]Plant Biology Laboratory, Salk Institute for Biological Studies, P.O. Box 85800, San Diego, California 92138, USA

Active defense of plants against fungal or bacterial pathogens often involves the rapid death of cells in intimate contact with the invading microorganism. Accompanying this so-called hypersensitive response (HR) are rapid localized changes in host metabolism which lead to the synthesis of potential defensive barriers ; these include the accumulation of low Mr antimicrobial compounds termed phytoalexins, the deposition of phenolic material and hydroxyproline-rich glycoproteins (HRGPs) in the host cell wall, and the synthesis of hydrolytic enzymes. The exact relationship between the HR and phytoalexin accumulation is still somewhat vague ; in some systems, hypersensitive cell death appears to be a pre-requisite for induction and accumulation of phytoalexins in neighboring healthy cells (Bailey 1982), whereas phytoalexin synthesis can be initiated in suspension cultured cells in the absence of serious effects on cell viability (Hamdan and Dixon 1986). It is, however, well documented that molecules from plant pathogenic fungi have the ability to induce both hypersensitive-type cell necrosis and/or phytoalexin accumulation in plant cells. These so-called elicitors often originate from the fungal cell wall.

The nature and activity of elicitors of fungal origin have been the subject of several reviews (Darvill and Albersheim 1984 ; Dixon 1986 : Lamb et al. 1989). It should be noted that race specificity (the ability of an elicitor to induce defense only on those host cultivars which exhibit resistance to the race of the pathogen from which the elicitor was isolated) has been demonstrated only in a limited number of cases (Dixon 1986). The exact role, if any, on non-race-specific elicitors in race-specific plant-pathogen interactions remains to be determined. Most

NATO ASI Series, Vol. H 47
Signal Perception and Transduction in Higher Plants
Edited by R. Ranjeva and A. M. Boudet

studies on the biochemistry and molecular biology of elicitation have, however, used such non-specific elicitors.

The simplest model for the induction of plant defense responses assumes that the elicitor binds to a specific receptor, probably located in the plant plasma membrane, and that this binding indirectly leads to changes in the transcriptional activity of genes involved in the elaboration of antimicrobial barriers by the host. The evidence for the presence of specific high-affinity elicitor binding sites in the soybean plasma membrane is now reasonably strong (Schmidt and Ebel 1987 : Cosio et al. 1988), although it is not yet known whether these are funtionally active. A large body of evidence has confirmed that elicitor-mediated changes in plant defense metabolism arise predominantly via transcriptional activation of defense response genes (Dixon and Harrison 1989 : Lamb et al 1989). The gap in our knowledge is therefore the link between elicitor binding and gene activation, i.e. the signal transduction pathway. No clear picture of the nature of this pathway has emerged from studies in which llevels of known response couplers in animal systems (e.g. polyamines, cyclic nucleotides, polyphosphoinositides) have been monitored in plant cells undergoing defense gene activation (reviewed in Dixon 1986). This may in part result from technical difficulties. A possible requirement for intracellular flow of calcium in elicitation has been suggested for soybean and carrot cell suspension cultures (Stab and Ebel 1987 ; Kurosaki et al. 1987), but the evidence is nevertheless indirect.

In the absence of any firm knowledge of the central components of defense-related signal transduction pathways, an alternative approach is to work backwards from the defense response pathways themselves. The present article approaches the problems of signal transduction in this context. It reviews (a), indirect evidence for the requirement for multiple signals based on differential expression patterns of defense reponse components and (b), the regulatory elements in defense response gene promoters which may bind specific *trans*-acting factors responsible for elicitor-mediated induction. Such *trans*-acting factors should be the final targets of the signal transduction pathway.

THE PHYTOALEXIN RESPONSE IN LEGUME CELLS : A MODEL SYSTEM FOR STUDING SIGNAL TRANSDUCTION

Cell suspension cultures of bean, soybean, alfalfa or chickpea respond to a range of elicitor molecules by accumulating their respective isoflavonoid phytoalexins (Dixon et al. 1983a ; Barz et al. 1988 ; Dalkin et al. 1989). This is preceeded by increases in the extractable activities of all or most of the enzymes involved in the biosynthesis of these phytoalexins. This in turn, where investigated, has been shown to result from increased transcriptional activation of the corresponding genes (Dixon and Harrison 1989). Such transcriptional activation is extremely rapid, occuring within less than five minutes of exposure of cells to elicitor (Lawton an Lamb 1987). It is therefore likely that the signal transduction pathway from elicitor binding to gene activation is relatively direct.

Measurement of transcript levels therefore provides a means of measuring the earliest results of signal transduction in elicited legume cells. As the enzymes and end products of the phytoalexin pathways are also readily assayable, it is possible to investigate whether different elicitors exert differential effects on the individual components of the pathways and thereby gain evidence for potential signal multiplicity. The availablity of a number of cloned genes for enzymes of these pathways (reviewed in Dixon and Harrison 1989) makes it possible to begin to dissect signal transduction at the level of elicitor-mediated events whose ultimate targets are the 5' upstream regions of these genes.

ELICITORS AS TOOLS FOR DEFINING THE COMPLEXITY OF SIGNAL TRANSDUCTION PATHWAYS

Although the physiological relevance of non-race-specific elicitors is open to debate, and several are relatively crude preparations, their use has revealed a number of interesting phenomena which may reflect the nature of signal transduction. One obvious feature is the ability of a particular elicitor to be effective in one species or cell type but not in others. As an example, crude polysaccharide preparations from cell walls of the bean pathogen _Colletotrichum lindemuthianum_ are effective elicitors of phytoalexin accumulation in cell suspension cultures of both bean and alfalfa. In contrast, the tripeptide glutathione is a strong and selective inducer of defense gene activation in bean cultures (Wingate et al. 1988), but is inactive towards alfalfa cultures growing under identical conditions (Dalkin et al 1989). The lack of effect of glutathione on alfalfa cell culture could be the result of lack of uptake into free intracellular pools (R. Edwards, unpubished results) ; alfalfa protoplasts are likewise unresponsive to glutathione, but are weakly elicited by the tripeptide in the presence of polyethylene glycol. Studies in electroporated alfalfa protoplasts have indicated that the promoter of the bean defense reponse gene chalcone synthase (CHS) is more inducible by glutathione than the endogenous alfalfa CHS, suggesting that the signalling pathway associated with glutathione action is operational in alfalfa but that promoters from the two organisms may show differential responsiveness (A.D. Choudhary and R.A. Dixon, unpublished results, see below).

Exposure of bean cells to crude or partially purified elicitor preparations from C. lindemuthianum results in differential effects with respect to the elicitor preparation used or the defense response gene under study. The following observations may relate to the potential complexity of elicitor-responsive signalling pathways.

(1) Increased transcription of phytoalexin biosynthetic genes (phenylalanine ammonia-lyase (PAL) and CHS) occurs almost immediately after addition of elicitor to cultured cells, whereas there is a lag of approximately 2 h before appearance of HRGP transcripts (Lawton and Lamb 1987). HRGP promoters are therefore probably activated by "secondary" signals.

(2) Elicitors from cell wall of C. lindemuthianum induce the early enzymes of phytoalexin biosynthesis and accumulation of isoflavonoid phytoalexins (Hamdan and Dixon 1986). In contrast, elicitors from the culture filtrate induce the same enzymes but no phytoalexin accumulation. It is not yet known whether the later enzymes are induced by culture filtrate elicitor ; if not, the pathway from phenylalanine to isoflavonoids may be activated by multiple signals.

(3) Fractionation of culture filtrate elicitor on concanavalin A Sepharose yields polysaccharide components, rich in mannose and galactose, which induce the extractable activity of CHS but do not induce PAL or chalcone isomerae (Hamdan and Dixon 1987b). This effect is dependent on the cell

line used. It is not known whether it reflects transcriptional events. Differential timing of PAL and CHS transcript accumulation has been reported in bean cotylons treated with different C. lindemuthianum elicitor fractions (Tepper et al. 1989).

(4) Individual bean CHS subunit isoforms (encoded by different members of the CHS multigene family) are induced in similar relative proportions in response to crude or fractionated C. lindemuthianum elicitors which differentially induce other polypeptides (Hamban and Dixon 1987a).

(5) Elicitor dose-response experiments have revealed complex concentration dependencies for induction of phytoalexin biosynthetic enzymes (Dixon et al), and differential effects on PAL synthesis and turnover at different elicitor concentrations (Lawton et al 1980).

(6) Pectic fragments (commercial preparations or of host origin) also induce phytoalexin accumulation in cultured bean cells, and the patterns of phytoalexins accumulating depend on the nature and concentration of the oligogalacturonide preparation (Dixon et al. 1989b). Synergism is observed between low concentrations of fungal elicitors and pectic fragments in the induction of phytoalexin biosynthesis in soybean and parsley (Davis et al 1986 ; Davis and Hahlbrock 1987), but not in bean cultures. The molecular level at which this synergism operates is not yet clear. Qualitative as well as quantitative changes in host response could imply co-cordination of different signalling pathways.

GENE EXPRESSION IN TRANSGENIC PLANTS : COMMONALITY OF ELICITOR-RESPONSIVE SIGNALLING PATHWAYS

Several studies to date have indicated that foreign genes may be expressed in transgenic plants in a tissue-specific and environmentally-regulated manner similar to that observed in the species of origin (Weising et al. 1988), implying that signal transduction pathways may be relatively conserved. This is well illustrated by the expression of legume (bean) defense gene promoters in a solanaceous plant (tobacco). A fragment of the bean CHS promoter (upstream to position -336 from the transcription start site) confers inducibility by wounding, UV irradiation, C. lindemuthianum elicitor, mercuric chloride or infection by Pseudomonas syringae, to the β-glucuronidase reporter gene in tobacco leaves (J. Schmidt, B. Stermer,

R.A. Dixon, C.J. Lamb, unpublished results) ; evidence to date suggests that the plant cell's own enzymes of flavonoid biosynthesis are not induced by microbial elicitors in solanaceous species (Dixon et al. 1988). A 1.1 kb fragment of the bean PAL promoter confers wound-inducibility to GUS in transgenic tobacco (Liang et al. 1989). Such studies provide the basis for functional analysis, by deletion, of regulatory sequences responsive to signals generated by elicitation or wounding (Thornburg et al. 1987 ; An et al. 1989 ; see below). In addition, measurement of GUS activity throughout transgenic tobacco leaves exhibiting a localized HR to P. syringae inoculated in a single area has indicated that the bean CHS promoter is activated both locally and at a distance from the HR site (B. Stermer, unpublished results).
This suggests that intercellular, transmissible signals may be conserved among plant species.

GENE EXPRESSION IN ELECTROPORATED PROTOPLASTS : FUNCTIONAL ANALYSIS OF CIS-ACTING REGULATORY SEQUENCES AND COMPLEXITY OF SIGNALLING PATHWAYS

Transient expression of gene promoter-reporter gene fusions is a relatively rapid method for the functional analysis of stimulus-responsive cis-acting elements. However, its use for elicitor studies is somewhat complicated due to the potential for elicitation to occur during the exposure of plant cells to the wall-degrading enzymes used for protoplast preparation. In soybean, protoplasts become elicited during preparation, but this declines on culturing and the protoplasts then become responsive to exogenously added elicitors (Dron et al. 1988). In alfalfa protoplasts, the extent of pre-elicitation depends on both the age (i.e. number of passages) of the cell culture from which the protoplasts are prepared, and the source of the wall-degrading enzymes (A.D. Choudhary and R.A. Dixon, unpublished results).
Expression of the bean CHS promoter fused to the bacterial chloramphenicol acetyltransferase (CAT) gene has been studied in both soybean and alfalfa protoplasts. In both cases, the promoter is responsive to C. lindemuthianum elicitor or glutathione ; in the case of alfalfa, the responsiveness to glutathione is not seen in the cells from which the protoplasts were isolated (see above). Deletion of a region of the bean

CHS promoter from -326 to -176 usptream of the transcriptional start site results in increased basal and elicitor-induced expression of the CAT gene in soybean protoplasts (Dron et al. 1988), and co-electroporation of this region in trans with the intact promoter-CAT construct likewise results in increased CAT expression. In contrast, in alfalfa protoplasts the deleted promoter supports approximately 2.5-fold less CAT expression, and preliminary data indicate that co-electroporation of the -326 to -173 region in trans results in increased basal CAT expression, a reduction below basal levels in response to C. lindemuthianum elicitor, and a markedly increased response to glutathione. These data suggest that the -326 to -173 region of the bean CHS promoter contains both negative (silencer) and positive (enhancer) regulatory elements, and that the expression patterns reflect the physiological status of the recipient protoplasts with respect to the relative level of trans-acting proteins which recognize these elements. They further suggest that elicitation by glutathione may be mediated by a signal pathway distinct in at least some feactures from that involved in fungal-elicitor-mediated gene activation, although elicitation by both ultimately involves factors which recognize the same region of the promoter.

Further deletion analysis of the CHS promoter has shown that a region between -130 and the TATA box is necessary for transcriptional activation by elicitors (Dron et al. 1988). This region, which has sequence homology to a region in the bean PAL promoter, has also been defined as an elicitor-responsive element on the basis of in vivo genomic footprinting of PAL genes in elicited parsley cells (Lois et al. 1989).

TRANS-ACTING DNA-BINDING PROTEINS

In vitro DNAse 1 footprinting and/or gel-retardation assays have shown that bean and alfalfa nuclear extracts contain proteins which bind specifically to the region of the bean CHS promoter ascribed a regulatory function on the basis of transient expression assays (Lawton et al. 1989 ; Dixon et al. 1989a). Three distinct protein-binding regions have been identified within the -326 to -173 fragment (Lawton et al. 1989) ; these can independently bind nuclear factors (L. Yu and M.J. Harrison, unpublished results). A nuclear protein of subunit Mr 55.000 has been observed by UV cross-linking to an oligonucleotide complementary to the strongest of the 3 binding sites (M.J. Harrison, unpublished results). Work is currently is progress to purify these trans-acting factors and to clone their genes.

The rapidity of elicitor-induced gene transcription, and studies on mammalian transcription factors, suggest that trans-acting factors are probably expressed constitutively in a given cell type and act as the last step in their signal transduction chain after stimulus-dependent post-translational modification. Thus, once these factors have been purified and cloned, they can be used as substrates for identifying the enzymes that modify them. These enzymes may in turn be activated by factors (e.g. elevated Ca^{2+} levels) closely associated with the initial effects of elicitor binding. Other additional signals may be necessary to target to the specific transduction loop for defense gene activation.

PLANT PHENOLICS AS SIGNAL MOLECULES

Recent results from a number of laboratories have revealed important biological activities for plant phenolic compounds, including gene activation (for Rhizobium nod genes and Agrobacterium vir genes), cell division promoting activity, regulation of hormone transport and transcriptional down-regulation of defense pathways (reviewed in Dixon and Lamb 1989). Although it is unlikely that phenolic compounds act in the initial stages of elicitor-mediated activation of plant defense response genes, they could act as secondary signals, either to modulate the expression of induced pathways or to transmit the elicitation response to surrounding cells. At present, evidence for such a role is only indirect. Cultured alfalfa cells exposed to C. lindemuthianum elicitor preparations release elicitor-active phenolic compounds into the culture medium. These compounds do not appear to accumulate intracellularly (R. Edwards, A. Harris, R.A. Dixon, unpublished results). Previous results have shown that bean cells release low Mr elicitors in response to macromolecular elicitors (Dixon et al. 1983b), and that the elicitor activity of autoclaved bean hypocotyl extracts does not appear to reside solely in polysaccharide or proteinaceous components (Dixon et al. 1989b). The very rapid induction of coniferyl alcohol dehydrogenase in elicited bean cells

(Walter et al. 1988) may be indicative of a function for cinnamyl alcohols other than as lignin precursors during plant defense ; this idea is attractive in view of the recent demonstration of the cell-division promoting activity of dehydrodiconiferyl glycosides (Lynn et al. 1987).
A larger body of evidence points to a possible role for the immediate product of the PAL reaction (Trans-cinnamic acid) as a signal molecule monitoring the flux through the phenylpropanoid pathway and acting to down-regulate the pathway at the transcriptional and post-translational levels. The basis for this model is summarized below for the bean cell culture system :

(1) Cinnamic acid selectively inhibits the appearance of transcripts encoding defense response enzymes in bean, while selectively inducing the appearance of other transcripts (Bolwell et al. 1988 ; M. Mavandad and R.A. Dixon, unpublished results).

(2) A proteinaceous PAL-inactivating factor is synthesized in response to cinnamic acid (Bolwell et al. 1986 ; M. Mavandad and R.A. Dixon, unpublished results).

(3) Inhibition of cinnamic acid production in vivo, by application of the PAL inhibitor AOPP, results in increased elicitor-induction of PAL and CHS transcripts (Bolwell et al. 1988) and, in the case of PAL, prevents the decline in enzyme activity which occurs after synthesis has ceased.

(4) The bean CHS promoter is an oder of magnitude more sensitive to transcriptional inactivation by cinnamic acid than is the cauliflower mosaic virus 35S promoter, as assessed by transient expression in electroporated alfalfa protoplasts (A.D. Choudhary and R.A. Dixon, unpublished results). Deletion analysis and co-electroporation experiments suggest the presence of cinnamic acid-responsive cis-acting sequences in the CHS promoter (M. Dron, unpublished results ; A.D. Choudhary, unpublished results).

(5) Transgenic tobacco plants over-expressing the bean PAL 2 gene have highly depressed PAL activity, very low soluble phenolic content and reduced lignin development, consistent with the dis-regulation of an endogenous cinnamic acid regulated signal pathway (Y. Elkind, R. Edwards, M. Mavandad, R.A. Dixon and C.J. Lamb, unpublished results).

It is likely that the next few years will witness a revival of interest in plant secondary metabolites as further examples of their regulatory roles are revealed.

CONCLUDING REMARKS

We still understand very little of the actual molecular events which link reception of microbial stimuli to activation of plant defense metabolism. It is likely that the pathways involved will be under subtle regulation as parts of a larger network of regulatory loops transducing signals involved in "housekeeping", developmental and environmental responses. The studies with elicitors described above have rather posed questions of complexity than defined mechanisms. New techniques for the molecular analysis of stimulus-responsive transcription factors (see Dixon and Harrison 1989), or genetic analysis in model systems such as Arabidopsis, should allow a more directed approach to the elucidation of components at the end of the signal transduction chain. This work will hopefully reveal links with transduction components defined by biochemical analysis.

ACKNOWLEDGEMENTS

We thank Scotty Mc Gill for preparation of the manuscript.

REFERENCES

An G, Mitra A, Choi MK, Costa MA, An K, Thornburg RW and Ryan, CA (1989). Functional analysis of the 3' control region of the potato wound-inducible proteinase inhibitor II gene. The plant Cell 1, 115-122.

Bailey JA (1982). Physiological and biochemical events associated with expression of resistance to disease. In : Active Defense Mechanisms in Plants, ed R.K.S. Wood, Plenum Press, New York, pp 39-65.

Barz W, Daniel S, Hinderer W, Jaques U, Kessmann H, Koster J, Otto C and Tiemann K (1988). Elicitation and metabolism of phytoalexins in plant cell cultures. In : Application of Plant Cell and Tissue Culture, Ciba Foundation Symp. No. 137, Wiley, Chichester, pp 178-198.

Bolwell GP, Cramer CL, Lamb CJ, Schuch W and Dixon RA (1986). L-phenylalanine ammonia-lyase from Phaseolus vulgaris : Modulation of the levels of active enzyme by trans-cinnamic acid. Planta 169, 97-107.

Bolwell GP, Mavandad M, Millar DJ, Edwards KJ, Schuch W and Dixon RA (1988). Inhibition of mRNA levels and activities by trans-cinnamic acid in elicitor-induced bean cells. Phytochemistry 27, 2109-2117.

Cosio EG, Popperl H, Schmidt W and Ebel J (1988). High-affinity binding of fungal β-glucan fragments to soybean (Glycine max L.) microsomal fractions and protoplasts. Eur. J. Biochem. 175, 309-315.

Dalkin K, Edwards R, Edington B and Dixon RA (1989). Defense responses in alfalfa (Medicago sativa L.) I. Elicitor induction of phenylpropanoid biosynthesis and hydrolytic enzymes in cell suspension cultures. Plant Physiol., submitted.

Darvill AG and Albersheim P (1984). Phytoalexins and their elicitors a defense against microbial infection in plants. Ann. Rev. Plant Physiol., 35, 243-298.

Davis KR, Darvill AG and Albersheim P (1986). Host-pathogen interactions XXXI. Several biotic and abiotic elicitors act synergistically in the induction of phytoalexin accumulation in soybean. Plant Mol. Biol., 6, 23-32.

Davis KR and Hahlbrock K (1987). Induction of defense responses in cultured parsley cells by plant cell wall fragments. Plant Physiol. 85, 1286-1290.

Dixon RA (1986). The phytoalexin response : elicitation, signalling and the control of host gene expression. Biol. Rev. 61, 239-291.

Dixon RA, Blyden ER, Robbins MP, van Tunen AJ and Mol JNM (1988). Comparative biochemistry of chalcone isomerases. Phytochemistry, 27, 2801-2808.

Dixon RA, Dey PM and Lamb CJ (1983a). Phytoalexins : enzymology and molecular biology. Adv. Enzymol. Rel. Areas Mol. Biol. 53, 1-136.

Dixon RA, Dey PM, Lawton MA and Lamb CJ (1983b). Phytoalexin induction in French bean. Intracellular transmission of elicitation in cell suspension cultures and hypocotyl sections of Phaseolus vulgaris. Plant Physiol. 71, 251-256.

Dixon RA, Dey PM, Murphy DL and Whitehead IM (1981). Dose response for Colletotrichum lindemuthianum elicitor-mediated enzyme induction in French bean cell suspension cultures. Planta 151, 272-280.

Dixon RA and Harrison MJ (1989). Activation, structure and organization of genes involved in microbial defense in plants. Adv. Genetics, in press.

Dixon RA, Harrison MJ, Lawton MA, Jenkins S and Lamb CJ (1989a). Defense gene transcription factors. J. Cell Biochem. Suppl. 13D, 247.

Dixon RA, Jennings AC, Davies La, Gerrish C and Murphy DL (1989b). Elicitor-active components from French bean hypocotyls. Physiol. Mol. Plant Pathol. 34, 99-115.

Dixon RA and Lamb CJ (1989). Regulation of secondary metabolism at the biochemical and genetic levels. Ann. Proc. Phytochem. Soc. Europe, in press.

Dron M, Clouse SD, Dixon RA, Lawton MA and Lamb CJ (1988). Glutathione and fungal elicitor regulation of a plant defense gene promoter in electroporated protoplasts. Proc. Natl. Acad. Sci. USA 85, 6738-6742.

Hamdan MAMS and Dixon RA (1986). Differential biochemical effects of elicitor preparations from Colletotrichum lindemuthianum. Physiol. Mol. Plant Pathol. 28, 329-344.

Hamdan MAMS and Dixon RA (1987a). Differential patterns of protein synthesis in bean cells exposed to elicitor fractions from Colletotrichum lindemuthianum. Physiol. Mol. Plant Pathol. 31, 105-121.

Hamdan MAMS and Dixon RA (1987b). Fractionation and properties of elicitors of the phenylpropanoid pathway from culture filtrates of Colletotrichum lindemuthianum. Physiol. Mol. Plant Pathol. 31, 91-103.

Kurosaki R, Tsurusawa Y and Nishi A (1987). The elicitation of phytoalexins by Ca^{2+} and cyclic AMP in carrot cells. Phytochemistry 26, 1919-1923.

Lamb CJ, Lawton MA, Dron M and Dixon RA (1989). Signals and transduction mechanisms for activation of plant defenses against microbial attack. Cell 56, 215-224.

Lawton MA, Dixon RA and Lamb CJ (1980). Elicitor modulation of the turnover of L-phenylalanine ammonia-lyase in French bean cell suspension cultures. Biochem. Biophys. Acta 633, 162-175.

Lawton MA, Kragh K, Jenkins SM, Harrison MJ, Yu L, Dron M, Dixon RA and Lamb CJ (1989). Multiple binding sites for nuclear factors are present in the silencer region of the promoter for the bean chalcone synthase gene. EMBO J., submitted.

Lawton MA and Lamb CJ (1987). Transcriptional activation of plant defense genes by fungal elicitor, wounding and infection. Mol. Cell Biol. 7, 335-341.

Liang X, Dron M, Schmidt J, Dixon RA and Lamb CJ (1989). Developmental and environmental regulation of a phenylalanine ammonia-lyase : β-glucuronidase gene fusion in transgenic tobacco plants. Proc. Natl. Acad. Sci USA, submitted.

Lois R, Dietrich A and Hahlbrock K (1989). A phenylalanine ammonia-lyase gene from parsley : structure, regulation and identification of elicitor and light-responsive cis-acting elements. EMBO J. 8, 1641-1648.

Lynn DG, Chen RH, Mannings KS and Wood HN (1987). The structural characterization of endogenous factors from Vinca rosea crown gall tumors that promote cell division of tobacco cells. Proc. Natl. Acad. Sci. USA, 84, 615-619.

Schmidt WE and Ebel J (1987). Specific binding of a fungal glucan phytoalexin elicitor to membrane fractions from soybean Glycine max. Proc. Natl. Acad. Sci. USA, 84, 4117-4121.

Stab MR and Ebel J (1987). Effects of Ca^{2+} on phytoalexin induction by fungal elicitor in soybean cells. Arch. Biochem. Biophys. 257, 416-423.

Tepper CS, Albert FG and Anderson AJ (1989). Differential mRNA accumulation in three cultivars of bean in response to elicitors from Colletotrichum lindemuthianum. Physiol. Mol. Plant Pathol. 34, 85-98.

Thornburg RW, An G, Cleveland TE, Johnson R and Ryan CA (1987). Wound-inducible expression of a potato inhibitor II - chloramphenicol acetyltransferase gene fusion in transgenic tobacco plants. Proc. Natl. Acad. Sci. USA, 84, 744-748.

Walter MH, Grima-Pettenati J, Grand C, Boudet AM and Lamb CJ (1988). Cinnamyl alcohol dehydrogenase, a molecular marker specific for lignin synthesis : cDNA cloning and mRNA induction by fungal elicitor. Proc. Natl. Acad. Sci. USA, 85, 5546-5550.

Weising K, Schell J and Kahl G (1988). Foreign genes in plants : transfer, structure, expression, and applications. Annu. Rev. Genet., 22, 421-477.

Wingate VPM, Lawton MA and Lamb CJ (1988). Glutathione causes a massive and selective induction of plant defense genes. Plant Physiol. 87, 206-210.

TRANSDUCTION OF THE GRAVITY SIGNAL IN PLANTS

A. Sievers
Botanisches Institut
Universität Bonn
Venusbergweg 22
D-5300 Bonn 1
Federal Republic of Germany

Graviperception in the root cap of Lepidium, a model system

The root cap of Lepidium, which has been characterized in detail by structural, physiological and electrophysiological evidences, offers the opportunity to formulate a sequence of events, how gravity perception occurs (Sievers et al. 1984; Sievers 1986).

Immediately upon gravistimulation of roots, the direction of their amyloplast pressure changes accordingly. Within seconds, i.e. clearly before the sedimenting amyloplasts reach a new stable position, the membrane potential of the lower flank statocytes transiently depolarizes (Behrens et al. 1985), while the membrane potential of the upper flank statocytes hyperpolarizes. This leads to an ion-flux towards the upper flank of the root cap. Such a phenomenon was measured after 30 s as an outwards directed electrical current (Behrens et al. 1982; Björkman and Leopold 1987a with maize; Iwabuchi et al. 1989).

Since it is known that endoplasmic reticulum (ER) vesicles from Lepidium roots are able to accumulate calcium actively (Buckhout 1983), it is supposed that a release of calcium from the ER occurs. Activation of plasmalemma located calcium and/ or other ion-pumps are possibly involved. The electrical coupling of root cells is more pronounced in longitudinal than transversal direction (electrical signal reduction at 390 μm in longitudinal vs. 140 μm in transversal direction; Behrens and Gradmann 1985), facilitating a differential ion-flow towards the elongation zone. Symplastic electrical coupling alone is

NATO ASI Series, Vol. H 47
Signal Perception and Transduction in Higher Plants
Edited by R. Ranjeva and A.M. Boudet

not sufficient for signal transmission, since the zone of maximal bending of cress roots is located 1-2 mm beyond the tip (Selker and Sievers 1987).

Which elements are responsible for the susception-perception transition, i.e. for the stimulus transduction proper, is still a matter of debate. Since it was shown (Wendt et al. 1987) that the distal ER-complex as an entity is not required for graviperception, the most reasonable candidate for stimulus transduction appears to be the ER-cisterna(e), located directly beneath the plasmalemma. This (these) ER-cisterna(e) is (are) anchored in its (their) position by microtubules (Hensel 1984), probably also by microfilaments (Hensel 1987). They may represent the functionally significant calcium-store, which releases and sequesters calcium according to the physical stimulus. Data on the sensitivity of roots in relation to the amount of distal ER will help in the future to define this function and assess its importance. The distal ER-complex may be regarded as an amplifier for graviperception.

The cell biological aspect: Statoliths and microfilaments in gravity perceiving plant cells

Most authors agree that the presence of sedimentable particles (statoliths) within the gravisensitive cells are responsible for graviperception (Volkmann and Sievers 1979; Wilkins 1984). In *Chara*, a green alga, barium-sulfate crystals filled small vacuoles had been identified as statoliths (Sievers and Schnepf 1981).

The most reasonable candidates to serve as statoliths in vascular plants are the starch filled amyloplasts (starch-statolith theory). Conclusive evidence for this was provided by Iversen (1969), who found that hormone-treated destarched roots of *Lepidium* were not able to respond gravitropically, while the growth of the roots remained widely unaffected. However, in addition to the hydrolysis of starch, a destruction of statocyte polarity was demonstrated to occur as an effect of hormone treatment (Busch and Sievers 1987).

Recently Caspar and Pickard (1989) demonstrated that the presence of starch in plastids is unnecessary for graviperception in a mutant of Arabidopsis. However, amyloplasts are necessary for full gravitropic sensitivity, and starchless, non sedimented plastids may also function as statoliths (Kiss et al. 1989). Here the question arises if the position and the translocation of statoliths in gravity perceiving cells is dependent on and influenced by elements of the cytoskeleton. In a first step actin filaments in two cell types which transduce the gravity stimulus are studied (Sievers et al. 1989).

First evidence for an involvement of microfilaments in stabilizing statoliths' position in the Chara rhizoid came from experiments with cytochalasin B (Hejnowicz and Sievers 1981) which causes sedimentation of the statoliths in apical direction. Recently, Bartnik and Sievers (1988) described a spherical aggregate of ER located beneath the statoliths. The ER aggregate disappears after treatment with cytochalasin B or phalloidin, followed by a sedimentation of the statolith, demonstrating that the integrity of the polar cell organization is actin-dependent. Microfilaments are made visible by labelling of actin with rhodamine conjugated phalloidin (Sievers et al. 1989). The rhizoid contains thick microfilament-bundles arranged longitudinally in its basal region. However, in the subapical and in the apical region, much thinner bundles exist. They contact the statoliths encircling them by a dense envelope.

The function of actin filaments in relation to positioning of statoliths in Chara rhizoids was studied under microgravity in the sounding rocket TEXUS (Volkmann, Sievers, Buchen: unpublished). After 6 min microgravity the statoliths are dislocated from their original site of sedimentation. The direction of dislocation is opposite to the direction of the originally acting gravity vector. The distance of the center of the statolith complex from the apical cell wall nearly doubled till the end of the microgravity period (i.e. during 6 min 55 s from 17 µm to 33 µm). The whole complex of statoliths changed its form drastically (Fig. 1). Under 1G condition, it is a flat disc mainly transversally oriented with respect to the geometry

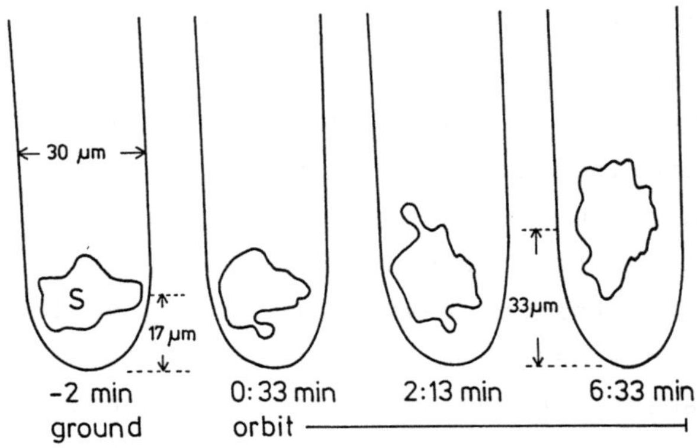

Fig. 1. A series of time-lapse drawings illustrating the dislocation of the statolith complex (S) in a Chara rhizoid under microgravity

of the rhizoid; however, under microgravity it becomes pear-shaped with its main axis longitudinally oriented.

The presence of microfilaments in root statocytes of Lepidium has also been claimed on account of experiments with cytochalasins (Hensel 1985; Wendt and Sievers 1986; Wendt et al. 1987) and proved by rhodaminyl-phalloidin labelling (Hensel 1989). Microfilaments play a role in stabilizing the positions of the nucleus and of the distal ER complex and are involved in ER-retransport after centrifugation. The relationship between statoliths (amyloplasts) and microfilaments in statocytes is demonstrated by the fact that cytochalasin B application results in an approximately three-fold increase of statolith sedimentation rates relative to controls (Sievers et al. 1989).

It is concluded from observations in both cells, the Chara rhizoid and the root statocyte of Lepidium, that microfilaments are involved in controlling the position of statoliths, and that under gravity a considerable tension on these microfilaments exists. In gravity perceiving plant cells statoliths may trigger the transduction mechanisms via actin filaments.

It is still an open question if the statoliths destroy the filament network during sedimentation when the organism is tilted out of the vertical direction, or if they sediment together with the surrounding and possibly attached microfilaments. Statoliths' movement may cause both, mechanical disruption followed by an active reorganization and/or stretching and destretching of filament bundles which possibly triggers graviperception. If the filament network is not destroyed by moving statoliths, actin filaments would be able in focussing and transmitting stress to their anchorage sites located at the plasma membrane and/or at ER-membranes. These sites may function as calcium pumps which exist in ER membranes (Sievers et al. 1984). Also stress activated ion-channels are made responsible for graviperception (Edwards and Pickard 1987). Wendt and Sievers (1989) recently demonstrated that the gravisensitivity of roots is dependent on the calcium concentration in statocytes. It should be mentioned that in both gravity sensing cell types, the _Lepidium_ root statocyte and the _Chara_ rhizoid, a special ER complex or an ER aggregate, respectively, is located beneath the statoliths, which in both cases may function as a calcium store.

The physiological aspect: Calcium and calmodulin in transduction of the gravity signal

In order to examine the role of calcium in graviperception, the calcium ionophore A23187 was used to elevate the concentration of free cytoplasmic calcium in statocytes of the roots of _Lepidium_ (Wendt and Sievers 1989). After a brief incubation (30 min) in a medium that contained 10 µM A23187 and 5.5 mM $CaCl_2$, 50 % of the roots bent gravitropically during a subseqent 2 h of horizontal exposure, with an angle of curvature that varied from 5° to 70°. The corresponding statocytes exhibited a polar arrangement of cell organelles as did the controls (Fig. 2). However, in statocytes from 50 % of the roots which were not curved after gravistimulation a portion of the distal ER complex was displaced in the direction of gravity within 30 min of horizontal exposure. After washing of the briefly treated roots for 24 h with 1 % dimethylsulfoxide the percentage of gravitropically bending roots increased to approx. 80 %, but the angle of curvature amounted to only 5°-10°. Longer treatment (2 h) with A23187 caused a complete loss of gravi-responsiveness which was accompanied by disintegration of sta-tocyte polarity. It is concluded from these results that calcium is involved in graviperception.

Calmodulin occurs in root apices in amyloplasts and in the cytoplasm (Dauwalder et al. 1986). Calmodulin antagonists, such as calmidazolium, compound 48/80 and chlorpromazine, inhibit the graviresponse of roots (Biro et al. 1982; Björkman and Leopold 1987b). In a light requiring gravitropic cultivar of maize, the activity of calmodulin increases upon illumination (Stinemetz et al. 1987). These data indicate a direct invol-vement of a calcium calmodulin mechanism in graviperception and in graviresponse. It should be mentioned that up to now contradictory data have been published about the microscopical localization of calcium in situ.

Nevertheless, calcium movement is involved in the percep-tion-response mechanism in roots. This was indicated by the reversible loss of gravitropic sensitivity after application of calcium chelators (Lee et al. 1983a). In root caps, calcium moves towards the lower flank of the tissue upon gravitropic stimulation. This seems to be in part an active process (Lee et al. 1983b). The latter was confirmed by Moore (1985a, b, 1986), who furthermore showed that such a downward movement of calcium was reduced in the amylomaize mutant of *Zea mays* and was lacking from non-responding roots of *Phaseolus* and *Allium*.

The polar movement of calcium appears to be directly linked together with the polar transport of auxin within the root cap, and as a consequence, with a differential transport of auxin into the physically lower root flank (Lee et al. 1984; Lee and Evans 1985; Evans and Hasenstein 1987).

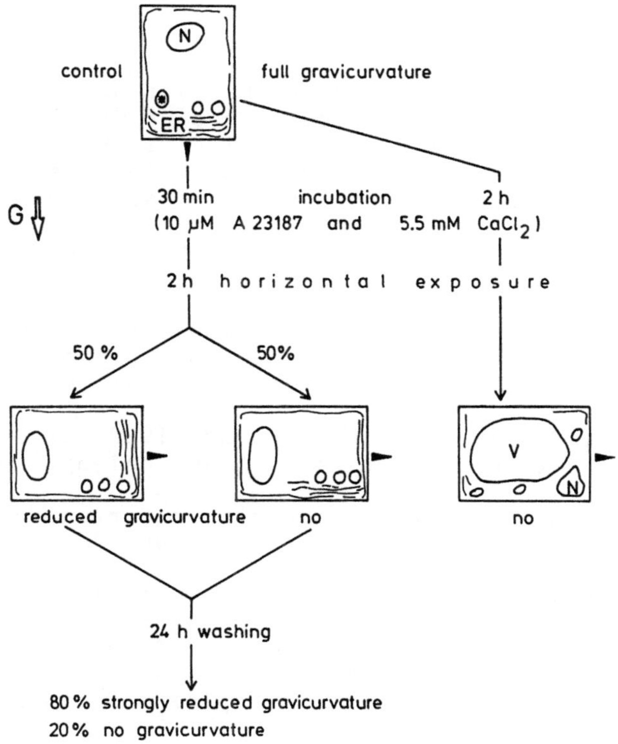

Fig. 2. Elevation of the concentration of free cytoplasmic calcium in statocytes of the roots of Lepidium causes destruction of structural polarity in statocytes and reduces gravicurvature of roots. Rectangles: central root statocytes demonstrating the arrangement of ER cisternae (streaks), amyloplasts (*), nucleus (N) and vacuole (V) after different treatment of the roots. G: direction of gravity. Small triangles: direction of the tip of the root cap

REFERENCES

Bartnik E, Sievers A (1988) In-vivo observations of a spherical aggregate of endoplasmic reticulum and of Golgi vesicles in the tip of fast growing Chara rhizoids. Planta 176:1-9

Behrens HM, Gradmann D (1985) Electrical properties of the vertically growing root tip of Lepidium sativum L. Planta 163:453-462

Behrens HM, Gradmann D, Sievers A (1985) Membrane-potential responses following gravistimulation in roots of Lepidium sativum L. Planta 163:463-472

Behrens HM, Weisenseel MH, Sievers A (1982) Rapid changes in the pattern of electric current around the root tip of Lepidium sativum L. following gravistimulation. Plant Physiol 70:1079-1083

Biro RL, Hale II CC, Wiegand OF, Roux SJ (1982) Effects of chlorpromazine on gravitropism in Avena coleoptiles. Ann Bot 50:735-747

Björkman T, Leopold AC (1987a) An electric current associated with gravity sensing in maize roots. Plant Physiol 84:841-846

Björkman T, Leopold AC (1987b) Effect of inhibitors of auxin transport and of calmodulin on a gravisensing-dependent current in maize roots. Plant Physiol 84:847-850

Buckhout TJ (1983) ATP-dependent Ca^{2+} transport in endoplasmic reticulum isolated from roots of Lepidium sativum L. Planta 159:84-90

Busch M, Sievers A (1987) Hormone treatment of roots causes reversible loss of structural polarity in statocytes. Abstr XIV Int Bot Congr.24 July-1 Aug 1987. Berlin, p 140

Caspar T, Pickard BG (1989) Gravitropism in a starchless mutant of Arabidopsis. Implications for the starch-statolith theory of gravity sensing. Planta 177:185-197

Dauwalder M, Roux SJ, Hardison L (1986) Distribution of calmodulin in pea seedlings: Immunocytochemical localization in plumules and root apices. Planta 168:461-467

Evans ML, Hasenstein K-H (1987) Stimulus-response coupling in the action of auxin and gravity on roots. In: Cosgrove DJ, Knievel DP (eds) Physiology of cell expansion during plant growth. The Amer Soc of Plant Physiol, p 202

Edwards KL, Pickard BG (1987) Detection and transduction of physical stimuli in plants. In: Wagner E, Greppin H, Millet B (eds) NATO ASI Series, vol 12. The cell surface in signal transduction. Springer, Berlin Heidelberg New York, p 41

Hejnowicz Z, Sievers A (1981) Regulation of the position of statoliths in Chara rhizoids. Protoplasma 108:117-137

Hensel W (1984) A role of microtubules in the polarity of statocytes from roots of Lepidium sativum L. Planta 162:404-414

Hensel W (1985) Cytochalasin B affects the structural polarity of statocytes from cress roots (Lepidium sativum L.) Protoplasma 129:178-187

Hensel W (1987) Cytodifferentiation of polar plant cells: Formation and turnover of endoplasmic reticulum in root statocytes. Exp Cell Res 172:377-384

Hensel W (1989) Tissue slices from living root caps as a model system in which to study cytodifferentiation of polar cells. Planta 177:296-303
Iversen T-H (1969) Elimination of geotropic responsiveness in roots of cress (Lepidium sativum) by removal of statolith starch. Physiol Plant 22:1251-1262
Iwabuchi A, Yano M, Shimizu H (1989) Development of extracellular electric pattern around Lepidium roots: its possible role in root growth and gravitropism. Protoplasma 148:94-100
Kiss JZ, Hertel R, Sack FD (1989) Amyloplasts are necessary for full gravitropic sensitivity in roots of Arabidopsis thaliana. Planta 177:198-206
Lee JS, Evans ML (1985) Polar transport of auxin across gravistimulated roots of maize and its enhancement by calcium. Plant Physiol 77:824-827
Lee JS, Mulkey TJ, Evans ML (1983a) Reversible loss of gravitropic sensitivity in maize roots after tip application of calcium chelators. Science 220:1375-1376
Lee JS, Mulkey TJ, Evans ML (1983b) Gravity-induced polar transport of calcium across root tips of maize. Plant Physiol 73:874-876
Lee JS, Mulkey TJ, Evans ML (1984) Inhibition of polar calcium movement and gravitropism in roots treated with auxin-transport inhibitors. Planta 160:536-543
Moore R (1985a) Movement of calcium across tips of primary and lateral roots of Phaseolus vulgaris. Am J Bot 72:785-787
Moore R (1985b) Calcium movement, graviresponsiveness and the structure of columella cells and columella tissues in roots of Allium cepa L. Ann Bot 56:173-187
Moore R (1986) Calcium movement, graviresponsiveness, and the structure of columella cells in primary roots of amylomaize mutants of Zea mays. Am J Bot 73:417-426
Selker JML, Sievers A (1987) Analysis of extension and curvature during the graviresponse in Lepidium roots. Am J Bot 74:1863-1871
Sievers A (1986) Transduction of the gravity stimulus in cress roots: A possible role of an ER-localized Ca^{2+} pump. In: Trewavas AJ (ed) Molecular and cellular aspects of calcium in plant development, Plenum, New York, p 225
Sievers A, Behrens HM, Buckhout TJ, Gradmann D (1984) Can a Ca^{2+} pump in the endoplasmic reticulum of the Lepidium root be the trigger for rapid changes in membrane potential after gravistimulation? Z Pflanzenphysiol 114:195-200
Sievers A, Kruse S, Kuo-Huang L-L, Wendt M (1989) Statoliths and microfilaments in plant cells. Planta in press
Sievers A, Schnepf E (1981) Morphogenesis and polarity of tubular cells with tip growth. In: Kiermayer O (ed) Cytomorphogenesis in plants. Springer, Wien New York, p 265
Stinemetz CL, Kuzmanoff KM, Evans ML, Jarrett HW (1987) Correlation between calmodulin activity and gravitropic sensitivity in primary roots of maize. Plant Physiol 84:1337-1342
Volkmann D, Sievers A (1979) Graviperception in multicellular organs. In: Haupt W, Feinleib ME (eds) Enc Plant Physiol NS vol 7. Springer, Berlin Heidelberg New York, p 573

Wendt M, Kuo-Huang L-L, Sievers A (1987) Gravitropic bending of cress roots without contact between amyloplasts and complexes of endoplasmic reticulum. Planta 172:321-329
Wendt M, Sievers A (1986) Restitution of polarity in statocytes from centrifuged roots. Plant Cell Environ 9:17-23
Wendt M, Sievers A (1989) The polarity of statocytes and the gravisensitivity of roots are dependent on the concentration of calcium in statocytes. Plant Cell Physiol in press
Wilkins MB (1984) Gravitropism. In: Wilkins MB (ed) Advanced plant physiology. Pitman, London, p 163

COMPARISON OF PLANT AND ANIMAL SIGNAL TRANSDUCING SYSTEMS

D. James Morré
Department of Medicinal Chemistry
Purdue University
West Lafayette, IN 47907 USA

A number of substances, ions, cyclic nucleotides and products of phospholipid hydrolysis, have been reported from animal systems to function as second messengers important to growth control. These substances function as parts of the same or different signal transducing systems and are produced in response to external stimuli transmitted across the plasma membrane to effect growth limiting reactions at the interior of the cell (Fig. 1).

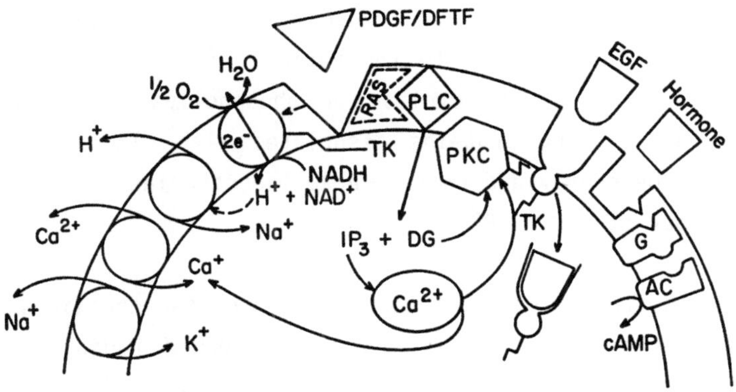

Figure 1. Sequence of events at the surface of a stimulated animal cell. Adapted and redrawn from Neumann (1985)

Despite the occurrence of similar molecules in both plants and animals, the signal transduction pathways of plants and animals for growth control may exhibit few parallels. This divergence is most apparent for the control of cell elongation by auxin where the overall mechanism may be unique to plants and apparently does not follow the same sequence of events as do the signal transduction paradigms of animal cells that respond to hormones and growth factors.

NATO ASI Series, Vol. H 47
Signal Perception and Transduction in Higher Plants
Edited by R. Ranjeva and A. M. Boudet

Hormone Receptors

Plant membranes, the plasma membrane included, bind auxins both specifically and reversibly (Hertel et al., 1972; Williamson et al., 1977; Ray et al., 1977b; Venis, 1986). However, other than high-affinity binding, there is little additional information upon which to classify the auxin-binding component of the plasma membrane as a hormone receptor. Auxin-binding proteins have been purified (Venis, 1980; Löbler and Klämbt, 1985) but little is known of their properties, function, or precise membrane location within the cell. However, the general animal mechanism of control by phosphorylation of tyrosine residues by means of a receptor-associated tyrosine protein kinase has already been largely excluded for the auxin receptor.

GTP-Binding (G) Proteins

In animals, GTP-binding proteins, for example the α-subunits of G-proteins, are frequently involved as part of transducing complexes involved in signal transduction from the receptor to the various effector systems diagrammed in Fig. 1 such as adenylate cyclase or phospholipase C (Gilman, 1987). Evidence for GTP-binding proteins associated with plant membranes (e.g. microsomes) has been provided from several laboratories (Dillenschneider et al., 1986; Drøbak et al., 1988).

Plasma membranes from etiolated hypocotyls of soybean contain a GTP-binding protein based on competitive displacement of bound radiolabeled GTP-γ-S (Fig. 2). The relative molecular weights of 28 and 24 kD correspond to none of the proteins of plasma membranes identified by Blum et al. (1988)

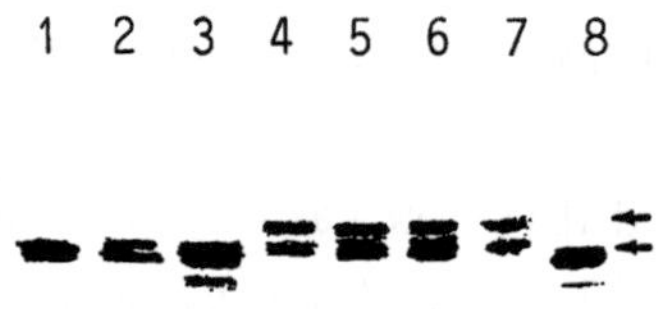

Fig. 2. Autoradiogram illustrating and comparing the major proteins binding [^{35}S]GTP-γ-S from plasma membranes isolated from soybean hypocotyls (lane 5), rat liver (lane 1), soybean root (lane 7) and rat adipocytes (lane 8), rat liver transitional endoplasmic reticulum (lane 3) and soybean hypocotyl microsomes (lane 4). Reference fractions include lower phase membranes from rat liver (lane 2) and soybean hypocotyl (lane 6). From a study with B. Zbell, University of Heidelberg, Federal Republic of Germany, results unpublished

using antisera to the highly conserved $G_{\alpha\text{-common}}$ peptide of animals. These proteins of soybean plasma membranes have been solubilized and partially purified. They bind [^{35}S]GTP-γ-S with high affinity (apparent kD of 10 nm). GTP is bound with lower affinity but not the other nucleotides (ATP, CTP, ITP, and UTP) tested, nor does ATP-γ-S prevent GTP-γ-S binding. GTP stimulates the release of Ca^{2+} from preloaded plant microsomes (Allan et al., 1989) as with animal cells but other functional roles for G proteins in plants and especially at the plasma membrane, remain to be discovered. Auxin responses by isolated plasma membranes thus far examined in our laboratory are either unaffected or variably inhibited by either GTP or GTP-γ-S (results unpublished).

Cyclic AMP/Adenylate Cyclase

One of the earliest examples to indicate that plants may be expected to have signal transducing pathways with characteristics different from those of animals came from studies with cAMP. Despite the fact that low levels of cyclic nucleotides may be found in some plant extracts (as products of microbial contaminants?), cyclic nucleotide-responsive proteins (e.g. protein kinases) are not (Hepler and Wayne, 1985). Auxin-responsive soybean tissues and membranes were shown early to lack both detectable cyclic AMP and adenylate cyclase (Yunghans and Morré, 1977) as determined by several sensitive techniques. By direct comparisons with the cyclase of rat liver plasma membranes, Yunghans and Morré (1977) were able to exclude both a cyclic AMP and an adenylate cyclase involvement from the auxin-response cascade. A cyclic GMP or a guanylate cyclase involvement may be excluded as well.

Protons

The induction of proton extrusion in plant stems and coleoptiles by auxin is a well established phenomenon (Cleland, 1976; Cleland and Rayle, 1978). Despite explanations such as increased deposition of cell wall matrix components, as an alternative to extrusion of protons to cause wall loosening in auxin-induced growth (Kutschera and Briggs, 1987b), there is still a possible second messenger role for proton extrusion both in response to auxin and to fusicoccin that may be unique to plants. In plants, the cytoplasmic acidification and proton pumping (membrane potential) responses are sinusoidal (Felle, 1989). They oscillate with the same periodicity but displaced from each other by $\lambda/2$. Thus, in addition to cell wall acidification in response to auxin, there is a balanced relationship between cell wall and cytoplasmic pH control.

One of the best studied examples of the regulatory effects of cytostolic pH in animal cells is activation of the sea urchin egg following fertilization (Busa and Nuccitelli, 1984). Proton extrusion by plasma membranes and attendant alkalinization of the cytoplasm occur in HeLa and other cultured animal cells, as well, where cytoplasmic acidification is related to activation of a Na^+/H^+ antiport (Sun et al., 1987). However, plasma membranes of plants appear to lack a Na^+/H^+ antiport although a Na^+/H^+ antiport has been described for tonoplast (Guern et al., 1989). Again, whereas proton extrusion may occur as part of the growth control paradigm in both plants and animals, the two processes appear to be fundamentally quite different.

Phospholipase C

In animal cells, the polyphosphoinositide PIP_2 present in the plasma membrane is cleaved by phospholipase C in response to external stimuli to produce the second messengers, inositol-1,4,5-trisphosphate (IP_3) and diacylglycerol (DAG) (Berridge and Irvine, 1984; Majerus et al., 1986; Berridge, 1987). DAG activates protein kinase C (Nishizuka, 1984) and IP_3 releases calcium from non-mitochondrial intracellular stores (Streb et al., 1983) to activate calcium-dependent enzymes.

Both PIP and PIP_2 are in the plasma membranes of higher plants (Wheeler and Boss, 1987; Irvine et al., 1989) and algae (Einspahr et al., 1988; Irvine et al., 1989). Additionally, plant plasma membranes contain phospholipase C activities both for phosphatidylinositol (Pfaffmann et al., 1987) and polyphosphoinositides (Melin et al., 1987). The relative activities for the two different substrates are dependent upon calcium concentration. The operation of a light-stimulated phospholipase C may be inferred for *Samanea saman* pulvini from increased levels of diacylglycerol produced (Morse et al., 1989). However, in the auxin hormone control of elongation growth, auxin has been without effect on phospholipase C both with isolated membrane vesicles and in situ with tissue segments (Pfaffmann et al., 1987). With PI as substrate, there was an insignificant (<3%) stimulation of the membrane-associated activity on incubation of isolated plasma membrane vesicles with 2,4-D and no effect on the soluble activity (Pfaffmann et al., 1987). A similar small but insignificant auxin effect (stimulation of about 4%) was obtained by M. Sommarin and A. S. Sandelius with the polyphosphoinositide-specific activity (private communication). Similarly with a synthetic water-soluble phospholipase C substrate, p-nitrophenylphosphorylcholine (pNPP-choline), the rate of hydrolysis with soybean membranes was increased 10 to 20% or less by addition of auxin to the assay.

Finally, evidence for phospholipid breakdown in situ by a phospholipase C-type mechanism was sought as well in the form of an auxin-induced increase in diacylglycerol (DAG). The pool of DAG is normally very small, ca. 4% of the total lipid, so that small increases could be easily detected and quantitated. With excised hypocotyl segments incubated with and without 2,4-D, steady state levels of DAG declined steadily over 4 h in the absence of auxin treatment (Morré et al., 1989). In the presence of 1 μM 2,4-D, the decline in DAG levels was less. However, when fractionated into membrane and soluble fractions, membrane-associated DAG was unchanged by auxin. The apparent auxin-induced increase was in the soluble fraction and presumably was unrelated to activation of a membrane-associated protein kinase. Isolated membrane vesicles showed no 2,4-D-induced accumulation of diglyceride despite extensive losses of phosphatidylcholine and other phospholipids (Table 1; Figure 3; Morré et al., 1989).

Table 1. Diacylglycerol content of soybean hypocotyls based on enzyme assay employing *E. coli sn*-1,2 DAG kinase and [^{32}P]ATP to convert DAG to [^{32}P] phosphatidic acid (Morré et al., 1989)

Treatment	DAG, nmol/g fresh weight Membranes	Supernatant
Initial	27.3±4.2	15.9±0.1
Control*	24.8±1.7	12.5±0.8
1 μM 2,4-D	25.6±0.1	13.8±0.4

*Determined after 4 h of incubation.

Similar results were obtained with 15 min of incubation with tissue and after 15 min of incubation with isolated membranes. In all examples the increase in DAG with auxin was small (less than 1% of the total lipids).

Fig. 3. Thin layer separation showing total phospholipids extracted from isolated plasma membranes incubated for 15 min in the absence (control) or presence of 1 μM 2,4-D

Calcium Ions

Calcium ions appear to be involved in the regulation of many cellular processes in plants (Hepler and Wayne, 1985; Roux and Serlin, 1987). They act, as well, as messengers in transducing external signals (Hepler and Wayne, 1985). As regulators of calcium mobilization in plants, inositol phosphates may be important (Ranjeva et al., 1988; Reddy and Poovaiah, 1987; Schumaker and Sze, 1987; Perdue et al., 1988) but it is not clear that auxin elicits their formation (see Phospholipase C) or what membrane components are involved. With mammalian cells, the inositol phosphates induce calcium release from endoplasmic reticulum. With plants, this release may be most prevalent from tonoplast (vacuole) vesicles (Schumaker and Sze, 1987; Ranjeva et al., 1988). IP_3 mobilized nearly 30% of the total in vitro accumulated Ca^{2+} in zucchini microsomes (Drøbak and Ferguson 1985), and release was proportional with IP_3 concentration from 10-30 μM. IP_3-induced Ca^{2+} release from purified endoplasmic reticulum from plants, however, has not been demonstrated.

There has been one report where exogenously added IP_3 causes a calcium efflux in plant cells. This was with protoplasts of cultured carrot cells (Rincon and Boss, 1987). Even if IP_3 is produced by phospholipase C action on PIP_2 in the plant plasma membrane, it still must reach the vacuole which is the primary IP_3-sensitive calcium store (Schumaker and Sze, 1987; Ranjeva et al., 1988) in plants in contrast to the endoplasmic reticulum for animal cells. Even if IP_3 is produced in selected tissues of higher plants (e.g. pulvini, Morse et al., 1987) or with *Catharanthus roseus* cells arrested in G phase in response to 2,4-D (Ettlinger and Lehle, 1988), it has not been possible to demonstrate that the IP_3 produced results in increased cytostolic calcium. Buckhout et al (1981a,b) reported a concentration-dependent release of bound calcium and other divalent ions from microsomes incubated with auxin in vitro. The membranes were equilibrated with $^{45}Ca^{2+}$ in the presence of the ionophore A23187 and the released calcium was from calcium-binding sites within the membrane as opposed to luminal Ca^{2+}. The release was in proportion to the logarithm of the auxin concentration to a maximum at 1 μM and paralleled the dose-response relationship of growth to auxin. IAA and 2,4-D but not 2,3-D produced the response. An actual loss of high affinity binding sites was determined from Scatchard analyses to be the basis for the calcium release rather than a change in binding affinity. Phospholipid changes were shown subsequently suggested to be responsible for the calcium release (Morré et al., 1984).

Other findings would argue against products of a PIP_2-specific phospholipase being responsible for calcium changes either in situ or with isolated membranes. Among these are the inability to detect IP_3 in stationary phase (growth arrested) *Catharanthus roseus* cells (Heim and Wagner, 1989) or carrot suspension culture cells labeled either with myo [2-^{3}H]inositol or $^{32}P_i$ (Rincon et al., 1989; Rincon and Boss, 1989). Calcium may have an important second messenger role in auxin control of plant growth, but unequivocal demonstration of a rise in cytosolic calcium in response to auxin addition in situ has not been possible to achieve nor has it been possible to show the generality of a rise in inositol phosphates (IP_3 or IP_2) potential effectors of calcium release in response to auxin.

Protein Kinase C

As with cyclic nucleotides, a regulatory role for diacylglycerols (DAG) in plants remains to be determined. There have been several reports that phospholipids (especially phosphatidylserine) plus DAG or phospholipids plus calcium or tumor promoting phorbol esters but not DAG plus calcium alone enhance a specific protein kinase activity (Elliott and Skinner, 1986; Olah and Kiss, 1986; Schafer et al., 1985; Favré and Turian, 1987). Morré et al. (1984) reported a nonspecific increase in phosphorylation of membrane proteins in the presence of DAG and calcium but not in response to auxin nor could a specific DAG response be demonstrated either in the presence or absence of auxin.

Protein kinases of plants appear to be immunologically related to animal protein kinase C (Elliot and Kokke, 1987). However, the plant kinases do not bind the phorbol ester tumor promoters nor do they show a calcium plus DAG dependence (see Harmon, 1989; Budde and Randall, 1989, for reviews).

Another important difference between protein phosphorylation in plants and animals concerns the phosphorylation site. In animals, a tyrosine protein kinase that is growth factor activated is often structured as part of the receptor. Tyrosine phosphorylation in plants (Torruella et al., 1986) has not been reproduced by others and there is no evidence from plants for tyrosine phosphorylation in response to growth factors. While phosphorylation sites other than tyrosine might well be involved in plants, it is clear that any signal transduction pathway in plants involving protein phosphorylation would be expected to exhibit major differences between the dominant tyrosine kinase control pathway at the membrane level that characterizes animal cell plasma membranes.

Phospholipase A_2

Phospholipase A_2 activity is present in all animal cells (van den Bosch, 1980). A soluble 70 kD form of phospholipase A_2 from mouse peritoneal macrophages requires Ca^{2+} and has been suggested to mediate the kinase C-related mobilization of arachidonic acid in macrophages (Wijkander and Sandler, 1989). The latter induces release of arachadonic acid from phosphatidylcholine, -ethanolamine, and -inositol. The A-type phospholipases cleave the fatty acids either in the *sn*-1 (phospholipase A_1) or *sn*-2 (phospholipase A_2) position. Phospholipase A_2 is an important source of arachadonic acid in many animal cells. Activation of phospholipase A_2 by increased cytosolic calcium has been implicated as an important step in the signalling pathway leading to the production of eicosanoids such as prostaglandins (Majerus et al., 1986; Berridge, 1984; Billah and Lapetina, 1983). The importance of phospholipase A_2 activity in signaling via release of arachadonic acid also must be missing in large measure from plants. Plants for the most part (mosses being an exception, Hartmann et al., 1986) lack arachadonic acid and most often are characterized by phospholipids with linoleic acid in the *sn*-2 position.

Phospholipase A_2 activity in dark adapted rod outer segments of bovine retina is activated by light in a transducin (G-protein) dependent mechanism (Jelsema, 1987). This and other evidence couple guanine nucleotide-binding proteins to the activation of phospholipase A_2 (Pelech and Vance, 1989).

Very little is known about A-type phospholipases in plants despite their presumed involvement in membrane phospholipid catabolism (Leshem, 1987). An important role for phospholipase A_2 activity in signal transduction in plants may be inferred in part from the action of the products of the enzyme on phospholipids, lysophospholipids, and free fatty acids, to modulate various growth-related components of the plant and animal plasma membrane. Importantly, both lysophospholipids and fatty acids have been shown to activate several plasma membrane enzymes in plants, including the vanadate-sensitive ATPase (Palmgren et al., 1988), and the auxin-stimulated NADH oxidase (Brightman et al., 1988; Fig 3). They inhibit the PI, PIP and DAG kinases (Wheeler and Boss, 1989).

The NADH oxidase of plasma membranes of rat liver is also stimulated by lysolipids and fatty acids (Table 2). The effect is seen in the presence of detergent and is not mediated through membrane fluidity or structure-linked latency (so-called detergent effect) (Table 2). It is given by palmitic and arachidonic acid as well as by linoleic and by either lysophosphatidylethanolamine or lysophosphatidylcholine. The concentrations of 30 µm are near optimal.

Fatty acids and lysolipids also activate protein kinases from plants (Scherer and Andre, 1989; Martiny-Baron and Scherer, 1988; Lucantoni and Polya, 1987; Klucis and Polya, 1987; Harmon, 1989). Stimulation was observed in a latency-free system to rule out a simple detergent effect. Plasma membranes when pretreated with detergent, still respond to lysophospholipids (Palmgren and Sommarin, in press). Additionally, the ATP-dependent accumulation of protons by plant plasma membrane vesicles has been shown recently to be stimulated by lysophospholipids as well (Palmgren and Sommarin, in press). Fatty acids stimulate ATPase but uncouple the proton pump when added to isolated vesicles.

Modulations of ATPases of animal membrane systems by lysophosphatidylcholine, free fatty acids or phospholipase A_2 has been reviewed by Palmgren et al. (1988). Included has been work from several groups (Swoboda et al., 1979; Taverna and Hanahan, 1980; Niggli et al., 1981; Sarkadi et al., 1982; Schmalzing and Kutschera, 1982; Swann, 1984; Kelly et al., 1986; Im et al., 1987). The H^+/K^+ ATPase of gastric mucose is stimulated by lysophosphatidylcholine (Im et al., 1987) as are the rates of Ca^{2+}k-ATPases (Taverna and Hanahan, 1980; Sarkadi et al., 1982; Scharff et al., 1983). The products produced by phospholipase A_2 may stabilize the Ca^{2+}-ATPase in a more reactive conformation (Scharff et al., 1983). A similar type of stabilization appears to be responsible for the lysophospholipid and fatty acid activation of the auxin-stimulated NADH oxidase of plants (Fig. 4). Lysophospholipids have been shown to regulate protein kinase C activity from pig brain (Oishi et al., 1988). Kuroda et

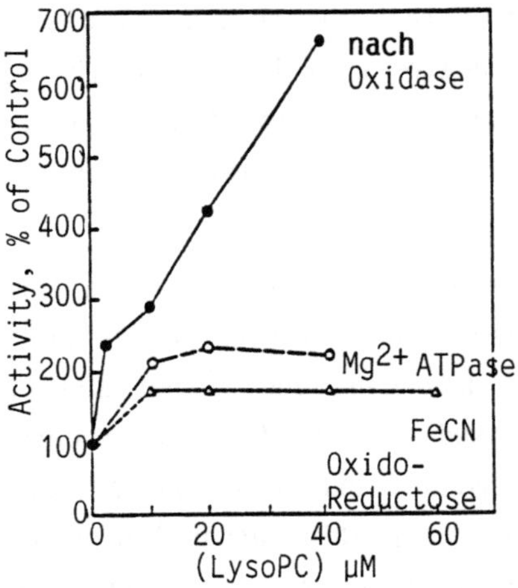

Fig. 4. Stimulation of NADH oxidase activity of soybean plasma membranes by lysophosphatidylcholine (Lyso PC) compared to the stimulation of Mg^{2+}ATPase and ferricyanide oxido-reductase (FeCN ORase) of the same membrane preparation. The stimulation of NADH oxidase by lyso PC is time dependent and maximum stimulations are usually obtained only after preincubations of approximately 10 min duration. Unpublished data of Andrew Brightman, Purdue University

al. (1987) found that LPIP rose in response to insulin in fat cells but attributed this rise in LPIP to a breakdown of PIP via phospholipase A_2. In addition to the effects of lysolipids and fatty acids on the activities noted above, lysolipids also influence the 1,3-ß-D-glucan synthetase (Kauss and Jeblick, 1986). The latter may relate to earlier reported effects of auxins on this activity with isolated plasma membrane vesicles (VanDerWoude et al., 1972). In animal cells, lysophosphatidylcholine stimulates membrane-associated sialyltransferase (Shier and Trotter, 1976) and galactosyltransferase activities (Kirschbaum and Bosmann, 1973). An auxin responsive phospholipase A_2 activity, however, has proven difficult to demonstrate for these auxin-responsive plasma membrane preparations from etiolated soybean hypocotyls.

Agonist-induced Phosphatidylcholine Breakdown

A number of reports have implicated the activation of a phosphatidylcholine-specific phospholipase C activity that is hormone responsive and not stimulated by calcium (Martin et al., 1987; see Pelech and Vance, 1989, for review). When isolated soybean microsomes (Morré et al., 1984) or purified plasma membranes (Fig. 3) are incubated for 10 to 15 min with or without 2,4-D, reductions in phosphatidylcholine as well as in phosphatidylinositol and other phospholipids result from the addition of 2,4-D but not growth inactive 2,3-D. With tissue preincubated with choline, the hydrolysis of choline-containing lipids is especially rapid. A plateau is reached within 15 min of incubation where up to 50% of the radioactivity lost from the membrane pellet appeared as soluble constituents in the supernatant (Morré and Drobes, 1987). The loss was accelerated only slightly by calcium or calmodulin alone or in various combinations with the 2,4-D. This hydrolysis is apparently not comparable to the

Table 2. Stimulation of NADH oxidase of isolated plasma membranes of rat liver by products of phospholipase A_2

Additions	0.1% Triton X-100	nmoles/min^{-1}/mg protein^{-1}
None	-	1.5
	+	1.6
Lysophosphatidyl choline (30 μM)	-	5.3
	+	5.8
Linoleic acid (30 μM)	-	3.4
	+	5.3

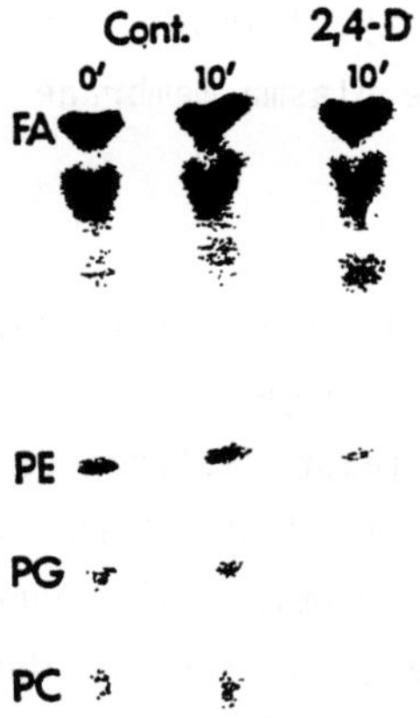

Fig. 5. Thin layer chromatographic analysis of the disappearance of radioactivity of plasma membranes from soybean hypocotyl isolated by aqueous two-phase partition. Membranes were prelabeled by exchange for 1 h at 25°C, pH 7.0, in the presence of 1-[^{14}C]linoleic acid to produce radiolabeled phosphatidylethanolamine (PE), phosphatidylglycerol (PG) and phosphatidylcholine (PC). FA = fatty acid. Unpublished data of F. E. Wilkinson and D. J. Morré, Purdue University.

phosphatidylcholine-specific phospholipase C of animal cells since there is no evidence for enhanced production of diacylglycerol (Table 1; Morré et al., 1989). It is unlikely to be a D-type activity since calcium does not stimulate appreciably and no accumulation of phosphatidic acid was observed to result from 2,4-D treatment. The nature of this activity is still under investigation. If the fatty acids of the phosphatidylcholine are labeled instead, a similar rapid loss of label from the membrane is observed (Fig. 5). These findings would be consistent with an auxin-stimulated phospholipase A_2 activity but the precise mechanism of hydrolysis remains to be established.

Platelet-activating Factor

Scherer and coworkers have shown that platelet-activating factor (PAF), an ether phospholipid (1-*O*-alkyl-2-acetyl-*sn*-glycero-3-phosphocholine) with hormone-like properties in animals (Hanahan, 1986), stimulates ATP dependent proton transport (Scherer, 1985; Scherer and Stoffel, 1987) as well as protein kinase activity (Martiny-Baron and Scherer, 1988) in microsomes from zucchini and from cultured soybean cells. A membrane lipid from plants similar to PAF was found that stimulated microsomal proton transport (Scherer et al., 1988). This latter activity, however, may be associated primarily with the tonoplast rather than the plasma membrane.

Summary

Plants respond to a variety of extracellular signals with products of lipid breakdown, calcium and protons serving as second messengers. For the plasma membrane and its response to growth hormones, major differences in signal transduction between plants and animals are evident. In some instances, responses attributed to plasma membrane in animal cells appear to be duplicated in plants by membranes of the vacuole (tonoplast). While animal studies are regarded as useful models for signal transduction pathways in plants, perhaps future encouragement should be given to searches for signal transduction pathways in growth control that may be unique to plants.

References

Allan E, Dawson A, Drøbak B, Roberts K (1989) GTP causes calcium release from a plant microsomal fraction. Cellular Signalling 1:23-29.

Berridge MJ (1984) Inositol triphosphate and diacylglycerol as second messengers. Biochem J 220:345-360.

Berridge MJ (1987) Inositol triphosphate and diacylglycerol: Two interacting second messengers. Annu Rev Biochem 56:159-163.

Berridge MJ, Irvine RF (1984) Inositol triphosphate: A novel second messenger in cellular signal transduction. Nature 312:315-321.

Billah MM, Lapetina EG (1983) Platelet-activating factor stimulates metabolism of phosphoinositides in horse platelets: Possible relationship to Ca^{2+} mobilization during stimulation. Proc Natl Acad Sci USA 80:965-968.

Blum W, Hinsch K-D, Schultz G, Weiller EW (1988) Identification of STP-binding proteins in the plasma membrane of higher plants. Biochem Biophys Res Commun 156:954-959.

Brightman AO, Barr R, Crane FL, Morré DJ (1988) Auxin-stimulated NADH oxidase purified from plasma membrane of soybean. Plant Physiol 86:1264-1269.

Buckhout TJ, Young KA, Low PS, Morré DJ (1981a) In vitro promotion by auxins of divalent ion release from soybean membranes. Plant Physiol 68:512-515.

Buckhout TJ, Young KA, Low PS, Morré DJ (1981b) Response of isolated plant membranes to auxins: Calcium release. Bot Gaz 141:418-421.

Budde RJA, Randall DD (1989) Protein kinases and future prospects. In: Morré DJ, Boss WF, Loewus FA (eds) Inositol metabolism in plants. New York, Allan R. Liss.

Busa WB, Nucitelli R (1984) Metabolic regulation via intracellular pH. Am J Physiol 246:409-438.

Cleland RE (1976) Kinetics of hormone-induced H^+ secretion. Plant Physiol 58:210-213.

Cleland RE, Rayle DL (1978) Auxin, H^+ excretion and cell elongation. Bot Mag Special Issue 1:1215-139.

Dillenschneider M, Hetherington A, Graziana A, Alibert G, Berta P, Haiech

J, Ranjeva R (1986) The formation of inositol phosphate derivatives by isolated membranes from *Acer pseudoplatanus* is stimulated by guanine nucleotides. FEBS Letts 208:413-417.
Drøbak BK, Ferguson IB (1985) Release of Ca^{2+} from plant hypocotyl microsomes by inositol-1,4,5-triphosphate. Biochem Biophys Res Commun 130:1241-1246.
Drøbak BK, Allen EF, Comerford JG, Roberts K, Dansson AP (1988) Presence of guanine nucleotide-binding proteins in a plant hypocotyl microsomal fraction. Biochem Biophys Res Commun 150:899-903.
Einspahr KJ, Peeler TC, Thompson GA Jr. (1988) Rapid changes in polyphosphoinsositide metabolism associated with the response of *Dunaliella salina* to hyposmotic shock. J Biol Chem 263:5775-5779.
Elliott DC, Kokke YS (1987) Partial purification and properties of a protein kinase C type enzyme from plants. Physochemistry 26:2929-2935.
Elliott DC, Skinner JD (1986) Calcium-dependent, phospholipid-activate4d protein kinase in plants. Phytochemistry 25:39-44.
Ettlinger C, Lehle L (1988) Auxin induces rapid changes in phosphatidylinositol metabolites. Nature 331:176-178.
Felle H (1989) pH as a second messenger in plants. In: Morré DJ, Boss WF (eds) Second Messengers in Plant Growth and Development. New York, Alan R. Liss pp. 145-166.
Favré B, Turian G (1987) Identification of a calcium and phospholipid-dependent protein kinase C in *neurospora crassa*. Plant Sci 49:15-21.
Gilman AG (1987) G proteins: Transducers of receptor-generated signals. Annu Rev Biochem 56:615-649.
Guern J, Mathieu Y, Kurkdjian A, Manigault P, Manigault J, Gillet B, Beloeil J-C, Lallemand J-Y (1989) Regulation of vacuolar pH of plant cells II: A ^{31} NMR study of the modifications of vacuolar pH in isolated vacuoles induced by proton pumping and cation/H^+ exchanges. Plant Physiol 89:27-36.
Hanahan DJ (1986) Platelet activating factor: a biologically active phosphoglyceride. Annu Rev Biochem 55:483-509.
Harmon AC (1989) Lipid activated Protein Kinases. In: Morré DJ, Boss WF, Loewus FA (eds) Inositol metabolism in plants. Alan R. Liss, New York.
Hartmann E, Beutelmann P, Vandekerkhove O, Euler R, Kohn G (1986) Moss cell cultures as sources of arachidonic and eicosapentaenoic acids. FEBS Letts 198:51-55.
Hartmann E, Pfaffmann H (1989) Mosses as a model system for involvement of phosphatidylinositol metabolism in signal transduction. In: Morré DJ, Boss WF, Loewus F (eds) Inositol Metabolism in Plants. Alan R. Liss, New York.
Heim S, Wagner KG (1989) Inositol phosphates in the growth cycle of suspension cultured plant cells. Plant Sci. In press.
Hepler PK, Wayne RO (1985) Calcium and plant development. Annu Rev Plant Physiol 36:397-439.
Hertel R, Thomson K-S, Russo VEA (1972) In vitro auxin binding to particulate cell fractions from corn coleoptiles. Planta 107:325-340.
Im WB, Blakeman DP, Davis JP (1987) Effect of lysophosphatidylcholine on K^+-transport in rat heavy gastric membranes enriched with (H^+-K^+)-ATPase. Biochem Biophys Res Commun 146:840-848.
Irvine RF, Letcher AJ, Lander DJ, Drøbak BK, Dawson AP, Musgrave A (1989) Phosphatidylinositol(4,5)biphosphate and phosphatidylinositol(4)-phosphate in plant tissues. Plant Physiol 89:888-892.
Jelsema CL (1987) Light activation of phospholipase A_2 in rod outer segments of bovine retina and its modulation by GTP-binding proteins. J Bio Chem 282:163-168.
Kauss H, Jeblick W (1986) influence of free fatty acids, lysophosphatidyl-

choline, platelet-activating factor, acylcarnitine, and echinocandin B on 1,3-ß-D-glucan synthase and callose synthesis. Plant Physiol 80:7-13.

Kelly RA, O'Hara DS, Mitch WE, Smith TW (1986) Identification of NaK-ATPase inhibitors in human plasma as momesterified fatty acids and lysophospholipids. J Bio Chem 261:11704-11711.

Kirschbaum BB, Bosman HB (1973) Lysolecithin enhancement of glycoprotein: glycosyltransferase activity. Fed Eur Biochem Soc 34:129-132.

Klucis E, Polya GM (1987) Calcium-independent activation of two plant leaf calcium-regulated protein kinases by unsaturated fatty acids. Biochem Biophys Res Commun 147:1041-1047.

Kuroda Y, Nakayama H, Ishibashi T, Aoki S, Tushima S, Nakagawa S (1987) A significant increase of lysophosphatidylinositol 4-phosphate with insulin in isolated rat fat cells. Fed Eur Biochem Soc 224:137-141.

Kutschera U, Briggs WR (1987b) Rapid auxin-induced stimulation of cell wall synthesis in pea internodes. Proc Natl Acad Sci USA 84:2747-2751.

Leshem YY (1987) Membrane phospholipid catabolism and Ca^{2+} activity in control of senescence. Physiol Plant 69:551-559.

Löbler M, Klämbt D (1985) Auxin-binding protein from coleoptile membranes of corn. I. Purification by immunoligical methods. J Biol Chem 260:9848-9853.

Lucantoni A, Polya GM (1987) Activation of wheat embryo calcium-regulated protein kinase by unsaturated fatty acids in the presence and absence of calcium. FEBS Lett 221:33-36.

Majerus PW, Connolly TM, Deckmyn H, Ross TS, Bross TF, Ishii H, Bansal VS, Wilson DB (1986) The metabolism of phosphoinositide-derived messenger molecules. Science 243:1519-1529.

Martin TW, Wysolmerski RB, Lagunoff D (1987) Phosphatidylcholine metabolism in endothelial cells: Evidence for phospholipase A and a novel Ca^{2+}-independent phospholipase C. Biochim Biohphys Acta 917:296-307.

Martiny-Baron G, Scherer GFE (1988) A plant protein kinase and plant microsomal H^+ transport are stimulated by the ether phospholipid platelet-activating factor. Plant Cell Rep 7:579-582.

Melin PM, Sommarin M, Sandelius AS, Jergil B (1987) Identification of Ca^{2+}-stimulated polyphosphoinositide phospholipase C in isolated plant membranes. Fed Eur Biochem Soc Lett 223:87-91.

Morré DJ, Drobes B (1987) A membrane-located, calcium-/calmodulin-activated phospholipase stimulated by auxin. In: Stumpf PK, Mudd JB, Nes WD (eds) The Metabolism Structure and Function of Plant Lipids. Plenum, New York, pp 229-231.

Morré DJ, Drobes B, Pfaffmann H, Wilkinson FE, Hartmann E (1989) Diacylglycerol levels unchanged during auxin-stimulated growth of excised hypocotyl segments of soybean. Plant Physiol. 90:275-279.

Morré DJ, Gripshover B, Monroe AL, Morré JT (1984) Phosphatidylinositol turnover in isolated soybean membranes stimulated by the synthetic growth hormone, 2,4-dichlorophenoxyacetic acid. J Biol Chem 259:15364-15368.

Morré DJ, Morré JT, Varnold RL (1984) Phosphorylation of membrane located proteins of soybean in vitro and response to auxin. Plant Physiol 75:265-268.

Morse MJ, Crain RC, Satter RL (1987) Phosphatidylinositol cycle metabolites in *Samanea saman* pulvini. Plant Physiol 83:640-644.

Morse MJ, Crain RC, Coté GG, Satter RL (1989) Light-stimulated inositol phospholipid turnover in *Samanea saman* pulvini. Increased levels of diacylglycerol. Plant Physiol 89:724-727.

Neumann P (1985) Events at the surface of the cell. Nature 317:380

Niggli V, Adunyah ES, Carafoli E (1981) Acidic phospholipids, unsaturated

fatty acids, and limited proteolysis mimic the effecto of calmodulin on the purified erythrocyte Ca^{2+}-ATPase. J Biol Chem 256:8588-8592.
Nishizuka Y (1984) Studies and perspective of protein kinase C. Science 225:1365-1368.
Oishi K, Raynor RL, Charp PA, Kuo JF (1988) Regulation of protein kinase C by lysophospholipids. J Biol Chem 263:6865-6871.
Olah Z, Kiss Z (1986) occurence of lipid and phorbol ester activated protein kinase in wheat cells. Fed Eur Biochem Soc Lett 195:33-36.
Palmgren MG, Sommarin M (In Press) Lysophosphatidylcholine stimulates ATP-dependent proton accumulation in isolated oat root plasma membrane vesicles. Plant Physiol.
Palmgren MG, Sommarin M, Ulvskov P, Jorgensen PL (1988) Modulation of plasma membrane H^+-ATPase from oat roots by lysophosphatidylcholine, free fatty acids and phospholipase A_2. Physiol Plant 74:11-19.
Pelech SL, Vance DE (1989) Signal transduction via phosphatidylcholine cycles. TIBS 14, 28-30.
Perdue DO, LaFavre AK, Leopold AC (1988) Calcium in the regulation of gravitropism by light. Plant Physiol 86:1276-1280.
Pfaffmann H, Hartmann E, Brightman A, Morré DJ (1987) Phosphatidylinositol specific phospholipase C of plant stems: Membrane associated activity concentrated in plasma membranes. Plant Physiol 85:1151-1155.
Ranjeva R, Carrasco A, Boudet AM (1988) Inositol trisphosphate stimulates the release of calcium from intact vacuoles isolated from *Acer* cells. FEBS Lett 230:137-141.
Ray PM, Dohrmann U, Hertel R (1977b) Specificity of auxin binding sites on maize coleoptile membranes as possible receptor sites for auxin action. Phant Physiol 60:585-591.
Reddy ASN, Poovaiah BW (1987) Inositol 1,4,5-triphosphate induced calcium release from corn coleoptile microsomes. J Biochem 101:569-573.
Rincon M, Boss WF (1987) Myo-inositol triphosphate mobilizes calcium from fusogenic carrot (*Daucus carota* L.) protoplasts. Plant Physiol 83:395-398.
Rincon M, Boss WF (1989) The second messenger role of phosphoinositides. In: Morré DJ, Boss WF, Loewus FA (eds) Inositol metabolism in plants. New York, Alan R. Liss.
Rincon M, Chen Q, Boss WF (1989) Characterization of inositol phosphates in carrot (*Daucus carota* L.) cells. Plant Physiol 89:126-132.
Roux SJ, Serlin BS (1987) Cellular mechanisms controlling light-stimulated gravitropism: Role of calcium. CRC Crit Rev Plant Sci 5:205-236.
Sarkadi B, Enyedi A, Nyers A, Gargdos G (1982) The function and regulation of the calcium pump in the erythrocyte membrane. Annu NY Acad Sci 402:329-348.
Schäfer A, Bygrave F, Matzenauer S, Marmé (1985) Identification of a calcium and phospholipid-dependent protein kinase in plant tissue. Fed Eur Biochem Soc Lett 187:25-28.
Scharff O, Foder B, Skibstead U (1983) Hysteretic activation of the Ca^{2+} pump revealed by calcium transients in human red cells. Biochim Biophys Acta 730:295-305.
Scherer GFE (1985) 1-O-alkyl-2-acetyl-*sn*-glycero-3-phosphocholine (platelet activating factor) stimulates plant H^+ transport in vitro and growth. Biochem Biophys Res Commun 133:1160-1167.
Scherer GFE, Andre B (1989) Auxin stimulates phospholipase A_2 in vivo and in vitro. Biochem Biophys Res Commun. submitted for publication.
Scherer GFE, Martiny-Baron G, Stoffel B (1988) A new set of regulatory molecules in plants: A plant phospholipid similar to platelet activating factor stimulates protein kinase and proton-translocating ATPase in membrane vesicles. Planta 175:241-253.

Scherer GFE, Stoffel B (1987) A plant phospholipid and platelet activating factor stimulate H^+ transport in isolated membrane vesicles. Planta 172:127-130.
Schmalzig G, Kutschera P (1982) Modulation of ATPase activities of human erythrocyte membranes by free acids or phospholipase A_2. J Membr Biol 69:65-76.
Schumaker KS, Sze H (1987) Inositol 1,4,5-trisphosphate releases Ca^{2+} from vacuolar membrane vesicles of oat roots. J Biol Chem 262:3944-3946.
Shier WT, Trotter JT III (1976) Stimulation of liver microsomal sialyltransferase activity by lysolecithin. Fed Eur Biochem Soc 62:165-168.
Streb H, Irvine RF, Berridge MJ, Schulz I (1983) Release of Ca^{2+} from a nonmitochondrial intracellular store in pancreatic acinar cells by inositol-1,4,5,-trisphosphate. Nature 306:67-68.
Sun IL, Garcia-Canero R, Liu W, Toole-Simms W, Crane FL, Morré DJ, Löw H (1987) Diferric transferrin reduction stimulates the Na^+/H^+ antiport of HeLa cells. Biochem Biophys Res Commun 145:465-473.
Sutherland EW, Robinson GA, Butcher RW (1968) Some aspects of the biological role of adenosine 3′,5′monophosphate (cAMP) Circulation 37:279-306.
Swann AC (1984) Free fatty acids and (Na^+, K^+)-ATPase: Effect on cation regulation, enzyme conformation, and interactions with ethanol. Arch Biochem Biophys
Swoboda G, Fritzsche J, Hasselbach W (1979) Effects of phospholipase A_2 and albumin on the calcium-dependent ATPase and the lipid composition of sarcoplasmic membranes. Eur J Biochem 95:77-88.
Taverna RD, Hanahan DJ (1980) Modulation of human erythrocyte Ca^{2+}/Mg^{2+}-ATPase activity by phospholipase A_2 and proteases: A comparison with calmodulin. Biochem Biophys Res Commun 94:652-659.
Toruella M, Casano LM, Vallejos RH (1986) Biology of cellular transducing signals '89. George Washington University Symposium '89. Abst# 226.
van den Bosch H (1980) Intracellular phospholipase A. Biochim Biophys Acta 604:191-246.
VanDerWoude WJ, Lembi CA, Morré DJ (1972) Auxin (2,4-D) simulation (*in vivo* and *in vitro*) of polysaccharide synthesis in plasma membrane fragments isolated from onion stems. Biochem Biophys Res Commun 46:245-253.
Venis MA (1980) Purification and properties of membrane-bound auxin receptors in corn. In: Skog F (ed) Plant Growth Substances, 1979. New York, Springer-Verlag, pp. 61-70.
Venis MA (1986) Receptors for plant auxin action and auxin transport. In: Conn PM (ed) The Receptors. Vol. IV. Academic Press, New York, pp 275-314.
Wheeler JJ, Boss WF (1987) Polyphosphoinositides are present in plasma membranes isolated from fusogenic carrot cells. Plant Physiol 85:389-392.
Wheeler JJ, Boss WF (1989) The presence of sn-1-palmitoyl-lysophosphatidylinotsitol monophosphate correlates positively with the fusion permissive state of the plasma membrane of fusogenic carrot cells grown in suspension culture. Biochim Biophys Acta. In press.
Wijkander J, Sandler R (1989) A phospholipase A_2 hydrolyzing arachidonylphospholipids in mouse peritoneal macrophages. FEBS Lett 244-51-56.
Williamson FA, Morré DJ, Hess K (1977) Auxin binding activities of subcellular fractions from soybean hypocotyls. Cytobiologie 16:63-71.
Yunghans WN, Morré DJ (1977) Adenylate cyclase activity not found in soybean hypocotyl and onion meristem. Plant Physiol 60:144-149.

CHEMICAL COMMUNICATION BETWEEN THE PLANT AND THE ENVIRONMENT

Jeffrey B. HARBORNE
School of Plant Sciences,
University of Reading,
Whiteknights,
Reading, RG6 2AS,
U.K.

INTRODUCTION

Most angiosperm plants in their natural habitats are in the unenviable position of having to both attract animals for the purposes of pollination and/or seed dispersal and at the same time repel them from feeding on their tissues. Chemical signalling by such plants can therefore be a highly complex process, involving many different chemicals and many different responses by animal visitors. In some cases, more than a hundred volatiles have been characterised in the flower scent or fruit odour of an individual species. Among these substances may well be those chemicals which will attract a particular pollinator or fruit eater, but equally there may be compounds which deter undesirable visitors, such as nectar thieves on flowers. Other tissues of the plant, such as the leaves, may well contain a further series of volatile chemical signals which will repel the majority of potential herbivores from feeding.

Chemical communication by a plant with its environment is, however, not restricted to the animal world, since it may interact with other plants or with micro-organisms. Plant-plant interactions which may involve the release of toxic chemicals into the soil are described as allelopathic, and may be with other members of the same species (autotoxic effects) or between plants of different species (e.g. a shrub-herb interaction). Plant-microbial interactions may be symbiotic, as in the familiar nitrogen-fixing interaction between legume and

NATO ASI Series, Vol. H 47
Signal Perception and Transduction in Higher Plants
Edited by R. Ranjeva and A. M. Boudet

Rhizobium or else may be hostile, as with pathogenic attack by infecting fungi. Chemical signals in such cases usually involve both low molecular weight and high molecular weight chemicals. In the legume-*Rhizobium* symbiosis, for example, both flavonoids and lectins are now recognised as signalling agents between plant and bacterium, necessary for the proper establishment of a highly specific infectious process.

Chemical signalling by plants is part of the subject of ecological biochemistry and much has been written on this topic in general terms (e.g. Harborne, 1988 and references therein). Chemical communication between plants and insects has been recently reviewed by Visser (1986) and by Labeyrie et al. (1987). Here, it is proposed to briefly consider recent developments in the biochemistry of plant signals, considering in turn volatile, tactile and visual signals. Such a division is of course an arbitrary one in that all three classes of signal may be involved in any particular plant-animal interaction.

VOLATILE SIGNALS

Volatile mono- and sesquiterpenoids have been identified in a number of lower plants, notably in fungi and liverworts, and their primary purpose would seem to be as defensive agents. Liverworts in particular are rarely eaten by insects or other herbivores. Some produce a turpentine-like odour, based on α- and β-pinene and camphene, while others develop a mushroom-like smell, due to the release of (+)-bornyl acetate and (-)-β-sabinene. Sesquiterpenoids are even more widespread in liverworts; some are responsible for the bitter and pungent tastes which are repulsive to herbivores (Asakawa, 1982). Fruiting bodies of basidiomycete fungi are also protected from animal attack by unpleasant odours or peppery tastes. Opposums, which are nocturnal feeders, occasionally eat toadstools, but they are careful to avoid those species which contain chemical repellents. The fly-agaric *Amanita muscaria*, which advertises its poisonous nature through its bright scarlet cap, may be tasted by the opposum. However, the hallucinogenic amine, muscimol, that is present in the cap is highly distasteful, inducing vomiting. The opposum learns to avoid feeding on _A. muscaria_ through the characteristic odour, octen-3-ol, which is released by the fungus but is itself quite innocuous (Camazine, 1985). It represents an olfactory cue released by the fly-agaric which acts as a warning signal to an animal to avoid further feeding.

Volatile terpenoids are widespread in the conifers, being released from the needles as well as the bark and oleoresin. Their defensive role is increased by their admixture with diterpene resin acids, also present in the same tissues. Their effectiveness as toxins is underlined by the fact that certain sawfly larvae 'borrow' these materials from their conifer host plants and use them for defence, discharging them as an oily effluent on ants and other predators (see Harborne, 1988). Adaptation to the toxic effects of pine terpenes can be seen in the successful exploitation of pine bark as feeding sites by various species of pine beetles. These animals use some of the dietarily-derived terpenes, directly or in modified form, as pheromones. Even these beetles find some monoterpenes such as limonene distasteful and particular monoterpene mixtures, present in some pine trees, can provide resistance to beetle attack. Furthermore, beetle feeding on pines can make the tree unpalatable due to the induction of a three-fold increase in monoterpene synthesis in the phloem around the feeding site (Miller et al., 1986).

The production of volatiles in plants only reaches its peak in the angiosperms when another role comes into play: flower scents are used to attract animals for carrying out cross pollination. The types of volatile released from petals are many and varied and besides the terpenoids, there are aliphatics, aromatics and aminoids (Table 1). Usually the odour released from a plant species will only be attractive to a small number

Table 1. Volatiles identified in flowers for attracting pollinators for references, see Harborne (1989)

Plant (family)	Flower volatiles	Pollinator(s)
ARACEAE		
Arum maculatum*) Dracunculus vulgaris)	indole, skatole, ammonia, isobutyl-amine	dung flies
D. canariensis	aliphatic esters	fruit flies
ASCLEPIADACEAE		
Araujia sericifolia	phenylacetaldehyde	moths
ORCHIDACEAE		
Catasetum spp.	trans-carvone epoxide	euglossine bees
Cypripedium calceolus	octyl and decyl acetates)	female bees
C. parviflorum	cis- and trans-β-ocimene)	
C. pubescens	1,3,5-trimethoxy-benzene)	
Platanthera chlorantha	methyl benzoate, monoterpenes	moths
RANUNCULACEAE		
Actaea spp.	geraniol, nerol and 14 other acyclic monoterpenes	nitulid beetles
Cimifuga simplex	methylanthranilate and isoeugenol	butterflies
ROSACEAE		
Rosa rugosa+	benzyl alcohol, pentan-1-ol, citronellol and farnesene	honey bees
SOLANACEAE		
Datura innoxia	tropane alkaloids (?)	hawkmoths
WINTERACEAE		
Exospermum spp.) Zygogynum spp.)	ethyl acetate, 2-methylpropyl acetate and isoprenoids	moths

*volatility of aminoid odour is increased by heating up of the spadix

+pollen has a chemical odour distinct from that of the petals of pollinators, usually of the same class, and it may act both as an attractant and as an isolating mechanism, ensuring sometimes that related species in the same genus are not pollinated by the same insect. Some odours are only released from the flowers during night time to attract nocturnal pollinators such as moths, rodents and bats.

How far floral odours, which attract one class of pollinator, repel others is not yet known in any detail. However, it is recognised (Faegri and van der Pijl, 1979) that bees are repelled by the aminoid odours released from _Arum maculatum_ to attract dung fly pollinators. Bees also have the ability to label the flowers they visit with odour traces, thus avoiding wasting energy on flowers already visited; hawkmoths, however, do not have this facility.

One of the more complex signalling systems discovered to date is that present in the orchids of the genus _Ophrys_. The attractiveness of the flowers to persuade male _Andrena_ bees to 'pseudocopulate' with them, pollinating them at the same time, is underlined by the presence of chemicals also produced by the female bees of the same genus in their pheromonal secretions. At least 150 volatiles have been characterised from _Ophrys lutea_; these include many aliphatic hydrocarbons, derived alcohols, aldehydes, ketones and esters, some 16 terpenoids and ten aromatics. In particular, geraniol, _E_, _E_-farnesol, nerol, citral and 6-methyl-5-hepten-2-one are present in both the orchid and female bee extracts. The first two of these compounds are highly attractive to male bees and are responsible for the odour-guided mating behaviour at the labellum of _Ophrys lutea_ flowers. Chemical mimicry is practiced by the plant to produce precisely the same chemicals that are pheromonal attractants in bee reproduction (Borg-Karlson and Tengo, 1986).

TACTILE SIGNALS

The most obvious form of chemical defence at the leaf surface is the presence of a sticky trap, which deters most insects from landing and prevents those which do land from eating the leaf. Sticky exudates may be released from glandular trichomes on the surface of the leaf or else a rubber-like exudate may be released onto the surface when a latex-bearing plant is damaged.

The biochemistry of sticky exudates has now been examined in a number of plants and a variety of different substances identified in this type of interaction. One of the more complex defences is present on the leaf surface of plants of the Solanaceae, where two different types of glandular trichome are present (Duffey, 1986). In Solanum, for example, type A trichomes release a phenol and phenolase enzyme system, which interact to give a liquid exudate, which becomes brown and hard with time. Type B trichomes, by contrast, release a volatile sesquiterpenoid signal, which acts as an alarm pheromone to aphids, together with a sticky material based on acylated sugar esters which forms an adhesive coating on the insect which lands on the leaf. A considerable range of different acylated glucose and sucrose esters have been detected in B type trichomes of different members of the Solanaceae (Table 2).

In species of Stylosanthes (Leguminosae), a slightly different mechanism operates. Here the sticky defence liberated from the trichome contains some α-pinene and this monoterpene kills the insects, such as cattle ticks, which are trapped at the surface (Sutherst and Wilson, 1986). In trichomes of Geranium x hortorum, the exudate toxins are o-pentadecenyl and o-heptadecenyl salicylic acid. The analysis of geranium cultivars

Table 2. Variation in sticky exudates from type B trichomes on leaves of Solanaceae

Sugar esters present*	Plant source[+]
2,3-Diacylated glucose 1,2,3-Triacylated glucose	Datura metel
3,4-Diacylated glucose 2,3,4-Triacylated glucose	non-tuberous Solanum spp.
3,4,6-Triacylated sucrose 3,4,6,3'-Tetracylated sucrose 2,3,4,3'-Tetracylated sucrose	tuberous Solanum spp.
2,3,4-Triacylated sucrose	Lycopersicon esculentum
2,3,4,6-Tetracylated sucrose	Petunia multiflora
2,3,4,6-Tetracylated glucose 2,3,4,6-Tetracylated sucrose	Nicotiana tabacum

* acylation involves a mixture of straight or branched chain organic acids. In Datura metel, there is a single acylating acid, hexanoic, but in non-tuberous Solanum, one of the esters that has been characterised has acetic, isobutyric and isocapric acids attached.

[+] see King et al. (1988) and references therein.

susceptible and resistant to spider mite attack indicates that the ability of the plant to resist the insect depends on a) having tall glandular trichomes at a high density relative to small squat trichomes and b) exuding the two toxins from the head of the trichome (Walters et al., 1989).

The production of latex, which has been detected in over 22 plant families and which is universal in such families as the Asclepiadaceae, is clearly a defensive chemical signal, since the milky secretion often contains poisonous terpenoids and alkaloids. Experimental evidence for a defensive role has been limited and until recently was only shown to deter ants. However, Dussoord and Eisner (1987) have established that many mandibular insects, in order to feed on latex-bearing plants, have responded to the latex defence by vein-cutting behaviour. This cutting of the veins blocks the flow of latex to the feeding sites and represents a counter-adaptation by the insect to the plant's defence. More direct experiments with droplets of the latex from the plant Asclepias syriaca showed that they repelled feeding by caterpillar of either the Monarch butterfly Danaus plexippus or the armyworm Spodoptera eridania.

Chemicals in the surface wax of the leaves can also act as important defence signals to herbivorous insects (Table 3). In some cases, chemicals originally secreted in the wax as a defensive barrier, are used by particular insects as tactile guides to feeding or oviposition. The most striking example of this is the female carrot fly, which responds to the toxic cocktail present on the surface of the carrot by ovipositing. The poisonous chemicals used in this way include three furanocoumarins and a polyacetylene, which in other circumstances can be shown to cause growth inhibition or death to plant herbivores (Stadler, 1986). It is interesting in this connection, that the female black swallowtail, Papilio polyxenes, which also lays her eggs on the carrot leaf, uses a relatively innocuous flavonoid and phenolic acid (Table 3) present in the vacuoles of the epidermal layers as a guide to oviposition.

In the case of the cabbage white butterfly which feeds on crucifer plants, the glucosinolate sinigrin acts both as a feeding stimulant to the larvae as well as an oviposition stimulant to the adult female. Rothschild et al. (1988) have recently shown that the adult female avoids laying her eggs on a potentially poisonous member of the Cruciferae by recognising the presence of small amounts of a cardiac glycoside based on strophanthidin which is located on the leaf surface. Thus, cardiac glycosides which occur together with glucosinolates in the wallflower Cheiranthus x allionii, can deter oviposition in this way.

Insects may also respond to the chemistry of the leaf wax,

Table 3. Surface chemicals eliciting insect responses

Plant species	Chemicals present*	Insect response(s)
Daucus carota (Umbelliferae)	methyleugenol, asarone, osthol, bergapten, xanthotoxin and falcarindiol	oviposition attraction by female carrot
	luteolin 7-(6"-malonylglucoside) and chlorogenic acid	oviposition attraction by the black swallowtail butterfly
Cheiranthus x *allionii* (Cruciferae)	strophanthidin glycoside	oviposition deterrence by cabbage white
Brassica campestris (Cruciferae)	sinigrin	feeding stimulus to caterpillar and oviposition attractant to female cabbage white
Sorghum bicolor	alkanes (C_{19}-C_{23}) and esters	feeding deterrent to locust
Morus nigra	*n*-hexacosanol and *n*-octacosanol	feeding stimulant to silkworm

* for references, see Harborne (1989); Woodhead and Chapman (1986).

either being deterred from or stimulated to feed by the alkanes and other hydrocarbon derivatives which make up the waxy layer (Woodhead and Chapman, 1986).

VISUAL SIGNALS

Flower colour in the angiosperms represents a form of visual signalling, advertising to the animal world the availability of nectar and pollen as a reward for pollinating visits. The colour has to be a contrast to pervasive green background of foliage, and almost every hue within the colour spectrum from white to purple black can be found in petals. The reason for the variation in colour and hue, which is often correlated with morphological variation, is usually explained on the basis of the colour preferences of different pollinators. Thus floral syndromes are recognised, in which bird-pollinated species have mainly bright red, long tubular petals and bee-pollinated species open flowers which are blue, purple, yellow or white but not red (Faegri and van der Pijl, 1979).

This traditional view that each flower colour acts as a signal to a particular class of pollinator has recently been questioned by Waser (1983), who has pointed out that there are very many exceptions to such generalisations. In particular, humming birds in central North America have been found to visit and pollinate plants with colours other than red or yellow. Colour preferences in birds can be learnt and varied and there is little doubt that the classic view of ornithophily has to be modified in the light of more recent findings.

Additionally, it is clear that other factors besides flower colour may determine which pollinators will visit a given plant. Many animals, including bees, can see in the ultraviolet and the appearance of the flower under such conditions - whether UV nectar guides are present or not - may determine the class of pollinator. Also, a significant number of species have the ability to signal to more than one pollinator within the period of flowering. For example, young flowers of Pedicularis densiflora are attractive to bumble bees, due to the magenta calyces and bracts, which are covered by UV reflective hairs. By contrast, the bright scarlet corollas of the older flowers attract humming birds for pollination. In the case of Ipomopsis aggregata (Polemoniaceae) there is a shift in flower colour from red to white during the flowering season, in order to compensate for the loss of humming bird pollination mid-season and in order to attract the attention of hawkmoths (Harborne, 1988).

The importance of blue flower colour has been explored in *Delphinium nelsonii*, plant populations of which contain small numbers of albino mutants. The white forms are under-visited compared to the blue; the reason seems to lie not in colour *per se* but in the fact that the blue forms have a contrasting honey guide in the lip and this allows the pollinators to collect the nectar more efficiently. The times required by the two pollinators, a bee and a humming bird, to collect nectar and travel between flowers are compared in Table 4, for blue and white forms respectively.

Table 4. Comparison of pollination efficiency on blue and white flowers of *Delphinium nelsonii*

	Time in secs taken between flowers	Time in secs taken to extract the nectar
BEES		
Blue flowers	2.1	2.7
White flowers	3.0	3.7
HUMMING BIRDS		
Blue flowers	0.38	0.72
White flowers	0.44	1.00

Data from Waser and Price (1983)

The evolutionary trend in the angiosperms towards a blue flower colour favoured particularly by bee pollinators, is clearly present in angiosperms. Blue flower colour is rare in primitive plant orders but is particularly well represented in many of the more highly advanced, herbaceous plant families, such as the Polemoniaceae, Hydrophyllaceae, Boraginaceae, Scrophulariaceae and Labiatae (Gottlieb, 1982). The chemical basis of blue flower colour has now been explored in some detail and a variety of mechanisms undoubtedly exist to provide true blueness (Table 5). Although the anthocyanidin delphinidin is commonly the basic pigment of such flowers, cyanidin and paeonidin derivatives can also on occasion undergo

Table 5. Chemical bases of blue flower colour

Anthocyanidin type	Modifying factors	Plant example
Delphinidin	flavone copigment	Fuchsia
Delphinidin	3-caffeoylquinic copigment	Hydrangea
Delphinidin	aromatic acylation	Gentiana, Platycodon, Tradescantia
Delphinidin	flavone copigment + magnesium	Commelina
Cyanidin	flavone copigment	Centaurea
Peonidin	flavone copigment	Ipomea

copigmentation to produce blues (Harborne, 1988).

Correlations between anthocyanidin type, flower colour and shape and pollinator have been established, when detailed studies are carried out within a given plant family. For example, in the Polemoniaceae, humming bird-pollinated species have either pelargonidin or cyanidin, bee-pollinated species delphinidin and lepidoptera-pollinated species mixtures of delphinidin and cyanidin. The anthocyanidins are thus still the major chemical basis for the provision of different colour types. Recently, Dukas and Schmida (1989) established a correlation between colour, size and shape with pollinator in the Cruciferae. Thus, violet flowers are large and tubular, being visited by large long-tongued bees, while yellow and white flowered species have small, open flowers, which are pollinated by small, short-tongued bees and flies. These authors conclude that flower colour in the Cruciferae is a communicative trait highly correlated with the mode of pollination.

Visual signalling is of considerable importance to herbivores consuming plant leaves and fruits but, in general, little biochemical work has been carried out on these systems. Most herbivorous insects are known to be sensitive to yellow colour (presumably of carotenoid origin) and it is possible to use this to advantage to trap aphids with yellow sticky traps (Prokopy and Owens, 1983). Whether they are sensitive to the greenness of leaves is less clear, though it is apparent that a green colour in a fruit is usually a signal to avoid feeding. This may be due to the fact that unripe green fruit is often sour or astringent to taste. Fruit ripeness is often advertised by a change in colour, e.g. from green to yellow, red or purple, but it is doubtful whether the colour per se is important. It is the contrast with the green background that allows the fruit to be seen and then eaten. Other chemical aspects of the fruit are probably more important in attracting the right animals to feed for successful seed dispersal and for discouraging the wrong dispersers (Janzen, 1978).

REFERENCES

Asakawa Y (1982) Chemical constituents of the Hepaticae. Fortschritte der Chem. organisch. Naturst. 42: 2-286.

Borg-Karlson AK, Tengo J (1986) Key substances in Ophrys lutea-Andrena pollination relationship. J Chem Ecol 12: 1928-1941.

Camazine S (1985) Olfactory aposematism: association of food toxicity with naturally occurring odour. J Chem Ecol 11: 1289-1296.

Duffey SS (1986) Plant glandular trichomes: their partial role in defence against insects. In: Juniper B, Southwood R (eds) Insects and the Plant Surface, Arnold, London, pp. 151-172.

Dukas R, Shmida A (1989) Correlation between the colour, size and shape of Israeli crucifer flowers and relationship to pollinators. Oikos 54: 281-286.

Dussoord DE, Eisner T (1987). Latex production as a defense against mandibular insects. Science 237: 898-901.

Faegri K, van der Pijl L (1979) Principles of pollination ecology, 3rd edn., Pergamon, Oxford.

Gottlieb OR (1982) Micromolecular evolution, systematics and ecology, Springer, Berlin.

Harborne JB (1988) Introduction to ecological biochemistry, 3rd edn., Academic, London.

Harborne JB (1989) Recent advances in chemical ecology. Natural Product Reports 6: 85-109.

Janzen DH (1978) Ecology and evolutionary biology of seed chemistry as related to seed predation. In: Harborne JB (ed) Biochemical Aspects of Plant and Animal Coevolution, Academic, London, pp. 163-206.

King RR, Calhoun LA, Singh RP (1988) 3,4-Di-O- and 2,3,4-tri-O-acylated glucose esters from the glandular trichomes of non-tuberous Solanum species. Phytochemistry 27: 3765-3768.

Labeyrie V, Fabres G, Lachaise D (eds) (1987) Insects-Plants, Dr W Junk, Dordrecht.

Miller RH, Berryman AA, Ryan CA (1986) Biotic elicitors of defence reactions in lodgepole pine. Phytochemistry 25: 611-612.

Prokopy RJ, Owens ED (1983) Visual detection of plants by herbivorous insects. Ann Rev Entomol 28: 337-364.

Rothschild M, Alborn H, Stenhagen G, Schoonhoven LM (1988) A strophanthidin glycoside in Siberian wallflower: a contact deterrent for the Large White butterfly. Phytochemistry 27: 101-108.

Stadler E (1986) Oviposition and feeding stimuli in leaf surface waxes. In: Juniper B, Southwood R (eds) Insects and the plant surface, Arnold, London, pp. 105-122.

Sutherst RW, Wilson LJ (1986) Tropical legumes and their ability to immobilise and kill cattle ticks. In: Juniper B, Southwood R (eds) Insects and the plant surface, Arnold, London, pp. 185-194.

Visser JH (1986) Host odour perception in phytophagous insects. Ann Rev Entomol 31: 121-144.

Walters D, Grossman H, Craig R, Mumma RO (1989) Geranium defensive agents. IV. Chemical and morphological bases of resistance. J Chem Ecol 15: 345-356.

Waser NM (1983) The adaptive nature of floral traits: ideas and evidence. In: Real L (ed) Pollination Biology, Academic, New York, pp. 242-286.

Waser NM, Price MV (1983) Pollinator behaviour and natural selection for flower colour in Delphinium nelsonii. Nature 302: 422-424.

Woodhead S, Chapman RF (1986) Insect behaviour and the chemistry of plant surface waxes. In: Juniper B, Southwood R (eds) Insects and the Plant Surface, Arnold, London, pp. 123-136.

SIGNAL PERCEPTION AND TRANSDUCTION IN HIGHER PLANTS: NOW AND THEN

Raoul Ranjeva and Alain Boudet

Centre de Physiologie Végétale de l'Université Paul Sabatier
URA CNRS n° 241, 118 Route de Narbonne F-31062 Toulouse Cédex

Plants are specific in that:

- they use light for conversion of inorganic carbon into biomass and for their morphogenesis is light dependent,
- they synthesize a large variety of secondary metabolites,
- they adapt themselves to variations of the environment without moving out,
- they may be transformed and/ or regenerated from single cells.

These characteristics make plants suitable for dissecting the process of signal perception and transduction. Advances in these topics become crucial since the main goal of modern agriculture is to improve the adaptation and the resistance of plants to different stresses and adverse conditions.

The ARW held in Toulouse (9th- 13th july, 1989) entitled " **Signal perception and transduction in higher plants**" gathered approximately fifty scientists. They included specialists of membrane proteins, patch-clamp, cell imaging, gene cloning ... all of them being in the fore-front of plant (or...animal) physiology.

Drawing a correct picture of all aspects of the workshop is not possible in a few words. However, it was very clear that plant physiology has emerged from the basic observations of the effects of different stimuli and is now applying almost routinely the concept and methodologies of cell biology molecular biology and biophysics. These new approaches have allowed to revisit old problems with new eyes.

INFORMATION IN PLANTS

From a formal point of view, plants basically use the same systems as other organisms for the translation of chemical or physical language into a biological response. A number of stimuli elicit changes in

NATO ASI Series, Vol. H 47
Signal Perception and Transduction in Higher Plants
Edited by R. Ranjeva and A. M. Boudet

enzyme activities as a result of the activation of the genome. Such a process is examplified in the case of morphogenetic events, adaptation to high or low temperatures or other climatic changes and responses to microorganisms. These physico-chemical signals promote the sequential activation of genes and synthesis of their corresponding translation products including : various stress proteins, PR proteins, enzymes of the phenolic pathway,...

At the present time, considerable progress has been made with simplified models and it is classically known that some plant-microorganism interactions may be simulated by substituting the microbe by oligosaccharides from the cell wall of either the host or the parasite.

The most important features are summarized as follows:

- a number of signal molecules are now obtained in a pure form and the structure/function relationship may be studied in detail with modified molecules,
- the host may be replaced in many cases by the corresponding cell-suspension cultures or protoplasts.

As a consequence, it is potentially possible to select or create cell lines with different degrees of sensitivity to either the signals or the elements of the transducing cascade. The recent design of suicide genes coupling the promoter of genes controlled by specific stimuli with coding regions corresponding to lethal proteins in a particular environments may help to select mutants altered in the transduction chains.

In addition to the signals that work at a short distance, long-distance communications within one organism or between different plants occur too. In this way, the propagation of an action potential takes place in response to mechanical stimuli. Furthermore chemical signals such as ethylene (a plant-growth substance) seems to play an important role in plant-plant communications.

SIGNAL PERCEPTION AND TRANSDUCTION

The structure of the photoreceptor phytochrome and the biological effects induced by its conformational changes are known for years. Very recently it has been shown that phytochrome may be a protein-kinase and is able to bind ATP. However, the precise subcellular localization of the photoreceptor and the primary site of light action are still unclear.

Different growth substances are able to bind soluble or particulate "receptors" or "acceptors"; in most cases, however, the biological significance of site occupancy remains obscure with the exception of a very few systems. Thus, binding of auxin to its membrane-bound receptor results in hyperpolarization that may be inhibited by antibodies raised against the purified receptor. Such an approach has been also developped for toxin receptors and reconstitution of isolated receptors in combination with various elements of the transduction chain is in progress.
In all the above-mentionned examples the selection of resistant or unsensitized cell-lines is of the utmost importance.
The interaction of a ligand and its receptor results in the generation of intracellular message among which are cytoplasmic pH and free calcium concentrations. Processes that have been shown to depend upon proton or/and calcium include: post-translationnal modifications of proteins (enzymes) transports and co-transports of solutes, secretion of proteins.... Here again, a large degree of homology with animal systems has been established including: the occurrence of calcium/calmodulin or phospholipd dependent processes, occurrence of cellular pools of calcium that may be mobilized by inositol tris - phosphate, characterization of G-proteins. However, some specificity of plant systems occurs also as for example the debate about phospholipase C and protein kinase C.

Concerning the final steps of the transduction chain it has to be mentioned that the characterization of trans activators involved in the control of gene expression is emerging in parallel with the isolation of plant gene promoters.

One of the newest approach during the ARW is the development of the strategy of predicting function from structural analogy. Such an approach was highlightened by the cloning of protein kinase genes by using conserved characteristics regions of animal counterpart. Such a strategy results in the demonstration that plant contains cAMP dependent kinases even if these enzymes have never been characterized by the biochemical approach.

In addition to the above-mentionned results there is a very impressive development in the biochemistry and biophysics of ion transport. Mineral ions are no more considered as simple nutrients but also as possible signals. In this connection the use of patch clamp on isolated organelles (namely vacuoles) or reconstituted transporters and channels has contributed to the understanding of molecular processes in osmoregulation and adaptation to salt stress.

SIGNAL INTEGRATION

If we have in hands all the ingredients involved in stimulus response coupling there is no experimental model, for which all the sequence of events starting from signal perception to the final biological response has been established. Therefore, we can foresee that the next workshop will deal with " Signal integration from membrane to nucleus"

If considerable progresses have been recently achieved on the characterization of different molecular events related to the perception and transduction of signals, these have to be now logically and sequentially ordered in an overall chain of reaction. The coordinated use of several approaches including "in vitro" reconstituted systems and mutants defective in some steps of the transduction chain should elicit further progress in this area.

Finally we must keep in mind that plants exhibit specific developmental strategies. If concepts arisen from animal physiology have been of the utmost importance the role of light in plant responses, the occurrence of walls, vacuoles and of volatile chemical messages emitted by plants are among some specific traits which can be related to new devices for stimulus response coupling in plants.

SUBJECT INDEX

NATO ASI Series H

Vol. 1: **Biology and Molecular Biology of Plant-Pathogen Interactions.** Edited by J. A. Bailey. 415 pages. 1986.

Vol. 2: **Glial-Neuronal Communication in Development and Regeneration.** Edited by H. H. Althaus and W. Seifert. 865 pages. 1987.

Vol. 3: **Nicotinic Acetylcholine Receptor: Structure and Function.** Edited by A. Maelicke. 489 pages. 1986.

Vol. 4: **Recognition in Microbe-Plant Symbiotic and Pathogenic Interactions.** Edited by B. Lugtenberg. 449 pages. 1986.

Vol. 5: **Mesenchymal-Epithelial Interactions in Neural Development.** Edited by J. R. Wolff, J. Sievers, and M. Berry. 428 pages. 1987.

Vol. 6: **Molecular Mechanisms of Desensitization to Signal Molecules.** Edited by T. M. Konijn, P. J. M. Van Haastert, H. Van der Starre, H. Van der Wel, and M. D. Houslay. 336 pages. 1987.

Vol. 7: **Gangliosides and Modulation of Neuronal Functions.** Edited by H. Rahmann. 647 pages. 1987.

Vol. 8: **Molecular and Cellular Aspects of Erythropoietin and Erythropoiesis.** Edited by I. N. Rich. 460 pages. 1987.

Vol. 9: **Modification of Cell to Cell Signals During Normal and Pathological Aging.** Edited by S. Govoni and F. Battaini. 297 pages. 1987.

Vol. 10: **Plant Hormone Receptors.** Edited by D. Klämbt. 319 pages. 1987.

Vol. 11: **Host-Parasite Cellular and Molecular Interactions in Protozoal Infections.** Edited by K.-P. Chang and D. Snary. 425 pages. 1987.

Vol. 12: **The Cell Surface in Signal Transduction.** Edited by E. Wagner, H. Greppin, and B. Millet. 243 pages. 1987.

Vol. 13: **Toxicology of Pesticides: Experimental, Clinical and Regulatory Perspectives.** Edited by L. G. Costa, C. L. Galli, and S. D. Murphy. 320 pages. 1987.

Vol. 14: **Genetics of Translation. New Approaches.** Edited by M. F. Tuite, M. Picard, and M. Bolotin-Fukuhara. 524 pages. 1988.

Vol. 15: **Photosensitisation. Molecular, Cellular and Medical Aspects.** Edited by G. Moreno, R. H. Pottier, and T. G. Truscott. 521 pages. 1988.

Vol. 16: **Membrane Biogenesis.** Edited by J. A. F. Op den Kamp. 477 pages. 1988.

Vol. 17: **Cell to Cell Signals in Plant, Animal and Microbial Symbiosis.** Edited by S. Scannerini, D. Smith, P. Bonfante-Fasolo, and V. Gianinazzi-Pearson. 414 pages. 1988.

Vol. 18: **Plant Cell Biotechnology.** Edited by M. S. S. Pais, F. Mavituna, and J. M. Novais. 500 pages. 1988.

Vol. 19: **Modulation of Synaptic Transmission and Plasticity in Nervous Systems.** Edited by G. Hertting and H.-C. Spatz. 457 pages. 1988.

Vol. 20: **Amino Acid Availability and Brain Function in Health and Disease.** Edited by G. Huether. 487 pages. 1988.

NATO ASI Series H

Vol. 21: **Cellular and Molecular Basis of Synaptic Transmission.** Edited by H. Zimmermann. 547 pages. 1988.

Vol. 22: **Neural Development and Regeneration. Cellular and Molecular Aspects.** Edited by A. Gorio, J. R. Perez-Polo, J. de Vellis, and B. Haber. 711 pages. 1988.

Vol. 23: **The Semiotics of Cellular Communication in the Immune System.** Edited by E. E. Sercarz, F. Celada, N. A. Mitchison, and T. Tada. 326 pages. 1988.

Vol. 24: **Bacteria, Complement and the Phagocytic Cell.** Edited by F. C. Cabello und C. Pruzzo. 372 pages. 1988.

Vol. 25: **Nicotinic Acetylcholine Receptors in the Nervous System.** Edited by F. Clementi, C. Gotti, and E. Sher. 424 pages. 1988.

Vol. 26: **Cell to Cell Signals in Mammalian Development.** Edited by S. W. de Laat, J. G. Bluemink, and C. L. Mummery. 322 pages. 1989.

Vol. 27: **Phytotoxins and Plant Pathogenesis.** Edited by A. Graniti, R. D. Durbin, and A. Ballio. 508 pages. 1989.

Vol. 28: **Vascular Wilt Diseases of Plants. Basic Studies and Control.** Edited by E. C. Tjamos and C. H. Beckman. 590 pages. 1989.

Vol. 29: **Receptors, Membrane Transport and Signal Transduction.** Edited by A. E. Evangelopoulos, J. P. Changeux, L. Packer, T. G. Sotiroudis, and K. W. A. Wirtz. 387 pages. 1989.

Vol. 30: **Effects of Mineral Dusts on Cells.** Edited by B. T. Mossman and R. O. Bégin. 470 pages. 1989.

Vol. 31: **Neurobiology of the Inner Retina.** Edited by R. Weiler and N. N. Osborne. 529 pages. 1989.

Vol. 32: **Molecular Biology of Neuroreceptors and Ion Channels.** Edited by A. Maelicke. 675 pages. 1989.

Vol. 33: **Regulatory Mechanisms of Neuron to Vessel Communication in Brain.** Edited by F. Battaini, S. Govoni, M. S. Magnoni, and M. Trabucchi. 416 pages. 1989.

Vol. 34: **Vectors as Tools for the Study of Normal and Abnormal Growth and Differentiation.** Edited by H. Lother, R. Dernick, and W. Ostertag. 477 pages. 1989.

Vol. 35: **Cell Separation in Plants: Physiology, Biochemistry and Molecular Biology.** Edited by D. J. Osborne and M. B. Jackson. 449 pages. 1989.

Vol. 36: **Signal Molecules in Plants and Plant-Microbe Interactions.** Edited by B. J. J. Lugtenberg. 425 pages. 1989.

Vol. 37: **Tin-Based Antitumour Drugs.** Edited by M. Gielen. 226 pages. 1990.

Vol. 38: **The Molecular Biology of Autoimmune Disease.** Edited by A. G. Demaine, J-P. Banga, and A. M. McGregor. 404 pages. 1990.

Vol. 39: **Chemosensory Information Processing.** Edited by D. Schild. 403 pages. 1990.

Vol. 40: **Dynamics and Biogenesis of Membranes.** Edited by J. A. F. Op den Kamp. 367 pages. 1990.

Vol. 41: **Recognition and Response in Plant-Virus Interactions.** Edited by R. S. S. Fraser. 467 pages. 1990.

NATO ASI Series H

Vol. 42: **Biomechanics of Active Movement and Deformation of Cells.**
Edited by N. Akkaş. 524 pages. 1990.

Vol. 43: **Cellular and Molecular Biology of Myelination.**
Edited by G. Jeserich, H. H. Althaus, and T. V. Waehneldt. 565 pages. 1990.

Vol. 44: **Activation and Desensitization of Transducing Pathways.**
Edited by T. M. Konijn, M. D. Houslay, and P. J. M. Van Haastert. 336 pages. 1990.

Vol. 45: **Mechanism of Fertilization: Plants to Humans.**
Edited by B. Dale. 710 pages. 1990.

Vol. 46: **Parallels in Cell to Cell Junctions in Plants and Animals.**
Edited by A. W. Robards, W. J. Lucas, J. D. Pitts, H. J. Jongsma, and D. C. Spray. 296 pages. 1990.

Vol. 47: **Signal Perception and Transduction in Higher Plants.**
Edited by R. Ranjeva and A. M. Boudet. 357 pages. 1990.

MIX
Papier aus verantwortungsvollen Quellen
Paper from responsible sources
FSC® C105338

If you have any concerns about our products,
you can contact us on
ProductSafety@springernature.com

In case Publisher is established outside the EU,
the EU authorized representative is:
Springer Nature Customer Service Center GmbH
Europaplatz 3, 69115 Heidelberg, Germany

Printed by Libri Plureos GmbH
in Hamburg, Germany